AF333052

Soil Biology

Series Editor: Ajit Varma

12

Volumes published in the series

Volume 1
Applied Bioremediation and Phytoremediation (2004)
A. Singh, O.P. Ward (Editors)

Volume 2
Biodegradation and Bioremediation (2004)
A. Singh, O.P. Ward (Editors)

Volume 3
Microorganisms in Soils: Roles in Genesis and Functions (2005)
F. Buscot, A. Varma (Editors)

Volume 4
In Vitro Culture of Mycorrhizas (2005)
S. Declerck, D.-G. Strullu, J.A. Fortin (Editors)

Volume 5
Manual for Soil Analysis – Monitoring and Assessing Soil Bioremediation
(2005)
R. Margesin, F. Schinner (Editors)

Volume 6
Intestinal Microorganisms of Termites and Other Invertebrates (2006)
H. König, A. Varma (Editors)

Volume 7
Microbial Activity in the Rhizosphere (2006)
K.G. Mukerji, C. Manoharachary, J. Singh (Editors)

Volume 8
Nucleic Acids and Proteins in Soil (2006)
P. Nannipieri, K. Smalla (Editors)

Volume 9
Microbial Root Endophytes (2006)
B.J.E. Schulz, C.J.C. Boyle, T.N. Sieber (Editors)

Volume 10
Nutrient Cycling in Terrestrial Ecosystems (2007)
P. Marschner, Z. Rengel (Editors)

Volume 11
Advanced Techniques in Soil Microbiology (2007)
Varma, Ajit; Oelmüller, Ralf (Editors)

Ajit Varma
Sudhir Chincholkar (Eds.)

Microbial Siderophores

With 24 Figures

 Springer

Prof. Dr. Ajit Varma
Director
Amity Institute of Microbial Sciences
and Vice Chairman (International)
Amity Science Technology Foundation
Amity University Uttar Pradesh
Sector-125
Noida 201303
India
E-mail: ajitvarma@aihmr.amity.edu

Prof. Sudhir B. Chincholkar
Director
School of Life Sciences
North Maharashtra University
P.O. Box 80
Jalgaon 425 001
India
E-mail: chincholkar_sb@hotmail.com

Library of Congress Control Number: 2007921586

ISSN 1613-3382
ISBN-978-3-540-71159-9 Springer-Verlag Berlin Heidelberg New York

Springer-Verlag is a part of Springer Science+Business Media
springeronline.com

Editor: Dr. Dieter Czeschlik, Heidelberg, Germany
Desk Editor: Dr. Jutta Lindenborn, Heidelberg, Germany
Cover design: WMXDesign GmhH, Heidelberg, Germany
Typesetting and production: LE-TEX Jelonek, Schmidt & Vöckler GbR, Leipzig, Germany
Printed on acid-free paper   SPIN 11372882   31/3100 YL 5 4 3 2 1 0 -

Preface

After the shift in paradigms of microbial technology, many more secondary metabolites have emerged as novel compounds. Some of them still remain unexplored. One of the secondary metabolites produced by microorganisms under iron starvation condition is recognized as a "Siderophore". To acquire iron, microbial species have to overcome the problems of iron insolubility and toxicity and practice several innovative mechanisms. To mobilize the iron metal in substrate, most fungi excrete ferric iron-specific chelators called Siderophores. These have the ability to chelate not only Fe^{3+} but also actinides, gallium, uranium, aluminum and copper. At present, because of their highest affinity towards Fe^{3+} these molecules are referred to as "Siderophores" but in the near future, we may expect new terminologies such as aluminophores, cuprophores, etc.

Several decades ago Dr. J.M. Meyer claimed that the parrot green color of pseudomonads fermentation broth is due to biosynthesis of Siderophores. Dr. R.Y. Stanier called this particular reaction remarkable. The detection of microbial Siderophores was simplified by the classical contribution of the "Universal Chemical Assay" developed by Dr. Joe Neilands, popularly known as the "Father of Ferruginous Facts". In due course, several hundreds of microbial Siderophores were characterized. However, the field of microbial Siderophores still remains largely unexplored and it attracts experts in the fields of microbiology, biochemistry, biotechnology, environment, biochemical technology, taxonomy, genetics and others. The microbial Siderophores have been exploited directly or indirectly in agriculture, medicine, industry and environment studies.

This volume contains 12 chapters authored by subject experts of great scientific repute. Pyoverdine is a yellow-green, water-soluble, major exogenous Siderophore of fluorescent pseudomonads. It is produced by iron starved cells and is a potent iron(III) scavenger and an efficient iron(III) transporter. Its importance as a ready marker for bacterial differentiation pyoverdines is discussed in detail in various chapters. Further chapters deal with synthesis, regulation, importance and involvement in the interactions of fluorescent pseudomonads with soil microflora and plants in the rhizosphere. Two chapters present the state of art of siderophore production and another chapter gives a glimpse of the present and futuristic applications of siderophores. Tools that will result in accurate, rapid and easy detection of bacteria for taxonomic purposes are in demand. One of the tools used for taxonomic purposes for Siderophore typing is referred as "siderotyping". Siderotyping is a tool to characterize, classify and identify fluorescent pseudomonads and non-fluorescent pseudomonas. Due to its major relevance in classification, siderotyping has also been discussed in this volume. The editors have included chapters that focus on ecological aspects and the role of siderophores in induced systemic resistance. Due to the significance of siderophore, diverse topics like microbial Siderophores in human and plant health-care and biotechnological production of Siderophores in symbiotic fungi are covered. Methodological approaches are also dealt with in one of the chapters. We are very happy and proud to present this volume for the benefit of researchers in general but students and teachers in particular.

We wish to thank Dr. Dieter Czeschlik and Dr. Jutta Lindenborn, Springer, Heidelberg, Germany for kind assistance and patience in finalizing the volume. The editors are grateful to the many people who have helped to publish this volume. This effort was successful due to encouragement given by Prof. R. S. Mali, former Vice-Chancellor, North Maharashtra University, Jalgaon. We also acknowledge Mr. Makarand Rane who helped in correcting and formatting the chapters. Finally, specific thanks go to our families, immediate, and extended, not forgetting the memory of those who have passed away, for their support or their incentives in putting everything together. Ajit Varma in particular is very grateful to Dr. Ashok K. Chauhan, Founder President, Ritnand Balved Education Foundation (an umbrella organization of Amity Institutions), New Delhi, for the kind support and constant encouragement received. Special thanks are due to my esteemed friend and well wisher Professor Dr. Sunil Saran, Director General, Amity Institute of Biotechnology and Adviser to Founder President, Amity Universe, all faculty colleagues of Amity Institute of Microbial Sciences Drs. Amit C. Kharkwal, Harsha Kharkwal, Shwet Kamal, Neeraj Verma, Atimanav Gaur and Debkumari Sharma and my Ph.D. students Ms. Aparajita Das, Mr. Ram Prasad, Ms. Manisha Sharma, Ms. Sreelekha Chatterjee, Ms. Swati Tripathi, Mr. Vipin Mohan Dan and Ms. Geetanjali Chauhan. The technical support received from Mr. Anil Chandra Bahukhandi is highly appreciated.

We hope that the volume will add definitively to the knowledge of students, researchers and faculty involved in the teaching of Life Sciences.

March 2007   *Ajit Varma*
                                                                                   Sudhir Chincholkar

Contents

**3   Siderotyping, a Tool to Characterize, Classify and Identify
      Fluorescent Pseudomonads** 67
Alain Bultreys

4   Siderophores of Symbiotic Fungi 91
Kurt Haselwandter and Günther Winkelmann

Contributors

Bakker, Peter A.H.M.
Institute of Biology, Section of Phytopathology, Utrecht University, PO Box 80084, 3508 TB Utrecht, The Netherlands

Bhatnagar, Kamya
Amity Institute of Herbal and Microbial Studies, Amity University, Sector 125, Noida, India

Bultreys, Alain
Département Biotechnologie, Centre Wallon de Recherches Agronomiques, Chaussée de Charleroi 234, B-5030 Gembloux, Belgium

Chaudhari, B.L.
School of Life Sciences, North Maharashtra University, Jalgaon, Post Box 80, Jalgaon 425001, Maharashtra, India

Chincholkar, S.B.
School of Life Sciences, North Maharashtra University, Jalgaon, Post Box 80, Jalgaon 425001, Maharashtra, India

Cornelis, Pierre
Department of Molecular & Cellular Interactions, Laboratory of Microbial Interactions, Flanders Interuniversity Institute of Biotechnology, Vrije Universiteit Brussel, Pleinlaan 2, 1050 Brussels, Belgium

Das, Aparajita
Amity Institute of Herbal and Microbial Studies, Amity University, Sector 125, Noida, India

Díaz de Villegas, María Elena
Cuban Institute for Research on Sugar Cane by Products (ICIDCA), Vía Blanca No 804 y Carretera Central, San Miguel Del Padrón, Ciudad de La Habana, Cuba

Faraldo-Gómez, José D.
Center for Integrative Sciences, University of Chicago, Chicago, USA

Giang, Pham Huong
International Centre of Genetic Engineering and Biotechnology (UNO, Triesta, Italy), New Delhi, India

Haselwandter, Kurt
Department of Microbiology, University of Innsbruck, Technikerstr. 25, A-6020 Innsbruck, Austria

Höfte, Monica
Laboratory of Phytopathology, Faculty of Bioscience Engineering, Ghent University, Coupure Links 653, B-9000 Gent, Belgium

Imperi, Franceso
Dipartimento di Biologia, Universita`di Roma Tre, Viale G. Marconi 446, 00146, Rome, Italy

Lamont, Iain L.
Department of Biochemistry, University of Otago, PO Box 56. Dunedin, New Zealand

Lemanceau, Philippe
INRA-CMSE, UMR 1229 INRA/Université de Bourgogne 'Microbiologie du Sol et de l'Environnement', 17 rue Sully, 21034 Dijon cedex, France

Matthijs, Sandra
Department of Molecular & Cellular Interactions, Laboratory of Microbial Interactions, Flanders Interuniversity Institute of Biotechnology, Vrije Universiteit Brussel, Pleinlaan 2, 1050 Brussels, Belgium

Mazurier, Sylvie
INRA-CMSE, UMR 1229 INRA/Université de Bourgogne 'Microbiologie du Sol et de l'Environnement', 17 rue Sully, 21034 Dijon cedex, France

Meyer, Jean-Marie
Département Microorganismes, Génomes, Environnement, Université Louis Pasteur-CNRS, UMR 7156, 28 rue Goethe, 67000 Strasbourg, France

Prasad, Ram
Amity Institute of Herbal and Microbial Studies, Amity University, Sector 125, Noida, India

Rane, M.R.
School of Life Sciences, North Maharashtra University, Jalgaon, Post Box 80, Jalgaon 425001, Maharashtra, India

Robin, Agnès
INRA-CMSE, UMR 1229 INRA/Université de Bourgogne 'Microbiologie du Sol et de l'Environnement', 17 rue Sully, 21034 Dijon cedex, France

Srivastava, Abhishek
International University of Bremen, School of Engineering and Sciences, Campus Ring 6, D-28759, Bremen, Germany

Vansuyt, Gérard
INRA-CMSE, UMR 1229 INRA/Université de Bourgogne 'Microbiologie du Sol et de l'Environnement', 17 rue Sully, 21034 Dijon cedex, France

Varma, Ajit
Amity Institute of Herbal and Microbial Studies, Amity University, Sector 125, Noida, India

Visca, Paolo
Dipartimento di Biologia, Universita`di Roma Tre, Viale G. Marconi 446, 00146, Rome, Italy

Winkelmann, Günther
Department of Microbiology and Biotechnology, University of Tübingen, Auf der Morgenstelle 28, D-72076, Germany

1 Fungal Siderophores: Structure, Functions and Regulation

Aparajita Das[1], Ram Prasad[1], Abhishek Srivastava[2],
Pham Huong Giang[3], Kamya Bhatnagar[1] and Ajit Varma[1]

1.1
Introduction

Fungi are eukaryotic, nonphotosynthetic organisms, and most are multicellular heterotrophs. Classically, the following groups of fungi have been considered:

1) **Slime molds** have a feeding phase of the life cycle (the trophic phase) that are motile and lack cell wall. Foods particles are ingested.
2) **Aquatic fungi** have cell wall and absorb nutrients rather than ingest them. The sex cells and spores of aquatic fungi are motile (zoospores).
3) **Terrestrial fungi** have cell wall and absorb nutrients rather than ingest them. The sex cells and spores are not motile (zoospores). Three major groups of fungi recognized are Zygomycetes (e.g., black bread mold, animal dung fungi), Ascomycetes (e.g., cup fungi, truffles) and Basidiomycetes (e.g., mushrooms or toadstools, puff balls, rusts, smuts). For details see Table 1.1; c.f. Giri et al. (2005)

1.1.1
Significance of Terrestrial Fungi

Fungi play important roles in the environment. Most fungi are either saprophytes or decomposers that break down and feed on decaying organic mate-

[1]Amity Institute of Herbal and Microbial Studies, Amity University, Sector 125, Noida, India,
email: ajitvarma@aihmr.amity.edu

[2]International University of Bremen, School of Engineering and Sciences,
Campus Ring 6, D-28759, Bremen, Germany

[3]International Centre of Genetic Engineering and Biotechnology (UNO, Triesta, Italy),
New Delhi, India

Soil Biology, Volume 12
Microbial Siderophores
A. Varma, S.B. Chincholkar (Eds.)

Table 1.1 Major groups of soil fungi

Group and representative members	Distinguishing characteristics	Asexual reproduction	Sexual reproduction
Zygomycetes *Rhizopus stolonifer* (black bread mold)	Multicellular, coenocytic mycelia	Asexual spores develop in sporangia on the tips of aerial hyphae	Sexual spores known as zygospores can remain dominant in adverse environment
Ascomycetes *Neurospora, Saccharomyces cerevisiae* (baker's yeast)	Unicellular and multicellular with septate hyphae	Common by budding, conidiophores	Involves the formation of an ascus on specialized hyphae
Basidiomycetes *Agaricus campestris* (meadow mushroom), *Cryptococcus neoformans*	Multicellular, uninucleated mycelia, group includes mushrooms, smuts, rusts that affect the food supply	Commonly absent	Produce basidiospores that are born on club shaped structures at the tips of hyphae
Deuteromycetes (Fungi Imperfecti) *Penicillium, Aspergillus*	A number of these are human pathogens	Budding	Absent or unknown

rial or dead organisms. Fungi obtain nutrients to absorb by secreting digestive enzymes onto the food source. The enzymes break down, or digest, the food. The breakdown, or digestion, of organic material can also be called decomposition. Any organism that causes decomposition can be called a decomposer. They are vital links in food webs, primarily as decomposers and pathogens of both plants and animals. They are excellent scavengers, breaking down dead plant and animal tissues, recycling elements back into food webs. Some fungi can establish mutualistic relationships with other organisms in nature. For example, some fungi form mycorrhizae with the roots of plants. The fungus supplies water and minerals and the plant provides carbohydrates and other organic compounds. Mycorrhizal fungi protect the root of plant against attack by parasitic fungi and nematodes. Other fungi grow with algae and cyanobacteria forming lichens. Lichens play an important role in soil formation.

1.1.2
Mycorrhiza

These are fungi which exhibit mutualistic relationships with plant roots; 90% of trees probably have them. Their presence significantly increases the roots' effective absorptive surface area and provides for a direct link between the process of decomposition (which yields raw materials) and the absorption of these materials by plants. Mycorrhizae are beneficial both in nature and agriculture. Plants colonized with them tend to grow better than those without them. Groups of mycorrhizae are given in Fig. 1.1.

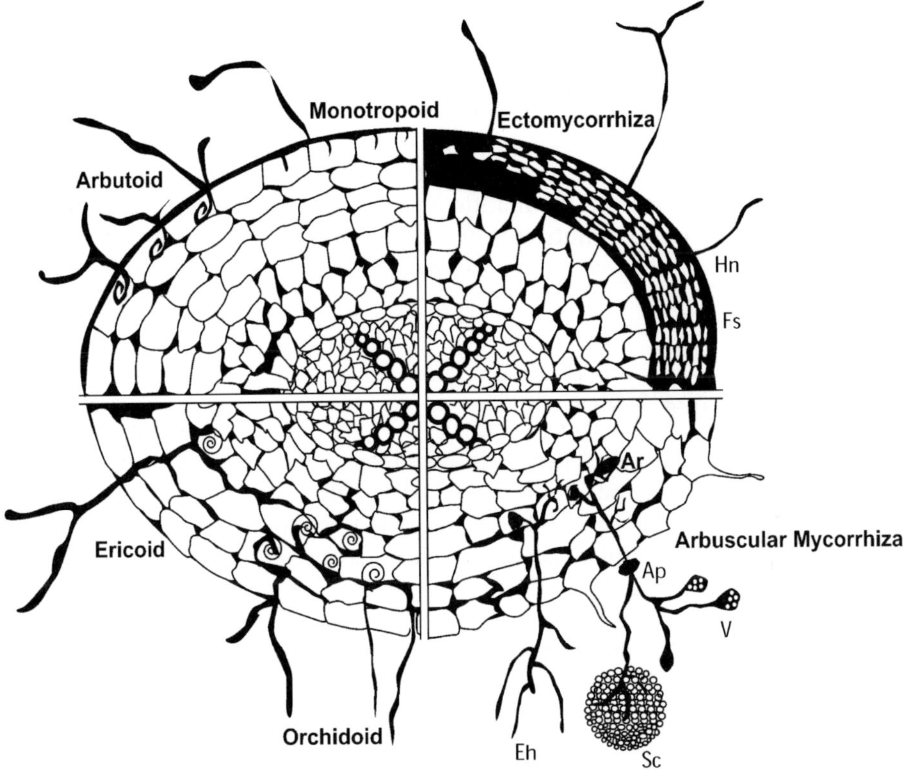

Fig. 1.1. Types of mycorrhizal fungi: Fs-fungal sheath; Eh-extramatrical hyphae; Hn-hartig's; V-vesicle; Sc-sporocarp; Ap-appresorium; Ar-arbuscule; Sc-sporocarp; Ap-appresorium

1.1.2.1
Types of Mycorrhizal Fungi

So far, seven types of mycorrhizae have come into general use over the years on the basis of morphology and anatomy but also of either host plant taxonomy or fungal taxonomy (Srivastava et al. 1996; Smith and Read 1997). These are: ectomycorrhiza, endomycorrhiza or arbuscular mycorrhiza, ericoid mycorrhiza, arbutoid mycorrhiza, monotropoid mycorrhiza, ect-endomycorrhiza and orchidaceous mycorrhiza.

1.1.2.1.1
Ectomycorrhiza (ECM)

Hyphae surround but do not penetrate the root cells. Ectomycorrhizae are commonly found in trees growing in temperate regions. The plant symbionts include both Gymnosperms and Angiosperms. Some have also been found in the tropics. The willow family (Salicaceae), birch family (Betulaceae), beech family (Fagaceae) and pine family (Pinaceae) have ectomycorrhizal associations. It possibly makes the trees more resistant to cold, dry conditions.

The hyphae grow in between the cortical and epidermal cells of the root forming a network called "Hartig's net". A mantle of hyphae covers the root surface, and mycelium extends from the mantle into the soil. It provides a large surface area for the interchange of nutrients between host and fungi. Most ectomycorrhizal fungi are basidiomycetes, but ascomycetes are also involved.

1.1.2.1.2
Arbuscular Mycorrhiza (AM)

The term refers to the presence of intracellular structures – vesicles and arbusculs – that form in the root during various phases of development. These mycorrhizae are the most commonly recorded group since they occur on a vast taxonomic range of plants, both herbaceous and woody species. The plant symbiont ranges from Bryophytes to Angiosperms. Aseptate hyphae enter the root cortical cells and form characteristic vesicles and arbuscules. The plasmalemma of the host cell invaginates and encloses the arbuscules. Arbuscular Mycorrhizal (AM) fungi belong to nine genera: *Gigaspora*, *Scutellospora*, *Glomus*, *Acaulospora*, *Entrophospora*, *Archaeospora*, *Gerdemannia*, *Paraglomus* and *Geosiphon*, the only known fungal endosymbiosis with cyanobacteria (Fig. 1.2).

1.1.2.1.3
Ericoid Mycorrhiza

In the Ericaceae, heather family, the ectomycorrhizal hyphae form a web surrounding the roots. The ericoid mycorrhizae are endomycorrhizae in the general sense, since the fungal symbiont penetrates and establishes into the cortical cells. Infection of each cortical cell takes place from the outer cortical wall; lateral spread from cell to cell does not occur. Infected cells appear to be fully packed with fungal hyphae. In the ericoid mycorrhizae, the host cell dies as the association disintegrates, thereby restricting the functional life (i.e. nutrient absorption) of these epidermal cells to the period prior to breakdown of the infected cell.

1.1.2.1.4
Arbutoid Mycorrhiza

The arbutoid mycorrhizae have characteristics which are found in both ECM and other endomycorrhizae. Intracellular penetration of cortical cells and formation of a sheath can occur, and a "Hartig's net" is present. A feature distinguishing

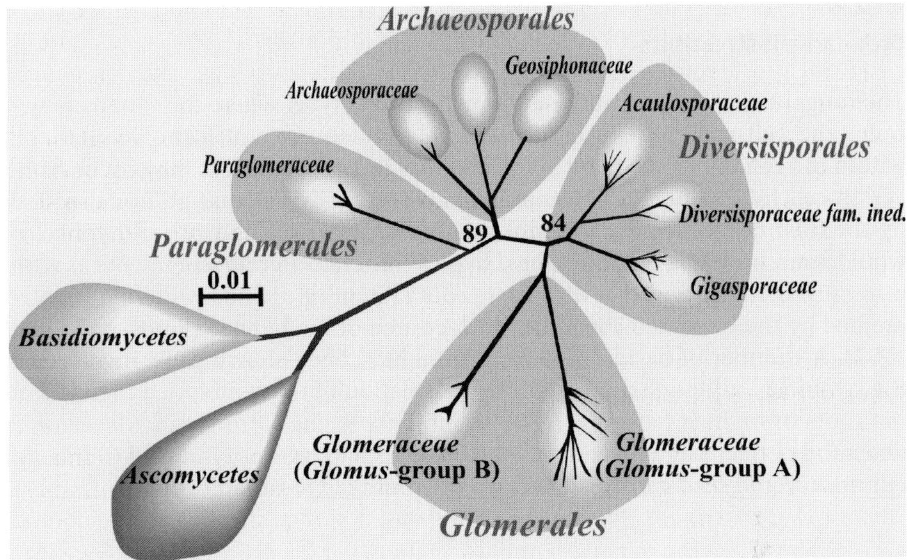

Fig. 1.2. Diagramatic representation of molecular taxonomic position of symbiotic fungi

them from ericoid mycorrhizae is the presence of dolipore septum in internal hyphae. Fungal associate in arbutoid mycorrhizae belong to basidiomycetes.

1.1.2.1.5
Monotropoid Mycorrhiza

This group of mycorrhiza is associated with the achlorophyllous plants in the family Monotropaceae. These mycorrhizae are very similar to the ECM and form a distinct sheath and "Hartig's net". However, they exhibit a distinctive type of intracellular penetration in cortical cells that is unlike other endomycorrhizal types. The fungus forms a peg into the cell wall.

1.1.2.1.6
Ect-endomycorrhiza

They are formed with the members of the Pinaceae. These Mycorrhizae form a "Hartig's net" in the cortex of the root but develop little or no sheath. Intracellular penetration of cortical cells takes place, and thus they are similar to the arbutoid type. Ectendomycorrhizae in Pinaceae seem to be limited to forest nurseries and are formed by a group of fungi called E-strain. These fungi are most likely to be the imperfect stage of ascomycetes; they may cause ect-endomycorrhizae in some tree species and ECM in other tree species.

1.1.2.1.7
Orchidaceous Mycorrhiza

The fungal association is of the endomycorrhizal type, where the fungus penetrates the cell wall and invaginates the plasmalemma and forms hyphal coil within the cell. Once the plant is invaded, spread of the fungus may occur from cell to cell internally. The internal hyphae eventually collapse and or digested by the host cell. Since the symbiosis forms an external network of hyphae, it would seem probable that the fungal hyphae function in nutrient uptake as with other mycorrhizae and that the coarse root system of orchids would be supplemented by the increased absorbing surface area of the hyphae (Smith and Read 1997). A number of basidiomycetes genera have been shown to be involved in the symbiosis, although many reports on isolation of the symbiotic fungus from the roots of orchids have placed the symbionts in the form genus *Rhizoctonia* when the perfect stage was not known or the isolate was not induced to fruit in culture. Orchid seed germinate only in the presence of suitable fungus.

1.1.3
Role of Fungi in Industry

1. Many fungi are valuable food source for humans. Yeast, such as *Saccharomyces*, is an important nutritional supplement because it contains vitamins, minerals, and other nutrients (Table 1.2).
2. Mushrooms are an important food. *Agaricus* (White Button), shiitake, and portabella mushrooms are often found in grocery stores.
3. In other places in the world, people prize the taste of Truffles and Morels, which are Ascocarps found near the roots of trees.
4. Many fungi are plant pathogens that attack grain and fruit. Wheat rust is a Basidiomycete that attacks wheat grains. Other fungi can attack food crops such as corn, beans, onions, squashes, and tomatoes.
5. Fungi are used to produce chemical compounds that are important to the food-processing industry such as citric and gluconic acid. Citric acid is used

Table 1.2. Food products and fungi

Type of Food	Fungus
Cheese: blue, Gorgonzola, Limburger, Roquefort, Camembert	*Penicillium* species
Beer, wine	*Saccharomyces carlsberbensis, Saccharomyces cerevisiae*
Soy products: miso (Japanese), soy sauce, tofu (Japanese)	*Aspergillus oryzae, Rhizopus* species, *Mucor* species
Nutritional yeast	*Saccharomyces* species
Breads	*Saccharomyces cerevisiae*

in soft drinks and candies. Gluconic acid is fed to chickens to enhance the hardness of eggshells.

6. *Ashbya gossypii* is a producer of Vitamin B2, an important nutritional supplement.

1.1.4
A Novel Endophytic Fungus *Piriformospora indica*

Symbiotic but cultivable fungus, *Piriformospora indica*, which is potential candidate to serve as biofertilizer, bioprotector, bioregulator, bioherbicide/weedicide, combats environmental stresses (chemical, thermal and physical) and is an excellent source for the hardening of the tissue culture raised crops/plants. *P. indica* tremendously improves the growth, overall biomass production and the synthesis of secondary metabolites (active ingredients) of diverse medicinal and plants of economic importance. This fungus also protects soil fertility and plant health as well (Fig. 1.3).

The properties of the fungus, *Piriformospora indica*, have been patented (Varma A and Franken P, 1997, European Patent Office, Muenchen, Germany. Patent No. 97121440.8-2105, Nov. 1998). The culture has been deposited at Braunsweich, Germany (DMS No.11827) and National Bureau of Agriculturally Important Microorganisms (NBAIM), Pusa, New Delhi.

Recently this fungus was found to enhance the anti-oxidants in *Baccopa monnieri* and Bacosides (secondary metabolites of *Bacopa monnieri*). A patent entitled '*Piriformospora indica* – a symbiotic fungus promotes biomass of *Bacopa monnieri*, enhances the antioxidant activity and bacoside concentration has

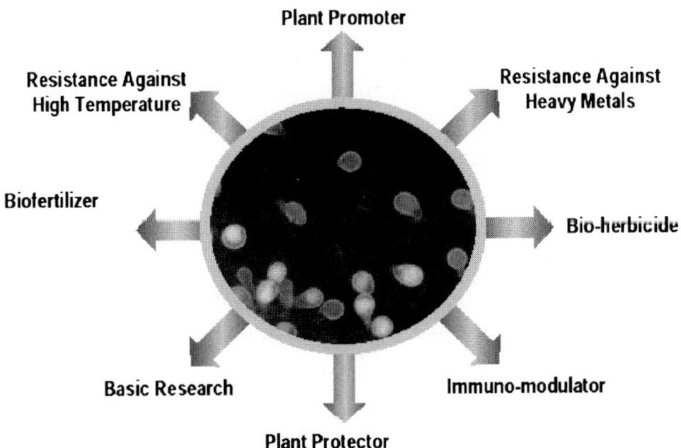

Fig. 1.3. Properties of *Piriformaspora indica*

been filed to seek the protection of this discovery from European and North American Countries. *Bacopa monnieri* is a plant of great economic importance and is known to be used as nerve tonic, adaptogen, antioxidant, antimicrobial, antifungal, sedative, cardio tonic, anticonvulsent, antidepressent and is useful for cancer patients experiencing pain. The fungus *P. indica* is documented to promote plant growth and protects the host against root pathogens and insects. Fungus root colonization promoted the plant growth and enhanced antioxidant activity and the active ingredient bacoside by several times. Symbiotic fungus, *P. indica*, enhances the plant biomass, antioxidant activity and bacoside concentration in *Bacopa monnieri*.

1.2
Siderophores

Bacteria and fungi, in response to low iron availability in the environment, synthesize microbial iron chelates called siderophores.

1.2.1
Bacterial Siderophores

There are a few reports of finding in vivo expression of siderophores by bacterial zoopathogens. For example, siderophores have been detected in sputum samples from the lungs of cystic fibrosis patients with infections due to *Pseudomonas aeruginosa* (Haas B et al. 1991) and enterochelin has been found in peritoneal washings of guinea pigs infected with *Escherichia coli* (Griffiths and Humphreys 1980). Immunoglobulins to siderophores have been detected in some instances. Such a host response is indicative of in vivo synthesis of iron chelators by some pathogens (Reissbrodt et al. 1997).

When we look more deeply into the large group of marine *Vibrios*, we notice that a broad range of structurally different siderophores is produced (Drechsel and Winkelmann 1997). Thus, catecholate siderophores have been detected in *Vibrio cholerae*, *Vibrio vulnificus* and *Vibrio fluvialis*. A mixed-type catecholate-thiazoline-hydroxamate siderophore, named anguibactin, has been isolated from *Vibrio anguillarun* and the citrate-based hydroxamate, aerobactin, has been described in certain marine vibrios. Also the occurrence of ferrioxamine G has been reported in vibrio species and we have recently identified the structurally related dihydroxamate, bisucaberin, in the fish pathogen *Vibrio salmonicida* (Winkelmann et al. 2002). This broad range of structurally different siderophores in the family Vibrionaceae may reflect the existence of a large pool of siderophore biosynthetic genes and may also indicate that the different genera of the family are more heterogeneous than previously assumed; vibrios are widespread in marine water, but this does not necessarily mean that they are really free-living bacteria. We may assume that most vibrios are somehow associated with particles in marine coastal water. Siderophores have also been isolated from several other marine bacteria, like *Alteromonas*, *Halomonas* and *Marinobacter*, indicating that siderophore production in the marine environment is widespread (Martinez et al. 2000).

1.2.1.1
Ecological Aspects

If we consider siderophore production within different microbial genera, we realize that catecholate siderophores predominate in certain Gram-negative genera, like the Enterobacteria and the genus *Vibrio*, but also in the nitrogen-fixing *Azotobacteria* and the plant-associated Agrobacteria. The reasons that these bacteria use catecholates may be manifold. However, lipophilicity, complex stability, high environmental pH and a weak nitrogen metabolism might favour catecholates. The Gram-positive streptomycetes produce hydroxamate-type ferrioxamines and the ascomycetous and basidiomycetous fungi synthesize ester- and peptide-containing hydroxamate siderophores that are acid-stable and well suited for environmental iron solubilization. Both the *Streptomycetes* and fungi show a versatile nitrogen metabolism with active N-oxygenases.

1.2.2
Fungal Siderophores

Two major responses to iron stress in fungi are a high-affinity ferric iron reductase and siderophore synthesis. Uptake of siderophores is a diverse process, which varies among the different classes of compounds. Three common classes – phenolates, hydroxamates, and polycarboxylates – are observed. Some phytopathogenic fungi produce unique compounds that function as phytotoxins but also chelate iron.

1.2.2.1
Historical Development of Fungal Siderophores

Research in this field began about five decades ago, and interest in it has accrued with the realization that most aerobic and facultative anaerobic microorganisms synthesize at least one siderophore. For details see Table 1.3.

Table 1.3. Development of siderophore research

Title	Authors	References	Year
Hydroxamate recognition during iron transport from hydroxamate-iron chelates	Haydon AH, Davis WB, Arceneaux JEL, Byers BR	J Bacteriol 115:912–918	1973
Siderophores in microbially processed cheese	Ong SA, Neilands JB	J Agric Food Chem 27:990–995	1979
The structure of the fungal siderophore, isotriornicin	Frederick CB, Bentley MD, Shive W	Biochem Biophys Res Commun 105:133–138	1982
Hydroxamate siderophore production by opportunistic and systemic fungal pathogens	Holzberg M, Artis WM	Infect Immun 40:1134–1139	1983
sid1, a gene initiating siderophore biosynthesis in *Ustilago maydis*: Molecular characterization, regulation by iron and role in phytopathogenicity	Mei B, Budde AD, Leong SA	Proc Natl Acad Sci USA 90:903–907	1993
urbs1, a gene regulating siderophore biosynthesis in *Ustilago maydis*, encodes a protein similar to the erythroid transcription factor GATA-1	Voisard C, Wang J, McEvoy JL, Xu P, Leong SA	Mol Cell Biol 13:7091–7100	1993
The role of ligand exchange in the uptake of iron from microbial siderophores by gramineous plants	Yehuda Z, Shenker M, Romheld V, Marschner H, Hadar Y, Chen Y	Plant Physiol 112:1273–1280	1996
Ferric rhizoferrin uptake into *Morganella morganii*: characterization of genes involved in the uptake of a polyhydroxycarboxylate siderophore	Kuhn S, Braun V, Koster W	J Bacteriol 178:496–504	1996
Double mutagenesis of a positive charge cluster in the ligand-binding site of the ferric enterobactin receptor, FepA	Newton SM, Allen JS, Cao Z, Qi Z, Jiang X, Sprencel C, Igo JD, Foster SB, Payne MA, Klebba PE	Proc Natl Acad Sci USA 94:4560–4565	1997
The C-terminal finger domain of Urbs1 is required for iron-mediated regulation of siderophore biosynthesis in *Ustilago maydis*	An Z, Zhao Q, McEvoy J, Yuan W, Markley J, Leon SA	Proc Natl Acad Sci USA 94:5882–5887	1997
The distal GATA sequences of the *sid1* promoter of *Ustilago maydis* mediate iron repression of siderophore production and interact directly with Urbs1, a GATA family transcription factor	An Z, Mei B, Yuan WM, Leong SA	EMBO J 16:1742–1750	1997
Siderophore-mediated iron transport: crystal structure of FhuA with bound lipopolysaccharide	Ferguson AD, Hofmann E, Coulton JW, Diederichs K, Welte W	Science 282:2215–2220	1998
Iron uptake in *Ustilago maydis*: tracking the iron path	Ardon O, Nudelman R, Caris C, Libman J, Shanzer A, Chen Y, Hadar Y	J Bacteriol 180:2021–2026	1998

Table 1.3. *(continued)* Development of siderophore research

Title	Authors	References	Year
Kinetics of iron complexing and metal exchange in solutions by rhizoferrin, a fungal siderophore	Shenker M, Hadar Y, Chen Y	Soil Sci Soc Am J 63:1681–1687	1999
Identification of a fungal triacetylfusarinine C siderophore transport gene (TAF1) in Saccharomyces cerevisiae as a member of the major facilitator superfamily	Heymann P, Ernst JF, Winkelmann G	BioMetals 12:301–306	1999
The *Aspergillus nidulans* GATA factor SREA is involved in regulation of siderophore biosynthesis and control of iron uptake	Haas H, Zadra I, Stoffler G, Angermayr K	J Biol Chem 274:46134619	1999
Siderophore-iron uptake in *Sacharomyces cerevisiae*. Identification of ferrichrome and fusarinine transporters	Yun CW, Tiedeman JS, Moore RE, Philpott CC	J Biol Chem 275:16354–16359	2000
Identification and substrate specificity of a ferrichrome-type siderophore transporter (Arn1p) in *Saccharomyces cerevisiae*	Heymann P, Ernst JF, Winkelmann G	FEMS Microbiol Lett 186:221–227	2000
A gene of the major facilitator superfamily encodes a transporter for enterobactin (Enb1p) in *Saccharomyces cerevisiae*	Heymann P, Ernst JF, Winkelmann G	Biometals 13:65–72	2000
Hydroxamate siderophore synthesis by *Phialocephala fortinii*, a typical dark septate fungal root endophyte	Bartholdy BA, Berreck M, Haselwandter K	Biometals 14:33–42	2001
SREA is involved in regulation of siderophore biosynthesis, utilization and uptake in *Aspergillus nidulans*	Oberegger H, Schoeser M, Zadra I, Abt B, Haas H	Mol Microbiol 41:1077–1089	2001
Siderophore uptake and use by the yeast *Saccharomyces cerevisiae*	Lesuisse E, Blaiseau PL, Dancis A, Camadro JM	Microbiology 147:289–298	2001
The role of the FRE family of plasma membrane reductases in the uptake of siderophore-iron in *Saccharomyces cerevisiae*	Yun CW, Bauler M, Moore RE, Klebba PE, Philpott CC	J Biol Chem 276:10218–10223	2001
Ferricrocin – an ectomycorrhizal siderophore of *Cenococcum geophilum*	**K. Haselwandter, G. Winkelmann**	BioMetals 15:73–77	2002
Identification of members of the *Aspergillus nidulans* SREA regulon: genes involved in siderophore biosynthesis and utilization	Oberegger H, Zadra I, Schoeser M, Abt B, Parson W, Haas H	Biochem Soc Trans 30:781–783	2002
Siderophore uptake by Candida albicans: effect of serum treatment and comparison with *Saccharomyces cerevisiae*	Lesuisse E, Knight SA, Camadro JM, Dancis A	Yeast 19:329–340	2002
Molecular genetics of fungal siderophore biosynthesis and uptake: the role of siderophores in iron uptake and storage	Haas H	Appl Microbiol Biotechnol 62:316–330	2003
The siderophore system is essential for viability of *Aspergillus nidulans*: functional analysis of two genes encoding l-ornithine-*N*-5-monooxygenase (sidA) and a non-ribosomal peptide synthetase (sidC)	Eisendle M, Oberegger H, Zadra I, Haas H	Mol Microbiol 49:359–375	2003

Table 1.3. *(continued)* Development of siderophore research

Title	Authors	References	Year
The siderophore system is essential for viability of *Aspergillus nidulans*: functional analysis of two genes encoding l-ornithine-*N*-5-mono-oxygenase (sidA) and a non-ribosomal peptide synthetase (sidC)	Eisendle M, Oberegger H, Zadra I, Haas H	Mol Microbiol 49:359–375	2003
Characterization of the *Aspergillus nidulans* transporters for the siderophores enterobactin and triacetylfusarinine C	Haas H, Schoeser M, Lesuisse E, Ernst JF, Parson W, Abt B, Winkelmann G, Oberegger H	Biochem J 371:505–513	2003
4'-Phosphopantetheinyl transferase-encoding npgA is essential for siderophore biosynthesis in *Aspergillus nidulans*	Oberegger H, Eisendle M, Schrettl M, Graessle S, Haas H	Curr Genet 44:211–215	2003
Siderophore biosynthesis but not reductive iron assimilation is essential for *Aspergillus fumigatus* virulence	Schrettl M, Bignell E, Kragl C, Joechl C, Rogers T, Arst HN Jr, Haynes K, Haas H	JEM 200:1213–1219	**2004**
Biosynthesis and uptake of siderophores is controlled by the PacC-mediated ambient-pH regulatory system in *Aspergillus nidulans*	Eisendle M, Oberegger H, Buttinger R, Illmer P, Haas H	Eukaryot Cell 3:561–563	2004
A putative high affinity hexose transporter, hxtA, of *Aspergillus nidulans* is induced in vegetative hyphae upon starvation and in ascogenous hyphae during cleistothecium formation	Wei H, Vienken K, Weber R, Bunting S, Requena N, Fischer R	Fungal Genet Biol 41:148–156	2004

1.2.2.2
Ecophysiological Functions of Iron

1.2.2.2.1
Importance of Iron

Iron is required by most living systems. To ensure a supply of the essential metal, pathogenic fungi use a great variety of means of acquisition, avenues of uptake, and methods of storage. Solubilization of insoluble iron polymers is the first step in iron assimilation.

Free iron is a devastating metal. In the Fenton reaction, Fe(II) reacts with H_2O_2 (a normal metabolite in aerobic organisms) to form a hydroxy radical, which binds to critical molecules found in living cells: sugars, amino acids, phospholipids, DNA bases, and organic acids (Byers and Arceneaux 1998; Fleischmann and Lehrer 1985). Microorganisms growing under aerobic conditions need iron for a variety of functions including reduction of oxygen for synthesis of ATP, reduction of ribotide precursors of DNA, for formation of heme, and for other essential purposes. A level of at least 1 µM iron is needed for optimum growth. Should the supply of Fe(II) dwindle, more is generated by the reduction of Fe(III) by superoxide anions (Byers and Arceneaux 1998; Fleischmann and Lehrer 1985). These environmental restrictions and biological imperatives have required that microorganisms form specific molecules that can compete effectively with hydroxyl ion for the ferric state of iron, a nutrient that is abundant but essentially unavailable.

The two methods most commonly used by microorganisms for solubilization of iron are reduction and chelation. The various means by which fungi acquire iron are listed in Table 1.4. Reduction of ferric iron to ferrous iron by enzymatic

Table 1.4. Mechanisms of iron acquisition by pathogenic fungi (Table adapted from Howard DH (1999) Acquisition, transport, and storage of iron by pathogenic fungi. Clin Microbiol Rev 12:394–404)

Mechanism	Examples	Reference(s)
Reduction of ferric to ferrous iron	*Candida albicans*	Morrissey et al. (1996)
	Cryptococcus neoformans	Jacobson et al. (1998)
	Geotrichum candidum	Mor et al. (1988)
	Saccharomyces cerevisiae http://cmr.asm.org/cgi/content/full/12/3/394/T1 - TF1-b[b]	Lesuisse and Labbe (1994)
Siderophore acquisition of ferric iron		
Hydroxamates (families)		
Rhodotorulic acid	*Epicoccum purpurescens*[a]	Frederick et al. (1981)
	Histoplasma capsulatum[a]	Burt (1982)
	Stemphilium botryosum	Manulis et al. (1987)
Coprogens	*Curvularia lunata*	Van der Helm and Winkelmann (1994)
	Epicoccum purpurescens	Frederick et al. (1981)
	Fusarium dimerum	Van der Helm and Winkelmann (1994)
	Histoplasma capsulatum	Burt (1982)
	Neurospora crassa	Van der Helm and Winkelmann (1994)
	Stemphilium botryosum	Manulis et al. (1987)
Ferrichromes	*Aspergillus* spp	Charlang et al. (1981)
	Epicoccum purpurescens	Frederick et al. (1981)
	Microsporum spp.	Bentley et al. (1986)
	Neurospora crassa	Van der Helm and Winkelmann (1994)
	Trichophyton spp.	Mor et al. (1992)
	Ustilago maydis	Ardon et al. (1997, 1998)

[a] Examples given are those discussed in the review; both zoopathogens and phytopathogens are included. In some examples, general references are given in which additional reports about the listed fungus are cited
[b] Some strains are reported to be human pathogens (Kwon-Chung and Bennett 1992)

Table 1.4. *(continued)* Mechanisms of iron acquisition by pathogenic fungi

Mechanism	Examples	Reference(s)
Fusarinines	*Aspergillus* spp.	Van der Helm and Winkelmann (1994)
	Epicoccum purpurescens	Frederick et al. (1981)
	Fusarium spp.	Van der Helm and Winkelmann (1994)
	Histoplasma capsulatum	Burt (1982)
	Paecilomyces spp.	Van der Helm and Winkelmann (1994)
Unidentified in report referenced	*Absidia corymbifera*[c]	Holzberg and Artis (1983)
	*Candida albicans*http://cmr.asm.org/cgi/content/full/12/3/394/T1 - TF1-c[c]	Holzberg and Artis (1983); Ismail et al. (1985)
	Madurella mycetomatis http://cmr.asm.org/cgi/content/full/12/3/394/T1 - TF1-c[c]	Mezence and Boiron (1995)
	Pseudallescheria boydii	de Hoog et al. (1994)
	Rhizopus arrhizus http://cmr.asm.org/cgi/content/full/12/3/394/T1 - TF1-c[c]	Holzberg and Artis (1983)
	Rhizopus oryzae http://cmr.asm.org/cgi/content/full/12/3/394/T1 - TF1-c[c]	Holzberg and Artis (1983)
	Scedosporium prolificans http://cmr.asm.org/cgi/content/full/12/3/394/T1 - TF1-d[d]	de Hoog et al. (1994)
	Sporothrix schenckii http://cmr.asm.org/cgi/content/full/12/3/394/T1 - TF1-c[c]	Holzberg and Artis (1983)
Polycarboxylates (rhizoferrin)	Zygomycetes	Van der Helm and Winkelmann (1994)
Phenolates-catecholates (chemical structures not identified)	*Candida albicans* http://cmr.asm.org/cgi/content/full/12/3/394/T1 - TF1-e[e]	Ismail et al. (1985)
	Wood-rotting fungi[f]	
Miscellaneous iron resources		
Hemin	*Candida albicans*	Moors et al. (1992)
	Histoplasma capsulatum	Worsham and Goldman (1988)
β-Keto aldehydes (phytotoxins	*Stemphylium botryosum*	Barash et al. (1982)
Acidification and mobilization	*Neurospora crassa*	Winkelmann (1979)
	Saccharomyces cerevisiae	Lesuisse and Labbe (1994)

[c] Exact chemical structures not provided. Identified as hydroxamates by the Neilands method (Holzberg and Artis 1983)

[d] Assay based on color formation on CAS medium; no further work reported (de Hoog 1994)

[e] Exact chemical structures not provided; phenolates identified by the Arnow assay (Ismail et al. 1985)

[f] Exact chemical structures not given in report; siderophores detected by CAS assay (Fekete et al. 1989) and identified as phenolates by paper chromatography (Fekete et al. 1983)

or nonenzymatic means is a common mechanism among pathogenic yeasts. Under conditions of iron starvation, many fungi synthesize iron chelators known as siderophores. The word siderophore (from the Greek for "iron carriers") is defined as relatively low molecular weight, ferric ion specific chelating agents elaborated by bacteria and fungi growing under low iron stress with a very high affinity for iron. The role of these compounds is to scavenge iron from the environment and to make the mineral, which is almost always essential, available to the microbial cell.

It must be stressed that not all microbes require iron, and siderophores can be dispensed with in these rare cases. Some lactic acid bacteria are not stimulated to greater growth with iron, they have no heme enzymes, and the crucial iron-containing ribotide reductase (Ardon et al. 1998) has been replaced with an enzyme using adenosylcobalamin as the radical generator. Other microbes need iron but grow anaerobically on Fe(II). While nearly all fungi make siderophores, both budding and fission yeast appear to be exceptions (An et al. 1997a, b).

In *S. cerevisiae*, the action of an Fe(III) reductase is followed by that of an Fe(II) oxidase, which generates Fe(III) for uptake. The Fe(III) would tend to polymerize to an insoluble form were it not prevented from so doing (Byers and Arceneaux 1998). Therefore, some method of iron storage must be used to escape the toxic carnage that would result from the occurrence of free iron and to prevent repolymerization of the iron transported into the fungal cell.

Among the alternative means of assimilating iron is surface reduction to the more soluble ferrous species, lowering the pH, utilization of heme, or extraction of protein-complexed metal. Siderophores appear to be confined to microbes and are not products of the metabolism of plants or animals, which have their own pathways for uptake of iron.

Since free iron is toxic, it must be stored for further metabolic use. Polyphosphates, ferritins, and siderophores themselves have been described as storage molecules.

1.2.2.2.2
Ecological Aspects

Siderophores are also involved in mycorrhizal symbiosis, as found in all terrestrial plant communities. One of the major types of mycorrhizae is the ectomycorrhiza, typically formed by almost all tree species in temperate forests. So far, only a few siderophores have been described due to the difficulties with cultivating the mycorrhizal fungi in pure culture under iron limitation. However, siderophores from three ericoid mycorrhizal fungal species, *Hymenoscyphus ericae*, *Oidiodendron griseum* and *Rhodothamnus chamaecistus*, and an ectendomycorrhizal fungus *Wilcoxina* and an ectomycorrhizal fungus *Cenococcusm geophilum*, have been isolated which all produce hydroxamate siderophores of the ferrichrome and fusigen class (Haselwandter and Winkelmann 2002). The ectomycorrhizal fungus *Cenococcum geophilum* was grown in low-iron medium and the excreted siderophores were extracted, purified and analyzed by HPLC. The principal hydroxamate siderophore produced in *Cenococcum geophilum* was identified as ferricrocin and is confirmed by analytical HPLC, FAB-mass spectrometry and 1H- and 13C-NMR spectra. Although the occurrence of ferricrocin has been shown earlier to occur in the ericoid mycorrhizal fungi, Haselwandter and Winkelmann were the first to report ferricrocin in a true ectomycorrhizal fungus which is taxonomically related to the ascomycetes (Haselwandter and Winkelmann 2002). The siderophore production of various isolates of *Phialocephala fortinii* (a typical dark septate fungal root endophyte) was assessed quantitatively as well as qualitatively in batch assays under pure culture conditions at different pH values and iron(III) concentrations (Bartholdy et al. 2001).

Zygomycetes produce solely aminocarboxylates based on citric acid and amines, which show optimal iron-binding activity at a weakly acidic pH (Drechsel et al. 1992). Although phenolate and catecholate pigments have been detected in higher fungi, defined structures of catecholate-based siderophores have never been reported in fungi. The concomitant production of organic acids by most fungi probably prevents the use of ferric catecholates that are unstable at acidic pH, while ferric hydroxamates are generally stable down to pH 2. Characterization of siderophore classes based on microbial groups, however, is not always possible. The phylogenetic distance between catechol- and hydroxamate-producing genera can be very small and occasionally both siderophore types have been observed in the same genus, and indeed in at least one case in a single siderophore (Carrano et al. 2001). There are reports that both catecholate- and hydroxamate-type siderophores have been isolated from the *Erwinia/Enterobacter/Hafnia* group, representing closely related genera of the family of Enterobacteriaceae (Deiss et al. 1998; Reissbrodt et al. 1990). Fluorescent pseudomonads and the related non-fluorescent Burkholderia group are well-known producers of linear peptide siderophores (Ongena et al. 2002). These groups of microbes might profit from the generally neutral environment of soil, where acid stability is of minor importance. The alternating d- and l-configuration of peptidic amino acids also makes these siderophores very resistant to microbial proteases.

1.2.2.3
Storage Molecules

The need for iron storage is universal. Soon after uptake, iron is found in the vacuoles of *S. cerevisiae*, where it is perhaps bound to polyphosphates (Lesuisse and Labbe 1994). The vacuoles serve as "major" storage compartments for iron in the yeast (Lesuisse and Labbe 1994). The reduced iron within the compartment is kept in the ferrous form and serves as the substrate for ferrochelatase, which is the enzyme involved in the insertion of iron into heme (Labbe-Bois and Camadro 1994). It is believed that intracellular movement of iron could be effected by intracellular citric and malic acids (Lesuisse and Labbe 1994).

Polyphosphate ferrous iron storage has also been revealed in the low-molecular-mass iron pool of *Escherichia coli* (Böhnke and Matzanke 1995).

Iron-rich proteins have been discovered in animals (ferritin), plants (phyto-ferritins), and bacteria (bacterioferritins). However, surprisingly few such molecules have been described in fungi. Those that have been are found in members of the Zygomycota (Matzanke 1994a). Three types of ferritins have been described: (i) mycoferritin, which resembles mammalian ferritins; (ii) zygoferritin, a unique form of ferritin found only in the zygomycetes; and (iii) a bacterioferritin found in Absidia spinosa (Carrano et al. 1996).

Ferritin-like molecules have not been discovered among members of the phyla Ascomycota and Basidiomycota. Work on the ascomycetes has focused on *Neurospora crassa* and *Aspergillus ochraceus*. *N. crassa* forms the two predominant hydroxamate siderophores coprogen and ferrichrocin. The major extracellular siderophore formed under conditions of iron depletion is desferricoprogen. Desferriferricrocin is found mostly intracellularly. After uptake of coprogen (the iron-charged desferricoprogen), iron is released and transferred to desferricrocin by ligand exchange (which is not necessarily enzymatically mediated (Matzanke et al. 1987), and the desferricrocin thereby becomes the ferricrocin and serves as the main intracellular iron storage compound (Matzanke et al. 1988). Ferrocrocin also serves as a long-term iron storage siderophore in *A. ochraceus* (Van der Helm and Winkelmann 1994).

The basidiomycetes that have been studied are *Rhodotorula minuta* and *Ustilago sphaerogena*. Biosynthesis of rhodotorulic acid is characteristic of the heterobasidiomycetous yeasts. Other members of the Ustilagnaceae produce ferrichrome type siderophores. Rhodotorulic acid (RA) forms a complex with iron, $Fe_2(RA)3$ that is not transported across the membrane; rather, the iron is transferred by ligand exchange to an internal pool of rhodotorulic acid that then functions as the iron storage molecule (Matzanke 1994b). It should be noted that energy-dependent ligand exchange occurs at the membrane. It is not a simple exchange between $Fe_2(RA)3$ and internal desferrirhodotorulic acid; an additional mediator is required (Matzanke 1994b). In *U. sphaerogena*, iron is transported by ferrichrome A, which does not accumulate. The iron storage compound is ferrichrome (Matzanke 1994b). In the ascomycetes and basidiomycetes that have been studied, hydroxamate-type siderophores are the iron storage molecules.

In summary, iron storage ferritin-like compounds have been found only in the zygomycetes. The iron storage function in ascomycetes and basidiomycetes is performed by polyphosphates and hydroxamates.

1.2.2.4

Production of Siderophores

Under conditions of extreme iron stress, fungi produce low-molecular-weight (Mr<1500) ferric iron chelators known collectively as siderophores (Guerinot

1994; Höfte 1993; Riquelme 1996; Telford and Raymond 1996). Most of the fungal siderophores are hydroxamates. However, the zygomycetes form iron-regulated polycarboxylates and there are well-documented reports of phenolate-catecholates in species of the wood-rotting fungi. The compounds were identified by comparison to known phenolic siderophores separated by paper chromatography (Fekete et al. 1983, 1989). There is also an unconfirmed report of phenolates being formed by *Candida albicans*, but this report was based solely on color reactions (Ismail et al. 1985). Siderophores are named on the basis of their iron-charged forms, while the deferrated form (the one that gathers iron) is called deferri-siderophore or desferri-siderophore (Fekete et al. 1983, 1989; Holzberg and Artis 1983; Ismail et al. 1985).

1.2.2.5
Detection of Fungal Siderophores

Detection methods have been covered in detail in several monographs (Fekete 1993; Neilands and Nakamura 1991; Payne 1994a), but since some of the ensuing discussion involves methods employed by investigators, a brief summary is given here. The siderophores are produced in large excess by fungi in iron-starved cultures and turn red upon capture of Fe(III). Thus, a limited-iron medium for the fungus of our choice needs to be designed (Cox 1994); the fungus is grown in the limited-iron medium, periodically iron salts are added to a centrifuged sample, and, when a sample turns red, the culture medium is harvested for further processing.

1.2.2.6
Chemical Types of Siderophores

Two classes of compounds that function in iron gathering are commonly observed: hydroxamates and polycarboxylates.

1.2.2.6.1
Hydroxamates

The hydroxamates all contain an *N*-hydroxyornithine moiety. Descriptions of chemical structures are those given by Höfte (1993). More complete consideration of the chemistry of the hydroxamates has been presented on a number of occasions (Jalal and Van der Helm1991; Matzanke 1991; Telford and Raymond 1996).

There are four recognized families of hydroxamates, as well as other hydroxamates that are reported to be produced by certain fungi.

1. Rhodotorulic acid – the diketopiperazine of N-acetyl-l-N-hydroxyornithine. It has been found mainly in basidiomycetous yeasts (Van der Helm and Winkelmann 1994). A derivative of rhodotorulic acid is a dihydroxamate named dimerum acid; 3 mol of dimerum acid (DA) is combined with 2 mol of iron (Fe$_2$(DA)3) to form the iron-bearing ligand. The binding of iron is weaker than that with the other three-hydroxamate families. Dimerum acid is produced by some phytopathogens (e.g., *Stemphylium botryosum* and *Epicoccum purpurescens* (Frederick et al. 1981; Manulis et al. 1987) and by *H. capsulatum* (Burt 1982) and *Blastomyces dermatitidis* among the medically important fungi.

2. Coprogens – these are composed of 3 mol of N-acyl-N-hydroxy-l-ornithine, 3 mol of anhydromevalonic acid, and 1 mol of acetic acid (Fig. 1.4). Unlike the situation with the rhodotorulic acid family, 1 mol of iron combines with one ligand in the coprogen, ferrichrome, and fusarinine families (Van der Helm and Winkelmann 1994). The coprogens are produced by a number of plant pathogens (Höfte 1993; Manulis et al. 1987), for example *H. capsulatum* (Burt 1982), *B. dermatitidis* and the occasional human pathogens *Fusarium dimerum* and *Curvularia lunata* (Van der Helm and Winkelmann 1994).

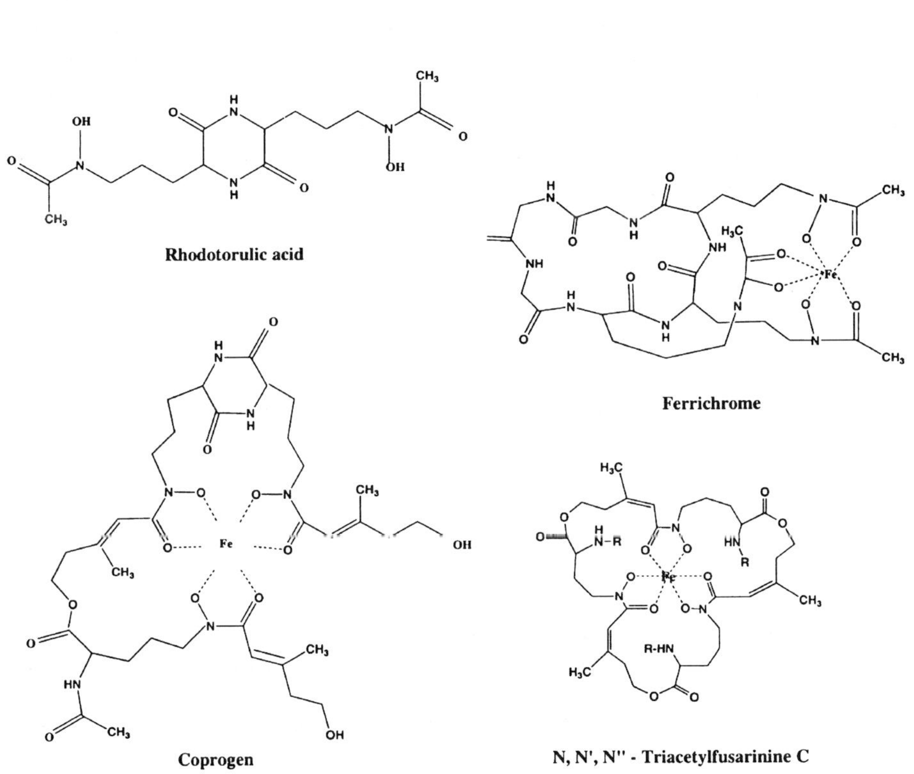

Fig. 1.4. Representative hydroxamate siderophores (adapted from Winkelmann 1993)

The coprogens may be considered trihydroxamate derivatives of rhodotorulic acid with a linear structure, and by such reckoning the number of hydroxamate families would be reduced to three. It is known that the hydrolysis of the ester group of coprogen B results in 1 mol of dimerum acid and 1 mol of *trans*-fusarinine (Winkelmann 1993). In the original report on DA (referred to as "dimerumic acid" in that report) from *H. capsulatum* (Burt 1982), the title implied that it was a degradation product of coprogen B. In fact, hydrolysis of the ester bond of coprogen gives rise to DA and trans-fusarinine (Winkelmann 1993), but DA is the predominant hydroxamate in liquid shake cultures of *H. capsulatum* (Burt 1982).

3. Ferrichromes – cyclic peptides containing a tripeptide of N-acyl-N-hydroxyornithine and combinations of glycine, serine, or alanine (Fig. 1.4). Among the pathogenic fungi, ferrichromes are produced by some phytopathogenic fungi (Höfte 1993) and by *Microsporum* sp. (Bentley et al. 1986; Mor et al. 1992), *Trichophyton* sp. (Mor et al. 1992), and *Aspergillus* spp. including the important pathogen *A. fumigatus* (Leong and Winkelmann 1998; Van der Helm and Winkelmann 1994). Another function of ferrichromes is the intracellular storage of iron.

4. Fusarinines – also called fusigens, these may be either linear or cyclic hydroxamates. Fusarinine is a compound in which N-hydroxyornithine is N-acylated by anhydromevalonic acid. Among zoopathogens, various fusarinines are found in *Fusarium* spp., *Paecilomyces* spp., and *Aspergillus* spp. (Van der Helm and Winkelmann 1994). Compounds identified as *trans*-fusarinine and an unidentified monohydroxamate were observed in culture filtrates of *Histoplasma capsulatum*, but they did not have biological activity (i.e., they did not stimulate growth of the fungus) (Burt 1982). The work on siderophores of *H. capsulatum* was initiated because of the well-known fact that the fungus does not clone in its yeast cell phase of growth on most culture media in vitro (Burt 1982). Burt used hydroxamates isolated from culture filtrates to relieve the growth restriction) (Burt 1982). The same strategy, i.e., growth stimulation, was used by Castaneda et al. (1988) in a study of *Paracoccidioides brasiliensis*. The siderophores coprogen B and DA isolated from *B. dermatitidis* were used to improve the plating efficiency of *P. brasiliensis* (Castaneda et al. 1988).

Unidentified hydroxamates – there are unconfirmed reports of formation of hydroxamates by *C. albicans* (Holzberg and Artis 1983; Ismail et al. 1985; Sweet and Douglas 1991). No exact chemical structures were presented, and it has been suggested by others (Van der Helm and Winkelmann 1994) that long-term cultivation (20 days) of a fungus will result in materials that react positively in tests based on color formation. Cutler and Han (1996) have reported their failure to display siderophores in *C. albicans*. The synthesis of hydroxamate siderophores by *C. albicans* has not been proven by chemical characterization.

Hydroxamates have been reported in other zoopathogens: *Absidia corymbifera, Madurella mycetomatis, Pseudallescheria boydii, Rhizopus arrhizus, R. oryzae, Scedosporium prolificans*, and *Sporothrix schenckii*. The report of hydroxamates from the zygomycetes *A. cormybifera*, *R. arrhizus*, and *R. oryzae* is unconfirmed (Van der Helm and Winkelmann 1994). It appears that an altogether different class of siderophores, the polycarboxylates, is formed by zygomycetes unconfirmed (Van der Helm and Winkelmann 1994). The report on *P. boydii* and *S. prolificans* was based solely on a positive CAS reaction. No studies of siderophore function or chemical characterization were performed (de Hoog et al. 1994). The report on *M. mycetomatis* was based on the CAS reaction (Mezence and Boiron 1995). The report on *S. schenckii* was based on a reaction with iron salts and the use of the Neilands method for hydroxamates (Holzberg and Artis 1983).

1.2.2.6.2
Polycarboxylates

Although there is one incomplete report of hydroxamates among the zygomycetes (Holzberg and Artis 1983), these fungi have not been observed by others to synthesize siderophores of the hydroxamate sort. Instead, a citric acid-containing polycarboxylate called rhizoferrin has been isolated from *Rhizopus microsporus* var. rhizopodiformis (Van der Helm and Winkelmann 1994). The molecule contains two citric acid units linked to diaminobutane (Fig. 1.5). Rhizoferrin is widely distributed among the members of the phylum Zygomycota, having been observed in the order Mucorales (families Mucoraceae, Thamndidiaceae, Choanephoraceae, and Mortierellaceae) and in the order Entomophthorales (Van der Helm and Winkelmann 1994).

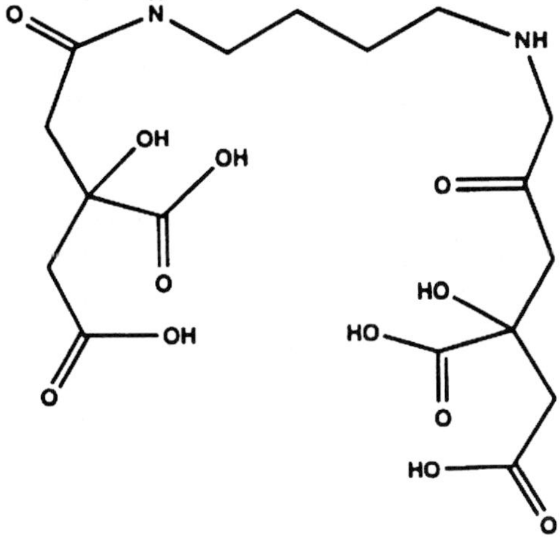

Fig. 1.5. The polycarboxylate rhizoferrin (adapted from **Winkelmann** 1993)

1.2.2.6.3
Phenolates-Catecholates

Most investigators consider that the phenolate-catecholate class of siderophores are not produced by fungi (Höfte 1993; Van der Helm and Winkelmann 1994). Only two exceptions have been reported; the first exception is *C. albicans* (Ismail et al. 1985) and this study was flawed by a lack of chemical structure studies. Phenolate siderophores were not found by two other groups of investigators (de Hoog et al. 1994; Holzberg and Artis 1983) who did, however, report hydrox-amates in *C. albicans*. The other exception involves certain wood-rotting fungal species (Fekete et al. 1983, 1989). The siderophores were detected by the CAS dye and identified as phenolates by comparison to standard phenolates (e.g., entero-chelin), and both the siderophores and the standard phenolates were separated by paper chromatography and visualized by UV fluorescence (Fekete et al. 1983). Further chemical structure studies were not reported, and the scientist reported that the compounds "...appear to be phenolate in character"(Jellison et al. 1991).

1.2.2.7
Synthetic Pathways of Siderophores

In this part we describe the various synthetic pathway of siderophores, their structural aspects and their mechanism of iron transport.

1.2.2.7.1
Diversity in Siderophore Synthesis

Diversity in siderophore synthesis is observed. Sometimes this diversity is with regard to a number of representatives in a single family, and at other times it is reflected in the synthesis of a number of representatives in several families. Fred-erick et al. (1981) provided evidence for synthesis of siderophores belonging to all four families of hydroxamates by *Epicoccum purpurescens* and Höfte (1993) reported that *Aspergillus ochraceus* "...can produce up to 10 or more different siderophores...". The reason for diversity is thought to be the ability of organisms to adapt to a wide variety of environmental situations (Höfte 1993; Winkelmann 1993). This idea is especially interesting for pathogens for which a host immune response may be involved (see "Siderophores as Pathogenic Factors" below).

1.2.2.7.2
Biosynthetic and Structural Aspects

The biosynthetic pathways of siderophores are tightly connected to aerobic me-tabolism involving molecular oxygen activated by mono-, di- and N-oxygenases and the use of acids originating from the final oxidation of the citric acid cycle, such as citrate, succinate and acetate. Moreover, all siderophore peptides are synthesized by non-ribosomal peptide synthetases and in the case of fungal siderophores are mainly built up from ornithine, a non-proteinogenic amino acid. Thus, siderophore synthesis is largely independent from the primary metabolism. Most siderophores contain one or more of the following simple bidentate ligands as building blocks: (1) a dihydroxybenzoic acid (catecholate) coupled to an amino acid, (2) hydroxamate groups containing N-5-acyl-N-5-hydroxyornithine or N-6-acyl-N-6-hydroxylysine and (3) hydroxycarboxylates consisting of citric acid or β-hydroxyaspartic acid.

Besides being precursors, most of the monomeric bidentates may also act as functional siderophores after excretion. The iron-binding affinity of bidentate siderophores, however, remains low compared with hexadentate siderophores. A phylogeny of siderophore structures is difficult to delineate. However, starting from simple precursors of each class, one can imagine that extended siderophore structures have been favoured during evolution, resulting in hexadentate siderophores possessing higher stability constants (chelate effect) compared with their monomeric precursors. Thus, higher denticity seems to have a selective advantage in siderophore evolution.

A further aspect of siderophore evolution is the optimization of chelate conformation. Although linear di-, tetra- and hexadentate siderophores have been found in all siderophore classes, there is a tendency for cyclization in the final biosynthetic end products (Fig. 1.6). Examples are enterobactin or corynebactin in the catecholate class, fusigen, triacetylfusarinine and ferrioxamines E and G, as well as the ferrichromes and asperchromes, in the hydroxamate class. Cyclization enhances complex stability, chemical stability and improves resistance to degrading enzymes. Cyclization is regarded as a common feature of secondary metabolism and is found in microbial peptides, polyketides, macrocyclic antibiotics and other bioactive compounds. Cyclization might also be advantageous for diffusion-controlled transport processes across cellular membranes. Moreover, due to a reduction of residual functional groups, the surface of the siderophores becomes non-reactive or inaccessible to modifying enzymes.

1.2.2.8
Transport of Siderophores

1.2.2.8.1
Transport Proteins

In bacterial systems, iron-regulated proteins occur and are expressed under conditions of iron depletion, often in concert with siderophore synthesis (Crosa 1997; Payne 1994b, c). Proteins from the plasma membrane of *N. crassa* that had been grown in iron-replete and iron-depleted media were compared (Van der Helm and Winkelmann 1994). There were no significant differences in the

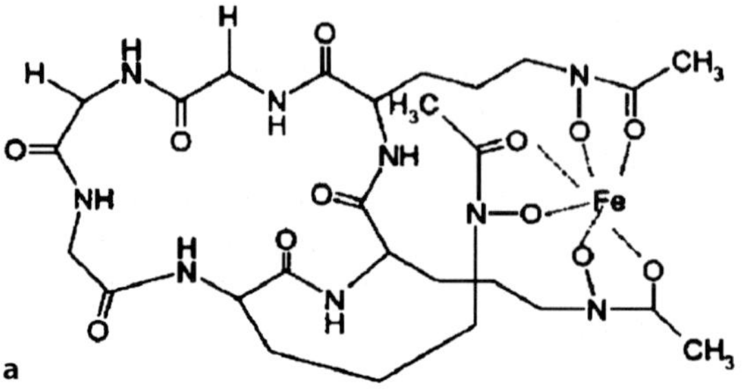

Fig. 1.6. Cyclic siderophores (adapted from Winkelmann 2001)

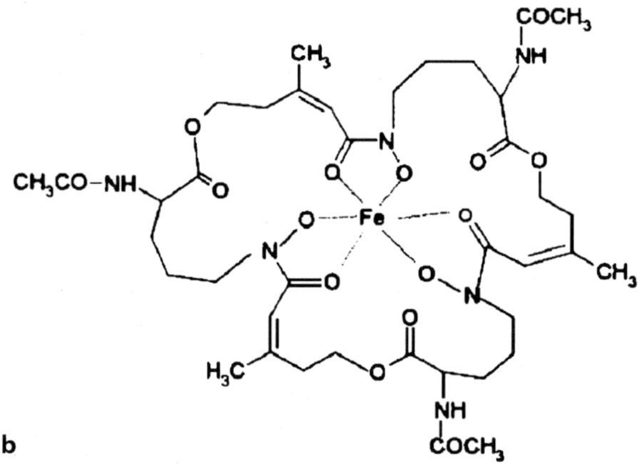

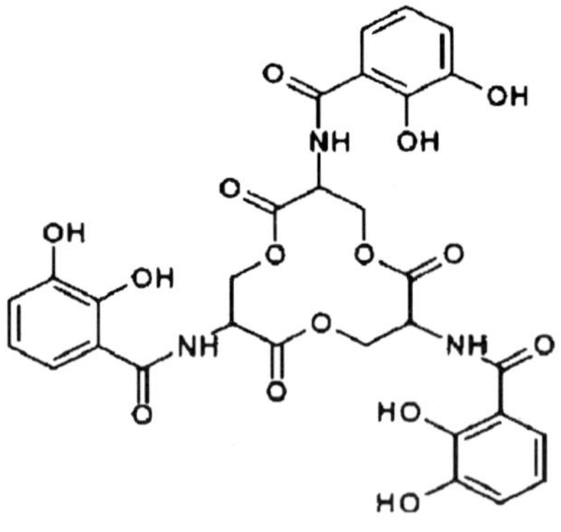

a

b

c

sodium dodecyl sulfate SDS-gel electrophoretic profiles of the two sources of proteins. There was no observed over-production of membrane proteins like that seen in bacterial systems (Payne 1994b, c). Because at least a fivefold difference in transport rates was observed, the scientists (authors) suggested that the membrane siderophore transport system is constitutively expressed (Van der Helm and Winkelmann 1994). In contrast to the work with *N. crassa*, iron-regulated outer membrane proteins have been found in the mycetoma-causing fungus, *Madurella mycetomatis* (Mezence and Boiron 1995). One unique protein and three amplified ones were observed under iron-limiting conditions (Mezence and Boiron 1995). The situation with other zoopathogens has not been explored.

1.2.2.8.2
Mechanism of Hydroxamate-Iron Transport

There are four described mechanisms of siderophore iron uptake across the cytoplasmic membrane of fungi. The following summary is modified from descriptions given by others (Carrano and Raymond 1978; Chung et al. 1986; Matzanke 1994a; Van der Helm and Winkelmann 1994).

1. Shuttle Mechanism. The intact siderophore-iron complex is taken into the cell. The iron is released by a reductase or by direct ligand exchange in which the recipient siderophore becomes the storage molecule. The gathering ligand is released to capture another iron molecule. This mechanism is the one used for uptake by siderophores of the coprogen and ferrichrome families (Matzanke 1994a; Winkelmann and Zahner 1973).
2. Direct-Transfer Mechanism. Iron is taken up without entrance of the ligand into the cell. The iron transfer is not a membrane-reductive event (Müller et al. 1985) but is a membrane-mediated exchange between the gathering siderophore and an internal chelating agent (Carrano and Raymond 1978). The transfer mechanism may be by ligand exchange (nonenzymatic) to an internal pool of the chelating agent, which then serves as the storage compound (Matzanke 1994a,b). This type of transfer has been reported with the rhodotorulic acid family of siderophores.
3. Esterase-reductase Mechanism. The esterase-reductase mechanism was shown to operate with the ferric triacetylfusarinine C (Winkelmann 1991, 1993). The ester bonds of the iron ligand are split after uptake of the ligand, the fusarinine moieties are excreted, and the ferric iron is reduced and stored (by an unknown mechanism).
4. Reductive Mechanism. Another reductive mechanism appears to be involved in the transport of some ferrichromes, which were shown not to enter cells but, rather, to give up ferric iron by reduction with transport of the ferrous iron (Winkelmann 1991, 1993). The storage mechanisms are not known. Fungi utilizing siderophores that they, themselves, do not synthesize commonly express this sort of transport.

1.2.2.8.3
Mechanism of Carboxylate Transport

The uptake of rhizoferrin has been studied Transport of the entire ligand is observed, and a Km value of 8 mM has been determined (Van der Helm and Winkelmann 1994). It is interesting that both ferrioxamines B and C are taken up at similar rates by *Rhizopus microsporus* var. rhizopodiformis, the species used to study rhizoferrin. This observation probably accounts for the cases of mucormycosis seen in patients treated with Desferal (desferrioxamine mesylate) (Tilbrook and Hider 1998; Boelaert et al. 1993). Phenolate transport has not been studied in either of the instances in which it was reported to occur in fungi (Fekete et al. 1989; Ismail et al. 1985).

1.3
Functions of Siderophores

Although their main function is to acquire iron from insoluble hydroxides or from iron adsorbed onto solid surfaces, siderophores can also extract iron from various other soluble and insoluble iron compounds, such as ferric citrate, ferric phosphate, Fe-transferrin, ferritin or iron bound to sugars, plant flavone pigments and glycosides or even from artificial chelators like EDTA and nitrilotriacetate by Fe(III)/ligand-exchange reactions. Thus, even if siderophores are not directly involved in iron solubilization, they are required as carriers mediating exchange between extracellular iron stores and membrane-located siderophore-transport systems.

The efficiency of siderophores in microbial metabolism is based mainly on three facts. (1) Siderophores contain the most efficient iron-binding ligand types in Nature, consisting of hydroxamate, catecholate or α-hydroxycarboxylate ligands that form hexadentate Fe(III) complexes, satisfying the six co-ordination sites on ferric ions. Moreover, siderophores possessing three bidentates in one molecule (iron-to-ligand ratio=1:1) show increased stability due to the chelate effect. (2) Regulation of siderophore biosynthesis is an economic means of spending metabolic energy, but it also allows for the production of high local concentrations of siderophores in the vicinity of microbial cells during iron limitation. This kind of overproduction may also be operating in host-adapted bacterial and fungal strains, leading to increased virulence. (3) Besides their ability to solubilize iron and to function as external iron carriers, siderophores exhibit structural and conformational specificities to fit into membrane receptors and/ or transporters. This has been amply demonstrated by modifying siderophore chemical structure, i.e. using derivatives, enantiomers, metal-replacement studies or by genetic and mutational analysis of receptors and membrane transporters (Stintzi et al. 2000; Huschka et al. 1986; Ecker et al. 1988).

1.4
Siderophores as Pathogenic Factor

1.4.1
Activities

1.4.1.1
Phytopathogens

Many siderophores are produced by both bacterial and fungal plant pathogens, but their role in pathogenesis is largely unknown (Loper and Buyer 1991). The siderophore of the bacterial phytopathogen *Erwinia chrysanthemi*, chrysobactin, has been identified as a virulence factor (Riquelme 1996). However, the fungal phytopathogen *Verticillium dahlia* produces the hydroxamate siderophores coprogen B and DA, but efforts to detect these compounds in plants grown under iron-limited conditions was unsuccessful (Barash et al. 1993).

1.4.1.2
Phytotoxins

Only a single example of phytotoxins is discussed. Stemphylotoxin I and II are produced by *Stemphilium botryosum* f. sp. *lycopersici* (Barash et al. 1982; Manulis et al. 1984) and are ferric iron chelators. Their formation depends on iron concentration but is not as stringently iron regulated, as are the hydroxamates, and their siderophore activity is thus secondary to their primary phytotoxicity (Manulis et al. 1984).

1.4.1.3
Wood-Rotting Fungi

The wood-rotting fungus *Gloeophyllum trabeum* produces iron-binding compounds "that appear to be phenolate in character" (Jellison and Goodell 1988). Partially purified iron-binding compounds were conjugated with bovine serum albumin and injected into rabbits (Fekete 1993). The antisera were then used to immunolocalize siderophore molecules in slices of wood infected with *G. trabeum* (Jellison and Goodell 1988, 1989; Fekete 1993; Jellison et al. 1991). Siderophores are indeed antigenic, and the approach by these workers is appealing in its potential application to mammalian systems.

1.4.1.4
Zoopathogenic Fungi

Studies to reveal the in vivo elaboration of siderophores by zoopathogens have not been performed, but the methods for revealing them directly or indirectly (see the section "Host Response").

1.4.2
Host Response

The siderophores are known to be antigenic (e.g., see "wood-rotting fungi," above). In fact it has been suggested that siderophores may not be effective iron scavengers in vivo because they bind to serum proteins and elicit an immune response (Reissbrodt et al. 1997). Such immunogenicity might compromise the efficiency of siderophores in iron gathering. The immune response is, however, also evidence of in vivo synthesis. For example, enterochelin-specific immunoglobulins are found in normal human serum (Moore et al. 1980). This siderophore is produced by a number of enteric bacteria. Therefore, it is not certain whether the occurrence of anti-enterochelin activity relates to a previous infection (e.g., with *Salmonella*) or is a response to *E. coli* resident in the bowel.

1.4.3
Clinical Applications

As naturally occurring chelating agents for iron, siderophores might be expected to be somewhat less noxious for deferrization of patients suffering from transfusion-induced siderosis. A siderophore from *Streptomyces pilosus*, desferrioxamine B, is marketed as the mesylate salt under the trade name Desferal and is advocated for removal of excess iron resulting from the supportive therapy for thalassemia. The drug must be injected, however, and an oral replacement is needed (Bergeron and Brittenham 1994).

The potency of common antibiotics has been elevated by building into the molecules the iron-binding functional groups of siderophores (Watanabe et al. 1987). The objective here is to take advantage of the high affinity, siderophore-mediated iron uptake system of the bacteria.

1.5
Agricultural Interest

Fluorescent pseudomonads form a line of siderophores comprised of a quinoline moiety, responsible for the fluorescence, and a peptide chain of variable length bearing hydroxamic acid and hydroxy acid functions. Capacity to form these pseudobactin or pyoverdine type siderophores has been associated with improved plant growth either through a direct effect on the plant, through control of noxious organisms in the soil, or via some other route. Nitrogenase can be said to be an iron-intensive enzyme complex and the symbiotic variety, as found in *Rhizobium* spp., may require an intact siderophore system for expression of this exclusively prokaryotic catalyst upon which all life depends.

1.6
Utilization of Siderophores by Nonproducers

In a number of instances, pathogenic fungi have been shown to use siderophores even though they cannot synthesize them. Some examples are given below.

S. cerevisiae. Ferrioxamine B, ferricrocin, and rhodotorulic acid are iron sources for S. cerevisiae. Two mechanisms for utilization of ferrioxamine B have been described: (i) at relatively high concentrations (360 µM), the iron is made available by reductive dissociation, while (ii) at low iron concentrations (7 µM), yeasts transport the entire iron-bearing ligand with iron into the cells and reductively remove the iron internally (Lesuisse and Labbe 1994).

C. neoformans. Growth of *C. neoformans* was stimulated by ferrioxamine B (160 µM) added to an iron-depleted medium (Jacobson and Petro 1987). Although diffusion of the siderophore would dilute the concentration from that applied to the paper disc used, the level of siderophores close to the disc would probably be high enough to indicate a reductive mobilization of Fe^{2+} from the ligand (by analogy to *S. cerevisiae*). Studies on very low concentrations of the siderophore were not conducted, but ligand uptake is also possible.

C. albicans. Ferrichrome and several of its constitutive peptides stimulated the growth of *C. albicans* in an iron-depleted medium (Minnick et al. 1991). There is no information on whether utilization involves reductive mobilization or ligand uptake or concentration-dependent utilization of one or the other mechanisms for iron recruitment.

U. maydis. Studies of the phytopathogen *Ustilago maydis* have provided fascinating evidence for utilization of both synthesized and nonsynthesized siderophores. Under iron-limiting conditions, *U. maydis* produces ferrichrome and ferrichrome B. However, under laboratory conditions, it also uses the bacterial siderophore ferrioxamine B (Ardon et al. 1997, 1998). Ferrichrome was taken up by entire-ligand transport, while the ferrioxamine B was utilized by reductive removal of the Fe(III) and transport of the Fe(II) (Ardon et al. 1998). Nonenzymatic NADH- and flavin mononucleotide-dependent reduction of ferric siderophores has also been recorded (Adjimani and Owusu 1997). Although utilization of an extraneous bacterial siderophore would not affect in vivo recruitment of iron by the phytopathogenic fungus, its saprophytic existence in soil could be influenced. The interest in such an occurrence is heightened by the fact that the concentration of ferrioxamine B in the soil can be as high as 0.1 μM (Lesuisse and Labbe 1994). The same arguments can be raised with regard to zoopathogenic fungi with a saprophytic form of growth that constitutes the infectious phase of the pathogen.

Rhizopus spp. The occurrence of mucormycosis in patients being treated with Desferal (a methanesulfonate salt of desferrioxamine) clearly indicates the utilization of this bacterial siderophore by *Rhizopus* sp. (Boelaert et al. 1993).

P. brasiliensis. The growth of *Paracoccidioides brasiliensis* is stimulated by coprogen B and DA synthesized by *B. dermatitidis* (Castaneda et al. 1988). *P. brasiliensis* may synthesize its own siderophores under conditions of iron stress, but appropriate studies have not been conducted.

Host Iron Proteins. In a mammalian host, iron is bound to transferrin and lactoferrin or stored in ferritin. No fungus has been reported to be able to remove iron from transferrin or lactoferrin, although several bacteria and some animal parasites can do so (Payne 1993; Wilson et al. 1994).

1.7
Acidification and Mobilization

Many fungi can grow anaerobically and will acidify their growth medium. It has been suggested for *Neurospora crassa* that, under acidic conditions, iron could accumulate at the cell surface and be mobilized by excreted hydroxy acids to supply iron to the cell (Winkelmann 1979). This means of iron acquisition has been mentioned for *S. cerevisiae* in its ecological locations in nature (Lesuisse and Labbe 1994). However, no evidence was found for participation of citric acid or other polycarboxylic acids in iron uptake by *C. neoformans* (Jacobson and Petro 1987).

1.8
Piriformospora indica and Siderophore

In an independent study, the interaction of *Piriformospora indica* with *Pseudomonas fluorescens* was observed. It was found that *Pseudomonas fluorescens* had completely blocked the growth of the fungus (Fig. 1.7). Towards *Pseudomonas*, the radius never increased more than 0.6 cm.

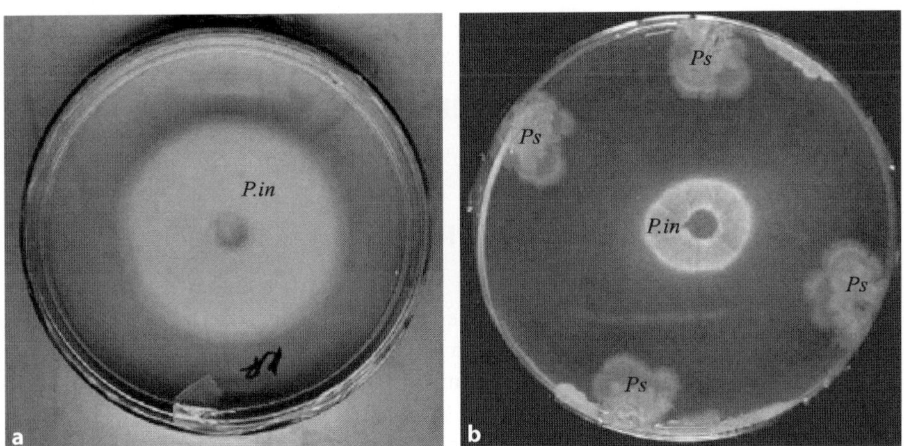

Fig. 1.7. a Axenic culture of *P. indica* on Aspergillus agar medium. **b** The interaction of *Piriformospora indica* with *Pseudomonas fluorescens* was observed in which *P. fluorescens* had completely blocked the growth of the fungus *P. indica*

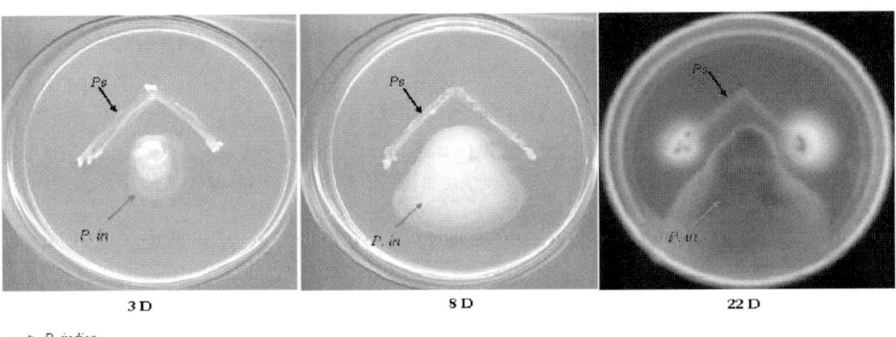

Fig. 1.8. Hyphal growth vs suppression assay showing the inhibitory effect of *Pseudomonas fluorescens* from Day-1 to Day-22 (release of fluorescent compound in the medium is clearly visible in Day-22 Petri plate)

In another set of experiments, *Pseudomonas fluorescens* colonies were raised in an angular manner on the petriplate and the fungus was placed in the centre. The growth of the later was completely blocked. After about three weeks, the synthesis of siderophore pyoverdine (and its derivatives) was detected It seems that the inhibition of fungus is multidirected (Fig. 1.8). In the beginning the inhibition was due to the synthesis of unidentified biomolecules and it was followed by production of siderophore. Identification of siderophore was based on the conventional spectrometry and spectrofluorimetry analysis.

1.9
Methods to Characterize Siderophore

1.9.1
Procedure for Detection of Siderophore

The iron salts procedure commonly used to detect siderophores is the method of Arkin et al. (Neilands and Nakamura 1991; Payne 1994a). The culture supernatants are mixed with a solution containing 5 mM $Fe(ClO_4)_3$ in 0.1 M $HClO_4$. The method is semiquantitative, and the amount of siderophore can be estimated by measurement of the optical density at 510 nm. The test may be used only as a rapid screening method because it lacks both specificity and sensitivity (Neilands and Nakamura 1991). Another colorimetric assay involves the use of the dye Chrome Asurol S (CAS) complexed with hexadecyltrimethylammonium bromide (HDTMA) (Neilands and Nakamura 1991; Payne 1994a; Fekete 1993). The removal of iron from CAS-HDTMA complex by a siderophore turns the dye yellow. The dye may also be included in a nutrient medium to measure the ability of an organism to remove iron from the dye complex (i.e., to make siderophores) by direct cultivation, a use that can be adapted to screening for non-siderophore-producing mutants (Neilands 1994). A caveat is, of course, that the medium must support the growth of the organism sometimes a bit tricky because of the potential toxicity of the HDTMA used to form the dye complex (Neilands and Nakamura 1991; Payne 1994b; Fekete 1993; Neilands 1994). Unfortunately, the CAS assay does have drawbacks. For example, phosphate strips iron off the dye, as also done by cysteine. Since there are several iron-binding substances that are not functional siderophores but nevertheless give positive color reactions with iron salts or CAS (Neilands and Nakamura 1991), the presence of hydroxamates or phenolates must be tested for; carboxylates are a different matter (Van der Helm and Winkelmann 1994). Hydroxamates may be detected by the Csáky assay (Neilands and Nakamura 1991; Payne 1994a), in which the end product detected is nitrite. The test is highly specific for hydroxamates (Neilands and Nakamura 1991). The Neilands assay (Neilands and Nakamura 1991) for hydroxamates identifies a *cis*-nitroso alkali dimer formed by periodic acid oxidation that is detected at 264 nm. The Neilands assay has a drawback of questionable specificity: the originator has said, "By working so deep in the ultraviolet it is difficult to have confidence that the material being measured is actually the desired dimer" (Neilands and Nakamura 1991). Phenolates-catecholates are identified by the Arnow assay (Neilands and Nakamura 1991; Payne 1994a), in which the centrifugate is treated with nitrous acid, molybdate, and alkali, which yields a pink compound that is detected at 515 nm. All of the color assays are subject to some uncertainties (Neilands and Nakamura 1991; Van der Helm and Winkelmann 1994) and chemical characterization is required to identify the assay reactivity as that of a siderophore (see the next section, on chemical structure).

The biological activity of a hydroxamate may be measured in a bioassay that employs a bacterium, *Aureobacterium* (*Arthrobacter*) *flavescens* JG.9 (ATCC 29091), which requires hydroxamate siderophores for mobilization of iron, as reflected by its growth (Neilands and Nakamura 1991; Fekete 1993). The bacterium uses a wide range of hydroxamates for its growth needs (Neilands and Nakamura 1991). Of course, *A. flavescens* may not have an appropriate receptor for a given compound. Thus, an assay procedure should eventually be developed that uses the siderophore-generating fungus to assess the biologic reactivity of the putative siderophore for the fungus from which it is isolated (Payne 1994b, c). There are bioassays for phenolates. These involve strains of *Salmonella typhimurium* or *Escherichia coli* defective in the biosynthesis of enterochelin (Neilands and Nakamura 1991).

Once it has been determined that a presumed siderophore is a hydroxamate (or phenolate), an excess of iron salts (often $FeCl_3$) is added to the harvested culture medium and the presence of red color is monitored through various procedures designed to isolate, purify, and characterize the compound(s) giving reactivity. This last step, chemical characterization, is essential. There are reports of siderophores for a given fungus that are based on the formation of a red color in the presence of iron salts or a yellow color in the presence of CAS. These initial color reactions for iron chelation may be followed by an assay for hydroxamates or phenolates that are again based on color formation, but no chemical structure studies have been presented It has been pointed out that "a variety of cellular materials from lysed cells may react positively in tests based on color formation" (Van der Helm and Winkelmann 1994). Thus, precise chemical characterization is required for further work with the putative siderophore. Such chemical characterization may be performed by thin-layer chromatography, high-performance liquid chromatography, nuclear magnetic resonance, and mass spectroscopy (Van der Helm and Winkelmann 1994).

1.9.2
Isolation of Siderophore

Since siderophores differ substantially in structure, no uniform procedure is available for their isolation. A preliminary examination by paper electrophoresis should reveal the charge profile as a function of pH, following which appropriate exchange resins can be applied for retention and elution of the compound(s). Most are water-soluble, and it is thus usually expedient to drive the siderophore into an organic solvent, such as benzyl alcohol or phenol-chloroform, in order to eliminate salt.

The siderophore may be isolated per se or as its iron chelate. The latter has the advantage of visual color, but the iron must be removed before any natural product can be characterized Vigorous hydrolysis in the presence of iron will destroy oxidizable moieties, and direct NMR analysis is ruled out by the paramagnetism of the ferric ion.

Structural characterization is best carried out by a combination of NMR and mass spectroscopy. Both of these techniques are sensitive and capable of providing absolute answers. Less than half of the known siderophores will crystallize; otherwise X-ray diffraction is the method of choice since it affords the configuration of those molecules containing a chiral center (Barash et al. 1993).

1.9.3
Cultivation of *Piriformospora indica*

The fungus is routinely cultivated on Aspergillus medium or Kaefer (Hill and Kaefer 2001). Composition of this medium is given in appendix. Fungal discs were prepared by using bottom of the sterile Pasteur pipette measuring about 4 mm in diameter and one disc was inoculated per petri-plate fortified with Kaefer medium containing 1% agar. These petri-plates were incubated at $28 \pm 2\,°C$ in dark. The growth normally commences on 3rd day and after 12 days the fungus completely covers the surface of the agar plate.

The broth culture of the fungus was made in Kaefer medium without agar. Five or more discs were transferred to 250-ml Erlenmeyer conical flasks containing 100 ml medium. The broth culture was incubated on Kühner rotary shaker at constant speed (150 rpm) and temperature $(28 \pm 2\,°C)$. The maximum growth was obtained after eight days of incubation.

1.10
Conclusions

Siderophores are common products of aerobic and facultative anaerobic bacteria and of fungi. Under conditions of iron starvation, many fungi synthesize low-molecular-weight iron chelators known as siderophores. Three common classes, phenolates, hydroxamates, and polycarboxylates, are examined. Some phytopathogenic fungi produce unique compounds that function as phytotoxins but also chelate iron.

Acknowledgments

The authors acknowledge the support and encouragement received from Dr. Ashok K Chauhan, Founder President, Ritnand Balved Education Foundation (an umbrella organization of Amity Institutions). The technical support received from Mr. Anil Chandra Bahukhandi is highly appreciated.

References

Adjimani JP, Owusu E (1997) Nonenzymatic NADH/FMN-dependent reduction of ferric siderophores. J Inorg Biochem 66:247–252

An Z, Mei B, Yuan WM, Leong SA (1997a) The distal GATA sequences of the sid1 promoter of Ustilago maydis mediate iron repression of siderophore production and interact directly with Urbs1, a GATA family transcription factor. EMBO J 16:1742–1750

An Z, Zhao Q, McEvoy J, Yuan WM, Markley JL, LongSA (1997b) The second finger of Urbs1 is required for iron-mediated repression of sid1 in Ustilago maydis. Proc Natl Acad Sci USA 94:5882–5887

Ardon O, Weizman H, Libman J, Shanzer A, Chen Y, Hadar Y (1997) Iron uptake in Ustilago maydis: studies with fluorescent ferrichrome analogues. Microbiology 143:3625–3631

Ardon O, Nudelman R, Caris C, Libman J, Shanzer A, Chen Y, Hadar Y (1998) Iron uptake in Ustilago maydis: tracking the iron path. J Bacteriol 180:2021–2026

Barash I, Pupkin G, Netzer D, Kashman Y (1982) A novel enolic -ketoaldehyde phytotoxin produced by Stemphylium botryosum f. sp. lycopersici. Plant Physiol 69:23–27

Barash I, Dori S, Mor H, Manulis S (1993) Role of iron in fungal phytopathologies, In: Iron chelation in plants and soil microorganisms. Academic Press, New York, pp 251–267

Bartholdy BA, Berreck M, Haselwandter K (2001) Hydroxamate siderophore synthesis by Phialocephala fortinii, a typical dark septate fungal root endophyte. BioMetals 14:33–42

Bentley MD, Anderegg RJ, Szaniszlo PJ, Davenport RF (1986) Isolation and identification of the principal siderophore of the dermatophyte Microsporum gypseum. Biochemistry 25:1455-1457

Bergeron RJ, Brittenham GM (eds) (1994) The development of iron chelators for clinical use. CRC Press, Boca Raton, FL

Böhnke R, Matzanke BF (1995) The mobile ferrous iron pool in Escherichia coli is bound to a phosphorylated sugar derivative. BioMetals 8:223–230

Boelaert JR, de Locht M, Van Cutsem J, Kerrels V, Cantinieaux B, Verdonck A, Van Landuty HW, Schneider YJ (1993) Mucormycosis during deferoxamine therapy is a siderophore-mediated infection. J. Clin. Investig 91:1979–1986

http://cmr.asm.org/cgi/external_ref?access_num=7647518&link_type=MEDBurt WR (1982) Identification of coprogen B and its breakdown products from Histoplasma capsulatum. Infect Immun 35:990–996

Byers BR, Arceneaux JEL (1998) Microbial iron transport: iron acquisition by pathogenic microorganisms. Metal Ions Biol Syst 35:37–66

Carrano CJ, Raymond KN (1978) Coordination chemistry of microbial iron transport compounds: rhodotorulic acid and iron uptake in Rhodotorula pilimanae. J Bacteriol 136:69-74

Carrano CJ, Böhnke R, Matzanke BF (1996) Fungal ferritins: the ferritin from mycelia of Absidia spinosa is a bacterioferritin. FEBS Lett 390:261–264http://cmr.asm.org/cgi/external_ref?access_num=8706873&link_type=MED

Carrano CJ, Jordan M, Drechsel H, Schmid DG, Winkelmann G (2001) BioMetals 14:119–125

Castaneda E, Brummer E, Perlman AM, McEwen JG, Stevens DA (1988) A culture medium for Paracoccidioides brasiliensis with high plating efficiency, and the effect of siderophores. J Med Vet Mycol 26:351–358

Charlang G, Ng B, Horowitz NH, Horowitz RM (1981) Cellular and extracellular siderophores of Aspergillus nidulans and Penicillium chrysogenum. Mol Cell Biol 1:94–100

Chung TDY, Matzanke BF, Winkelmann G, Raymond KN (1986) Inhibitory effect of the partially resolved coordination isomers of chromic desferricoprogen on coprogen uptake in Neurospora crassa. J Bacteriol 165:283–287

Cox CD (1994) Deferration of laboratory media and assays for ferric and ferrous iron. Methods Enzymol 235:315–329

Crosa JH (1997) Signal transduction and transcriptional and posttranscriptional control of iron-regulated genes in bacteria. Microbiol Mol Biol Rev 61:319–336

Cutler JE, Han Y (1996) Fungal factors implicated in pathogenesis. In: Esser K, Lemke PA (eds) The Mycota, vol. VI. Human and animal relationships. Springer, Berlin Heidelberg New York, pp 3–29

de Hoog GS, Marvin-Sikkema FD, Lahpoor GA, Gottschall JG, Prins RA, Guého E (1994) Ecology and physiology of the emerging opportunistic fungi Pseudallescheria boydii and Scedosporium prolificans. Mycoses 37:71–78

Deiss K, Hantke K, Winkelmann G (1998) BioMetals11:131–137

Drechsel H, Winkelmann G (1997) In: Winkelmann G, Carrano CJ (eds) Transition metals in microbial systems. Harwood Academic Publishers, Amsterdam, pp 1–49

Drechsel H, Jung G, WinkelmannG (1992) BioMetals 5:141–148

Ecker DJ, Loomis LD, Cass ME, Raymond KN (1988) J Am Chem Soc 110:2457–2464

Eisendle M, Oberegger H, Zadra I, Haas H (2003) The siderophore system is essential for viability of Aspergillus nidulans: functional analysis of two genes encoding l-ornithine N-5-monooxygenase (sidA) and a non-ribosomal peptide synthetase (sidC). Mol Microbiol 49:359–375

Eisendle M, Oberegger H, Buttinger R, Illmer P, Haas H (2004) Biosynthesis and uptake of siderophores is controlled by the PacC-mediated ambient-pH regulatory system in Aspergillus nidulans. Eukaryot Cell 3:561–563

Fekete FA (1993) Assays for microbial siderophores. In: Iron chelation in plants and soil microorganisms. Academic Press, New York, pp 399–417

Fekete FA, Spence JT, Emery T (1983) A rapid and sensitive paper electrophoresis assay for detection of microbial siderophores elicited in solid-plating culture. Anal Biochem 131:516–519

Fekete FA, Chandhoke V, Jellison J (1989) Iron-binding compounds produced by wood-decaying basidiomycetes. Appl Environ Microbiol 55:2720–2722

Ferguson AD, Hofmann E, Coulton JW, Diederichs K, Welte W (1998) Siderophore-mediated iron transport: crystal structure of FhuA with bound lipopolysaccharide. Science 282:2215–2220

Fleischmann J, Lehrer RI (1985) Phagocytic mechanisms in host response. In: Howard DH (ed) Fungi pathogenic for humans and animals. B. Pathogenicity and detection: II. Marcel Dekker, New York, pp 123–149

Frederick CB, Szaniszlo PJ, Vickrey PE, Bentley MD, Shive W (1981) Production and isolation of siderophores from the soil fungus Epicoccum purpurescens. Biochemistry 20:2432–2436

Frederick CB, Bentley MD, Shive W (1982) The structure of the fungal siderophore, isotriornicin. Biochem Biophys Res Commun 105:133–138

Giri B, Giang PH, Kumari R, Prasad R, Varma A (2005) In: Buscot F, Varma A (eds) Micro-organismsin soils: roles in genesis and functions. Springer, Berlin Heidelberg New York, pp 213–247

Griffiths E, Humphreys J (1980) Isolation of enterochelin from the peritoneal washings of guinea pigs lethally infected with Escherichia coli. Infect Immun 28:286–289

Guerinot ML (1994) Microbial iron transport. Annu Rev Microbiol 48:743–772

Haas B, Kraut J, Marks J, Zanker SC, Castignetti D (1991) Siderophore presence in sputa of cystic fibrosis patients. Infect Immun 59:3997–4000

Haas H (2003) Molecular genetics of fungal siderophore biosynthesis and uptake: the role of siderophores in iron uptake and storage. Appl Microbiol Biotechnol 62:316–330

Haas H, Zadra I, Stoffler G, Angermayr K (1999) The Aspergillus nidulans GATA factor SREA is involved in regulation of siderophore biosynthesis and control of iron uptake J Biol Chem 274:4613–4619

Haas H, Schoeser M, Lesuisse E, Ernst JF, Parson W, Abt B, Winkelmann G, Oberegger H (2003) Characterization of the Aspergillus nidulans transporters for the siderophores enterobactin and triacetylfusarinine C. Biochem J 371:505–513

Haselwandter K, Winkelmann G (2002) Ferricrocin – an ectomycorrhizal siderophore of Cenococcum geophilum. BioMetals 15:73–77

Haydon AH, Davis WB, Arceneaux JEL, Byers BR (1973) Hydroxamate recognition during iron transport from hydroxamate-iron chelates. J Bacteriol 115:912–918

Heymann P, Ernst JF, Winkelmann G (1999) Identification of a fungal triacetylfusarinine C siderophore transport gene (TAF1) in Saccharomyces cerevisiae as a member of the major facilitator superfamily. BioMetals 12:301–306

Heymann P, Ernst JF, Winkelmann G (2000) Identification and substrate specificity of a fer-richrome-type siderophore transporter (Arn1p) in Saccharomyces cerevisiae. FEMS Microbiol Lett 186:221–227

Heymann P, Ernst JF, Winkelmann G (2000) A gene of the major facilitator superfamily encodes a transporter for enterobactin (Enblp) in Saccharomyces cerevisiae. BioMetals 13:65–72

Hill TW, Kaefer E (2001) Improved protocols for Aspergillus medium: trace elements and minimum medium salt stock solution. Fungal Genet Newsl 48:20–21

Höfte M (1993) Classes of microbial siderophores. In: Barton LL (ed) Iron chelation in plants and soil microorganisms. Academic Press, New York, pp 3–26

Holzberg M, Artis WM (1983) Hydroxamate siderophore production by opportunistic and systemic fungal pathogens. Infect Immun 40:1134–1139

Howard DH (1999) Acquisition, transport,and storage of iron by pathogenic fungi. Clin Microbiol Rev 12:394–404

Huschka H, Jalal MAF, Van der Helm D, Winkelmann G (1986) J Bacteriol 167:1020–1024

Ismail A, Bedell GW, Lupan DM (1985) Siderophore production by the pathogenic yeast, Candida albicans. Biochem Biophys Res Commun 130:885–891

Jacobson ES, Petro MJ (1987) Extracellular iron chelation in Cryptococcus neoformans. J Med Vet Mycol 25:415–418

Jacobson ES, Goodner AP, Nyhus KJ (1998) Ferrous iron uptake in Cryptococcus neoformans. Infect Immun 66:4169–4175

Jalal MAF, van der Helm D (1991) Isolation and spectroscopic identification of fungal siderophores. In: Winkelmann G (ed) Handbook of microbial iron chelates. CRC Press, Boca Raton, Fl, pp 235–269

Jellison J, Goodell B (1988) Immunological detection of decay in wood. Wood Sci Technol 22:293–297

Jellison J, Goodell B (1989) Inhibitory effects of undecayed wood and the detection of Postia placenta using the enzyme-linked immunosorbent assay. Wood Sci Technol 23:13–20

Jellison J, Chandhoke V, Goodell B, Fekete FA (1991) The isolation and immunolocalization of iron-binding compounds produced by Gloeophyllum trabeum. Appl Biotechnol Microbiol 35:805–809

Kuhn S, Braun V, Koster W (1996) Ferric rhizoferrin uptake into Morganella morganii: characterization of genes involved in the uptake of a polyhydroxycarboxylate siderophore. J Bacteriol 178:496–504

Kwon-Chung KJ, Bennett JE (1992) Medical mycology. Lea & Febiger, Philadelphia

Labbe-Bois R, Camadro JM (1994) Ferrochelatase in Saccharomyces cerevisiae. In: Winkelmann G, Winge DR (eds) Metal ions in fungi, vol. 11. Marcel Dekker, New York, pp 413–453

Leong SA, Winkelmann G (1998) Molecular biology of iron transport in fungi. Metal Ions Biol Syst 35:147–186

Lesuisse E, Labbe P (1994) Reductive iron assimilation in Saccharomyces cerevisiae. In: Winkelmann G, Winge DR (eds) Metal ions in fungi, vol. 11. Marcel Dekker, New York, pp 149–178

Lesuisse E, Blaiseau PL, Dancis A, Camadro JM (2001) Siderophore uptake and use by the yeast Saccharomyces cerevisiae. Microbiology 147:289–298

Lesuisse E, Knight SA, Camadro JM, Dancis A (2002) Siderophore uptake by Candida albicans: effect of serum treatment and comparison with Saccharomyces cerevisiae. Yeast 19:329–340

Loper JE, Buyer JS (1991) Siderophores in microbial interactions on plant surfaces. Mol Plant-Microbe Interact 4:5–13

Manulis S, Kashman Y, Netzer D, Barash I (1984) Phytotoxins from Stemphylium botryosum: structural determination of stemphyloxin II, production in culture and interaction with iron. Phytochemistry 23:2193–2198

Manulis S, KashmanY, Barash I (1987) Identification of siderophores and siderophore-mediated uptake of iron in Stemphylium botryosum. Phytochemistry 26:1317–1320

Martinez JS, Zhang GP, Holt PD, Jung HT, Carrano CJ, Haygood MG, Butler A (2000) Science 87:1245–1247

Matzanke BF (1991) Structures, coordination chemistry and functions of microbial iron chelates. In: Winkelmann G (ed) Handbook of microbial iron chelates. CRC Press, Boca Raton, Fla, pp 15–64

Matzanke BF (1994a) Iron storage in fungi. In: Winkelmann G, Winge DR (eds) Metal ions in fungi, vol. 11. Marcel Dekker, New York, pp 179–214

Matzanke BF (1994b) Iron transport: siderophores. In: King RB (ed) Encyclopedia of inorganic chemistry, vol. 4, Iro-Met. Wiley, New York, pp 1915–1933

Matzanke BF, Bill E, Trautwein AX, Winkelmann G (1988) Ferricrocin functions as the main intracellular iron-storage compound in mycelia of Neurospora crassa. Biol Metals 1:18–25

Mei B, Budde AD, Leong SA (1993) sid1, a gene initiating siderophore biosynthesis in Ustilago maydis: molecular characterization, regulation by iron and role in phytopathogenicity. Proc Natl Acad Sci USA 90:903–907

Mezence MIB, Boiron P (1995) Studies on siderophore production and effect of iron deprivation on the outer membrane proteins of Madurella mycetomatis. Curr Microbiol 31:220–223

Minnick AA, Eizember LE, McKee JA, Dolence EK, Miller MJ (1991) Bioassay for siderophore utilization by Candida albicans. Anal Biochem 194:223–229

Moore DG, Yancey RJ, Lankford CE, Earhart CF (1980) Bacteriostatic enterochelin-specific immunoglobulin from normal human serum. Infect Immun 27:418–423

Moors MA, Stull TA, Blank KJ, Buckley HR, Mosser DM (1992) A role for complement receptor-like molecules in iron acquisition by Candida albicans. J Exp Med 175:1643–1651

Mor H, Pasternak M, Barash I (1988) Uptake of iron by Geotrichum candidum, a non-siderophore producer. BioMetals 1:99–105

Mor H, Kashman Y, Winkelmann G, Barash I (1992) Characterization of siderophores produced by different species of dermatophytic fungi Microsporum and Trichophyton. BioMetals 5:213–216

Morrissey JA, Williams PH, Cashmore AM (1996) Candida albicans has a cell-associated ferric reductase activity which is regulated in response to levels of iron and copper. Microbiology 142:485–492

Müller G, Barclay SJ, Raymond KN (1985) The mechanism and specificity of iron transport in Rhodotorula pilimanae probed by synthetic analogs of rhodotorulic acid. J Biol Chem 260:13916–13920

Neilands JB (1994) Identification and isolation of mutants defective in iron acquisition. Methods Enzymol 235:352–357

Neilands JB, Nakamura K (1991) Detection, determination, isolation, characterization and regulation of microbial iron chelates. In: Winkelmann G (ed) Handbook of microbial iron chelates. CRC Press, Boca Raton, Fla, pp 1–14

Newton SM, Allen JS, Cao Z, Qi Z, Jiang X, Sprencel C, Igo JD, Foster SB, Payne MA, Klebba PE (1997) Double mutagenesis of a positive charge cluster in the ligand-binding site of the ferric enterobactin receptor, FepA. Proc Natl Acad Sci USA 94:4560–4565

Oberegger H, Schoeser M, Zadra I, Abt B, Haas H (2001) SREA is involved in regulation of siderophore biosynthesis, utilization and uptake in Aspergillus nidulans. Mol Microbiol 41:1077–1089

Oberegger H, Zadra I, Schoeser M, Abt B, Parson W, Haas H (2002) Identification of members of the Aspergillus nidulans SREA regulon: genes involved in siderophore biosynthesis and utilization. Biochem Soc Trans 30:781–783

Oberegger H, Eisendle M, Schrettl M, Graessle S, Haas H (2003) 4'-Phosphopantetheinyl transferase-encoding npgA is essential for siderophore biosynthesis in Aspergillus nidulans. Curr Genet 44:211–215

Ong SA, Neilands JB (1979) Siderophores in microbially processed cheese. J Agric Food Chem 27:990–995

Ongena M, Jacques P, Delfosse P, Thonart P (2002) BioMetals15:1–13

Payne SM (1993) Iron acquisition in microbial pathogenesis. Trends Microbiol 1:66–68

Payne SM (1994) Detection, isolation, and characterization of siderophores. Methods Enzymol 235:329–344

Payne SM (1994) Effects of iron deprivation in outer membrane protein expression. Methods Enzymol 235:344–352

Payne SM (1994) Identification and isolation of mutants defective in iron acquisition. Methods Enzymol 235:352–356

Reissbrodt R, Rabsch W, Chapeaurouge A, Jung G, Winkelmann G (1990) BioMetals 3:54–60

Reissbrodt R, Kingsley R, Rabsch W, Beer W, Roberts M, Williams PH (1997) Iron regulated excretion of -keto acids by Salmonella typhimurium. J Bacteriol 179:4538–4544

Riquelme M (1996) Fungal siderophores in plant-microbe interactions. Microbiologia 12:537–546

Schrettl M, Bignell E, Kragl C, Joechl C, Rogers T, Arst HN Jr, Haynes K, Haas H (2004) Siderophore biosynthesis but not reductive iron assimilation is essential for Aspergillus fumigatus virulence. JEM 200:1213–1219

Shenker M, Hadar Y, Chen Y (1999) Kinetics of iron complexing and metal exchange in solutions by rhizoferrin, a fungal siderophore. Soil Sci Soc Am J 63:1681–1687

Smith SE, Read DJ (1997) Mycorrhizal symbiosis. Academic Press, San Diego, California, London, pp 605

Srivastava D, Kapoor R, Srivastava SK, Mukerji KG (1996) Vesicular arbuscular mycorrhiza: an overview. In: Mukerji KG (ed) Concepts in mycorrhizal research. Kluwer Academic Publishers, The Netherlands, pp 1–39

Stintzi A, Barnes C, Xu J, Raymond KN (2000) Proc Natl Acad Sci USA 97:10691–10696

Sweet SP, Douglas LJ (1991) Effect of iron concentration on siderophore synthesis and pigment production by Candida albicans. FEMS Microbiol Lett 80:87–92

Telford JR, Raymond KN (1996) Siderophores. In: Latwood J, Davis JED, MacNicol DD, Vogtle F (eds) Comprehensive supramolecular chemistry, vol. 1. Elsevier Science, Oxford, United Kingdom, pp 245–266

Tilbrook GS, Hider RC (1998) Iron chelators for clinical use. Metal Ions Biol Syst 35:691–730

Van der Helm D, Winkelmann G (1994) Hydroxamates and polycarboxylates as iron transport agents (siderophores) in fungi. In: Winkelmann G, Winge DR (eds) Metal ions in fungi, vol 11. Marcel Dekker, New York, pp 39–98

Voisard C, Wang J, McEvoy JL, Xu P, Leong SA (1993) urbs1, a gene regulating siderophore biosynthesis in Ustilago maydis, encodes a protein similar to the erythroid transcription factor GATA-1. Mol Cell Biol 13:7091–7100

Watanabe NA, Nagasu T, Katsu K, Kitoh K (1987) Antimicrob Agents Chemother 31:497–504

Wei H, Vienken K, Weber R, Bunting S, Requena N, Fischer R (2004) A putative high affinity hexose transporter, hxtA, of Aspergillus nidulans is induced in vegetative hyphae upon starvation and in ascogenous hyphae during cleistothecium formation. Fungal Genet Biol 41:148–156

Wilson ME, Vorhies RW, Anderson KA, Britigen BE (1994) Acquisition of iron from transferrin and lactoferrin by the protozoan Leishmania chagasi. Infect Immun 62:3262–3269

Winkelmann G (1979) Surface iron polymers and hydroxy acids. A model of iron supply in sideramine-free fungi. Arch Microbiol 121:43–51

Winkelmann G (1991) Specificity of iron transport in bacteria and fungi. In: Winkelmann G (ed) Handbook of microbial iron chelates. CRC Press, Boca Raton, Fla, pp 65–105

Winkelmann G (1993) Kinetics, energetics, and mechanisms of siderophore iron transport in fungi. In: Barton LL (ed) Iron chelation in plants and soil microorganisms. Academic Press, New York, pp 219–239

Winkelmann G (2001) Microbial siderophore-mediated transport. Biochem Soc Trans 30:691–696

Winkelmann G, Zahner H (1973) Stoffwechselprodukte von Mikroorganismen. 115, Mitteilung. Eisenaufnahme bei Neurospora crassa. I. Zur Spezifitat des Eisentransportes. Arch Mikrobiol 88:49–60

Winkelmann G, Schmid DG, Nicholson G, Jung G, Colquhoun DJ (2002) BioMetals 15:153–160

Worsham PL, Goldman WE (1988) Quantitative plating of Histoplasma capsulatum without addition of conditioned medium or siderophores. J Med Vet Mycol 6:137–143

Yehuda Z, Shenker M, Romheld V, Marschner H, Hadar Y, Chen Y (1996) The role of ligand exchange in the uptake of iron from microbial siderophores by gramineous plants. Plant Pysiology 112:1273–1280

Yun CW, Tiedeman JS, Moore RE, Philpott CC (2000) Siderophore-iron uptake in Sacharomyces cerevisiae. Identification of ferrichrome and fusarinine transporters J Biol Chem 275:16354–16359

Yun CW, Bauler M, Moore RE, Klebba PE, Philpott CC (2001) The role of the FRE family of plasma membrane reductases in the uptake of siderophore-iron in Saccharomyces cerevisiae. J Biol Chem 276:10218–10223

Appendix

Kaefer medium or Aspergillus Broth or Aspergillus Medium (Hill and Kafer 2001)

Chemicals	Composition
Glucose	20 g/L
Peptone	2 g/L
Yeast extract	1 g/L
Casamino acid	1 g/L
Vitamin stock solution	1 ml
Macroelements from stock	50 ml
Microelements from stock	2.5 ml
Agar	10 g
$CaCl_2$ 0.1 M	1 ml
$FeCl_3$ 0.1 M	1 ml
pH	6.5

Macroelements (major elements)	Stock g/L
$NaNO_3$	120.0
KCl	10.4
$MgSO_4.7H_2O$	10.4
KH_2PO_4	30.4

Microelements (Trace elements)	Stock g/L
Zn $SO_4.7H_2O$	22.0
H_3BO_3	11.0
$MnCl_2.4H_2O$	5.0
$FeSO_4.7H_2O$	5.0
$CoCl_2.6H_2O$	1.6
$CuSO_4.5H_2O$	1.6
$(NH_4)_6 Mo_7O_{27}.4H_2O$	1.1
Na_2EDTA	50.0

Vitamins	Percent
Biotin	0.05
Nicotinamide	0.50
Pyridoxal phosphate	0.10
Amino benzoic acid	0.10
Riboflavin	0.25

pH of the medium was adjusted to 6.5 with 1 N HCl
All the stocks were stored at 4 °C except vitamins, which are stored at −20 °C

2 Siderotyping and Bacterial Taxonomy: A Siderophore Bank for a Rapid Identification at the Species Level of Fluorescent and Non-Fluorescent Pseudomonas

Jean-Marie Meyer

2.1
Introduction

Bacteria belonging to the genus *Pseudomonas* are largely distributed in nature and can be isolated from most environments including soil, plant rhizosphere and phylosphere, or water. A few species are pathogens for animals, e.g., *Pseudomonas plecoglossicida* (Nishimori et al. 2000) or are, like the genus type-species *Pseudomonas aeruginosa*, opportunist human pathogens implicated in severe illnesses like cystic fibrosis. A greater number are plant pathogens, mainly found on the surfaces of plant leaves and stems such as *Pseudomonas syringae* and the related species *Pseudomonas amygdali*, *Pseudomonas avellanae*, *Pseudomonas cannabina*, *Pseudomonas ficuserectae*, *Pseudomonas meliae*, *Pseudomonas savastanoi*, *Pseudomonas tremae* and *Pseudomonas viridiflava* (Gardan et al. 1999). Others, e.g., *Pseudomonas palleroniana* and *Pseudomonas salomonii* (Gardan et al. 2002), or *Pseudomonas tolaasii* and *Pseudomonas costantinii* (Munsch et al. 2002), have been associated with serious crop damages affecting rice, garlic or mushrooms, respectively. However, most of the *Pseudomonas* spp. remain to be considered as non-pathogenic saprophytic bacteria, harboring for many of them behaviours of biotechnological interests such as chemical bioremediation, crop protection or plant growth promotion.

In soil, pseudomonads represent one of the most important Gram-negative genera among culturable aerobic bacteria usually found. According to student lab courses done on soil samples over many years in our laboratory, 1–10% of soil isolates correspond to bacteria easily recognized as fluorescent *Pseudomonas* thanks to the yellow-green, highly fluorescent halo existing around such colonies growing on the iron-poor Casamino acid (CAA)-agar medium. Some

Département Microorganismes, Génomes, Environnement, Université Louis Pasteur-CNRS, UMR 7156, 28 rue Goethe, 67000 Strasbourg, France,
email: meyer@gem.u-strasbg.fr

Soil Biology, Volume 12
Microbial Siderophores
A. Varma, S.B. Chincholkar (Eds.)
© Springer-Verlag Berlin Heidelberg 2007

of these bacteria have been recognized as efficient plant growth promoting rhizobacteria (PGPR), exercising a powerful biological control on soil-borne phytopathogens (Haas and Défago 2005). Iron competition was first thought to be the major mechanism responsible of the antagonistic effects of pseudomonads towards pathogenic microorganisms (Kloepper et al. 1980). However, antifungal antibiotic synthesis as well as plant elicitor production suggested that the pseudomonad PGPR effects were not restricted to siderophore-mediated iron competition (Haas and Défago 2005).

Diversity within the genus *Pseudomonas*, first mainly established on the basis of a restricted panel of phenotypic traits (Palleroni 1984; Bossis et al. 2000), has been largely developed since the last decade with more than 30 new fluorescent *Pseudomonas* species described thanks to the application of polyphasic taxonomy (Vandamme et al. 1996). Today, the definition of a novel *Pseudomonas* species or, more frequently, the characterization of a *Pseudomonas* isolate at the species level, requires, in order to be successful, the use of numerous phenotypic and genomic methods, which means much investment in time, labor, material and money. Such practical considerations should explain why some of the earliest defined and most important species like *Pseudomonas fluorescens* or *Pseudomonas putida* are still commonly in use, although well recognized by taxonomists as groups formed by a complex of species (Grimont et al. 1996; Yamamoto and Harayama 1998; Bossis et al. 2000). Siderotyping methods, in use in our laboratory for many years, proved to be rapid, accurate and inexpensive tools for pseudomonad characterization and identification (Meyer et al. 1997). It already helped in the characterization of 8 of these new *Pseudomonas* species (see below) and allowed the recognition of 9 sub-groups among a collection of 28 *P. fluorescens* strains belonging to the biovar I of the species (Meyer et al. 2002b). Interestingly, three of these sub-groups correlated with newly described *Pseudomonas* species, i.e., *Pseudomonas jessenii*, *Pseudomonas mandelii* and *Pseudomonas rhodesiae*, respectively, while the most important sub-group by number of isolates was considered as corresponding to the *P. fluorescens* sensu stricto species, since containing the type strain *P. fluorescens* ATCC 13525 among its representatives. Other sub-groups could be considered as potential new species which remain to be defined. Such an example well illustrates the efficiency and usefulness of siderotyping in *Pseudomonas* biodiversity studies as well as in taxonomy studies.

2.2
Siderophores of *Pseudomonas*

Siderotyping is primarily based on the characterization of the so-called siderophores (Neilands 1981), which are molecules excreted by numerous microorganisms growing under iron deficiency. Structurally, they are especially designed to chelate strongly traces of iron(III) ions present in the microbial environment and to transport them into the cells thanks to specialized membrane receptors. An overview on siderophores and iron transport in bacteria has recently been published to which the reader is referred (Crosa et al. 2004).

Table 2.1 illustrates the diversity encountered among the siderophores already recognized among pseudomonads. According to our own experience, it should represent only a small part of the siderophores produced by these bacteria, and many of them remain to be discovered and to be structurally characterized.

The most important ones, the pyoverdines, characterizing the so-called fluorescent *Pseudomonas*, are molecules made of three different parts which are i) a fluorescent chromophore based on a quinolin cycle responsible for the yellow-green color and bright fluorescence of pyoverdine; ii) a side-chain made of a carboxylic acid residue branched to an amino group of the chromophore and iii) a peptidic chain attached to a carboxyl group on the chromophore. The huge diversity which prevails within the pyoverdine family, with the description of so far close to 50 different compounds, comes from that third part which varies from one pyoverdine to another by the number (from 6 to 12) and types

Table 2.1. *Pseudomonas* siderophores

Siderophores	Producing strains	References
Fluorescent pseudomonads		
Pyoverdines	All	Budzikiewicz (2004)
Pyochelin	*P. aeruginosa*	Cox et al. (1981)
Salicylic acid	*P. fluorescens* CHA0 *Pseudomonas* spp. *P. fluorescens* AH2 *P. fluorescens* WCS 374	Meyer et al. (1992) Visca et al. (1993) Anthony et al. (1995) Mercado-Blanco et al. (2001)
Quinolobactin	*P. fluorescens* ATCC17400	Mossialos et al. (2000)
Pseudomonine	*P. fluorescens* AH2 *P. fluorescens* WCS 374	Anthony et al. (1995) Mercado-Blanco et al. (2001)
Pyridine-2,6-bis (monothio-carbo xylic acid) (PDTC)	*P. putida* DSM 3601 *P. putida* DSM 3602	Lewis et al. (2004)
Non-fluorescent pseudomonads		
Corrugatin	*P. corrugata*	Risse et al. (1998)
Desferriferrioxamine E	*P. stutzeri* ATCC 17488	Meyer and Abdallah (1980)
Catechol-type	*P. stutzeri* RC7	Chakraborty et al. (1990)
PDTC	*P. stutzeri* KC	Lewis et al. (2004)
Aerobactin	*Pseudomonas* sp.	Buyer et al. (1991)
Tropolone	*Pseudomonas* sp.	Lindberg et al. (1980)
Unnamed (pHi 7.2)	*P. plecoglossicida*	Meyer et al. (2002b)
Unnamed (pHi 5.9)	*P. graminis*	Meyer et al. (2002b)
Unnamed (pHi 3.9)	*P. frederiksbergensis*	Meyer et al. (2002b)

of amino acyl residues. For a comprehensive and detailed description of these molecules, see the recent review by Budzikiewicz 2004.

Besides pyoverdines, a few other structurally unrelated siderophores have been described in various fluorescent *Pseudomonas* strains, i.e., pyochelin found in almost if not all *P. aeruginosa* isolates, salicylic acid, quinolobactin and pseudomonine found in a few *P. fluorescens* strains, whereas pyridine-2,6-bis(monothiocarboxylic acid) (PDTC) was described in two *P. putida* isolates (Table 2.1). Concerning the non-fluorescent *Pseudomonas*, other molecules have usually been described. However, many investigations remain to be done in the field since only 4 of the 20 presently well defined species have been extensively analyzed for siderophore production: corrugatin, an original lipopeptidic siderophore, has been described in several *Pseudomonas corrugata* isolates (Risse et al. 1998; Meyer et al. 2002b), whereas three still structurally unknown compounds, presently defined by their respective pHi values, characterized the *Pseudomonas plecoglossicida*, *Pseudomonas graminis* and *Pseudomonas frederiksbergensis* species, respectively (Meyer et al. 2002b). The few other siderophores cited in Table 2.1 were described as produced by strains of *Pseudomonas stutzeri*, a notorious heterogeneous complex of species (Cladera et al. 2004), or by uncharacterized isolates.

2.3
Siderotyping Methods

Table 2.2 gives an overview of the different methods which could be used to characterize these siderophores or any siderophore produced by a given microorganism grown under iron starvation. These methods can be roughly subdivided into two major groups, i.e., the analytical and the biological methods. Amongst the first group, methods which have been used to differentiate siderophores based on their physico-chemical properties are mainly isoelectrofocusing (IEF), high performance liquid chromatography (HPLC) or HPLC coupled with electrospray mass spectrometry (HPLC/ES-MS). The latter has the advantage to determine the molecular mass of the siderophore molecules and to give also some structural information. Other analytical methods concern a quantitative analysis of the amino acid content of siderophores (limited, indeed, to those siderophores containing a peptidic chain, mainly the pyoverdines and to a less extent the ornibactins) and a recently introduced and promising molecular biology method based on the recognition of siderophore-related specific DNA sequences (presently limited to the few fully sequenced pseudomonads).

The biological methods are mainly based on the use of siderophores as iron-transporters in experiments directly measuring the resulting incorporation of labeled iron into iron-starved cells, or in growth experiments where the siderophore-mediated iron feeding is detected thanks to its growth-stimulating effects. In order to easily detect such growth stimulations, the presence in the biological

Table 2.2. Advantages and disadvantages of the main siderotyping methods

Siderotyping methods	Principes	Advantages (+)/ disadvantages (−)	References
Analytical methods			
IEF	Comparison of the isoelectro-phoretic pattern of siderophores	(+): easy, fast, inexpensive; up to 20 strains analyzed/gel (−): very close IEF patterns in some cases; requires standardized culture conditions	Koedam et al. (1994) Meyer et al. (1997) Meyer (1998) Fuchs et al. (2001)
Amino acid analysis	Quantification of the aa content of the siderophore peptidic part by hydrolysis and aa analysis	(+): partial structure determination (−): expensive; requires highly puri-fied siderophores; limited to aa-contain-ing siderophores	Cornelis et al. (1989) Bultreys and Gheysen (2000)
HPLC	Comparison of the HPLC pattern of siderophores	(+): easy; sensitive (−): expensive, time consuming; use of ferrisiderophores	Meyer et al. (1992) Meyer et al. (1995) Bultreys et al. (2003)
HPLC/ES-MS	Characterization of siderophores (pyoverdines) in culture broth by HPLC coupled with electrospray mass spectrometry	(+): molecular mass determination and partial struc-tural information (−): expensive; use of ferripyoverdines	Kilz et al. (1999) Fuchs et al. (2001)
RAPD/PCR	Fingerprinting by DNA amplifica-tion using specific siderophore-related DNA sequences	(+): easy; fast; sensitive (−): expensive; restricted to sid-erotypes of DNA-sequenced strains	DeVos et al. (1997) Smith et al. (2005)
Biological methods			
Cross-feeding	Stimulation of bacterial growth by siderophore sup-plementation on plate or in liquid growth medium containing an iron-chelator	(+): easy, fast, inexpensive (−): strain sensitiv-ity to iron-chelator, false-positifs due to heterologous and/or inducible iron uptake systems	Cornelis et al. (1989) Meyer et al. (1997, 1998)
Fe-siderophore-mediated uptake	Comparison of bacterial sidero-phore-mediated iron uptake specificity as determined after incubation of cells with labeled iron-siderophore complex	(+): easy, fast, sensitive (−): radioactivity handling, cross-up-take with structur-ally closely related siderophores	Hohnadel and Meyer (1988) Meyer et al. (1997) Meyer (1998)

Table 2.2. *(continued)* Advantages and disadvantages of the main siderotyping methods

Siderotyping methods	Principes	Advantages (+)/ disadvantages (−)	References
Biological methods			
Receptor immunoblotting	Specific immunological detection of the ferrisiderophore outer membrane receptor	(+): sensitive (−): time-consuming and expensive (OM purification, SDS-PAGE, Western blot, antisera...)	Cornelis et al. (1989) Meyer et al. (1990, 1997)

assays (on agar plates or in liquid cultures) of a strong iron chelator like ethylene-diamino-dihydroxyphenyl acetic acid (EDDHA) is highly recommended. It chelates the residual (contaminant) iron present in all culture media and, not being usable by the bacteria, it increases the iron-deficiency of the cells, thus lowering or fully inhibiting, depending on the concentration, the bacterial growth. Unfortunately, the purity grade of the iron-free EDDHA is no longer commercially available, although the iron-complex form could still be found on the phytosanitizing market, sold as a fertilizer (purification and especially complete removal of the complexed iron remain, however, to be done). The indirect recognition of siderophores through the immunological detection of their respective outer membrane receptors is also cited in Table 2.2, although it has been used for a very limited number of siderophores only (i.e., the three pyoverdines characterizing the *P. aeruginosa* species), the method being particularly time-consuming and expensive. A precise technical description of each of these different siderotyping methods will not be detailed in the present chapter and could be found in the references cited in Table 2.2. Emphasis will be given, however, on the two methods which have been selected in the author's laboratory as the most convenient ones, namely IEF and siderophore-mediated (^{59}Fe)-iron uptake.

2.3.1
Siderotyping Through Isoelectrofocusing

Molecules able to complex mineral ions like Fe^{3+} have electric charges which result in their migration in an electric field and therefore allow the characterization, thanks to isoelectrophoresis, of their respective isoelectric pH values (pHi or pI values). The isoelectrofocusing method was first applied to siderophores in a work focused on pyoverdines (Koedam et al. 1994). Such siderophores are easily detected on a polyacrylamide gel by their strong and characteristic fluorescence when illuminated under UV light at 365 nm. Interestingly, the electrophoresed gel could be also covered by a Chrome-Azurol S (CAS)-containing agarose overlay which specifically reacts with siderophore molecules (Schwyn and Neilands 1987). Thus, the IEF method could be extended to all siderophores and, therefore, with regard to the *Pseudomonas*, can also be applied to the study of the non-fluorescent species.

Figure 2.1 shows different IEF patterns obtained by analyzing through isoelectrophoresis the pyoverdines produced by 15 fluorescent *Pseudomonas* strains. Several fluorescent bands are visualized for each deposit, representing the different pyoverdine isoforms of an otherwise identical molecule (Budzikiewicz 2004). The pHi values, the number as well as the intensity of the bands, vary from one strain to another, depending on the type of pyoverdine they produce. It is of interest to note that the IEF patterns of pyoverdines cover the full experimental pH range (3.9 to 9.2) developed in the polyacrylamide gel containing a pH 3–10 ampholine mix. Some patterns present acidic bands exclusively (e.g., deposits 13 and 15), others are characterized by one or more acidic band(s) and one supplementary band in the middle of the gel (neutral pH; ex: deposits 1 to 3). A third type of IEF patterns is composed of band(s) with neutral and alcaline pHi (deposits 5 and 7–12), while some other less frequent pyoverdines dispatch bands within the three pH zones, as illustrated in Fig. 2.1 by deposit 6. As can be expected, these differences within pyoverdine IEF patterns reflect differences within pyoverdine structures, at the level of the different peptidic part of the molecules, as well as the level of the carboxylic side chains. Conversely, strains showing an identical pyoverdine-IEF pattern, as for deposits 7–12 or 13 and 15 in Fig. 2.1, were demonstrated to produce an identical pyoverdine (Fuchs et al. 2001).

Indeed, some pyoverdines could present so close IEF patterns that their differentiation becomes problematic. This is illustrated in Fig. 2.1 when comparing deposits 2 and 3: the presence of a supplementary band at acidic pH is visible for deposit 3, while the upper band in deposit 2 is very faint compared to the same in deposit 3. Do these small differences reflect changes at the structure level? To answer the question, another siderotyping method is required.

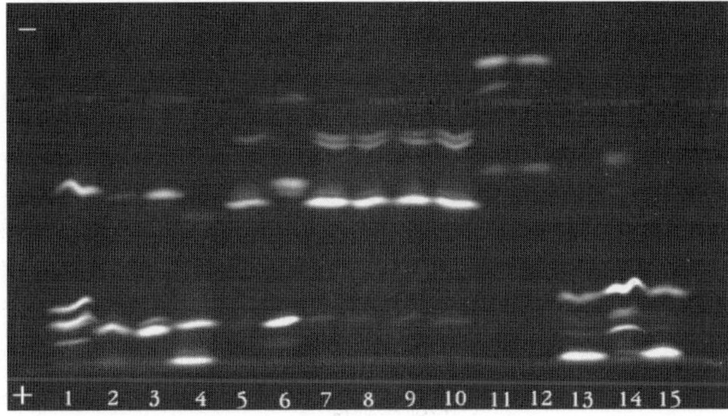

Fig. 2.1. Pyoverdine-isoelectrofocusing pattern of 15 fluorescent pseudomonad strains

2.3.2
Siderotyping Through Siderophore-Mediated Iron Uptake

A strict specificity concerning siderophore-mediated iron transport is usually observed among *Pseudomonas* strains and their respective siderophores. This general rule has been well established for pyoverdines (Hohnadel and Meyer 1988; Meyer et al. 1997), as well as for other siderophores produced by fluorescent or non-fluorescent *Pseudomonas* (Meyer et al. 1992, 2002b). Thus, it can be deduced that the siderophore-mediated iron uptake capacity of one bacterial strain towards a siderophore of foreign origin can give informations concerning the structure of the compound: the bacteria will use it only if it is identical to their own siderophore. Deciding whether two pyoverdines with closely related IEF patterns like deposits 2 and 3 in Fig. 2.1, are identical or not could thus be resolved very easily by determining their cross-reactivity in iron uptake experiments involving the two compounds and their respective producer strains.

The procedure is very rapid, requiring only a 20-min incubation of iron-starved cells with the labeled iron-siderophore complex, followed by a quick filtration step allowing the radioactivity incorporated into the cells during the incubation time to be counted. Just 1 ml of bacterial culture is usually enough for several assays. Thus, cells can be harvested and washed by eppendorf-sized centrifugation. Furthermore, there is no need at this stage to purify the siderophore: the labeled iron-siderophore complex is prepared de novo by mixing the label source with culture supernatant. This is possible particularly when working with pyoverdines. A detailed procedure has been recently described (Meyer and Geoffroy 2004).

As shown in Table 2.3 describing 25 strains and their structurally different pyoverdines, 14 of them corresponded to strains presenting a strict specificity of recognition towards their own pyoverdines. Five others were able to use a supplementary pyoverdine, while six strains used two pyoverdines of foreign origin. It is clear from Table 2.4 that cross-reactivity is explained by a strong structural similarity at the peptide level (with common amino acid residues marked in bold characters) in between the pyoverdines of concern. Moreover, these partial identities could tell us which part of the peptide chain could be the recognition site towards the ferripyoverdine outer membrane (OM) receptor. As argued in Meyer et al. 2002a, the glycine-allo-threonine motif present at the same position in the three pyoverdines of *P. fluorescens* PL7, *P. fluorescens* PL8 and *P. aeruginosa* ATCC 27853, is very likely the recognition site between these pyoverdines and their respective OM receptors.

Table 2.3. Structures of 25 pyoverdines detected by siderotyping as novel pyoverdines and their respective iron-uptake specificity

Strains	Pyoverdine peptide chain[a]	Crossreactivity with pyoverdine of strain	References
P. chlororaphis D-TR133	Asp-FOHOrn-Lys-(Thr-D/LAla-L/DAla-FoOHOrn-Ala)	Pfl CHA0	Barelmann et al. (2003)
P. costantinii CFBP 5705	Ser-AOHOrn-Gly-aThr-Thr-Gln-Gly-Ser cOHOrn	Pfl A6	Fernandez et al. (2001)
P. fluorescens 9AW	Ser-Lys-OHHis-aThr-Ser cOHOrn	None	Budzikiewicz et al. (1997)
P. fluorescens 51W	Ala-Lys-Gly-Gly-OHAsp-Gln-Ser-Ala-Gly-aThr-cOHOrn	None	Voss et al. (1999)
P. fluorescens 1.3	Ala-Lys-Gly-Gly-OHAsp-Gln-Dab-Gly-Ser-cOHOrn	Pp gwose, Pfl ATCC 17400	Georgias et al. (1999)
P. fluorescens PL7	Ser-AOHOrn-Ala-Gly-aThr-Ala-cOHOrn	Pfl PL8, Pa ATCC 27853	Barelmann et al. (2002)
P. fluorescens PL8	Lys-AOHOrn-Ala-Gly-aThr-Ser-cOHOrn	Pfl PL7, Pa ATCC 27853	Barelmann et al. (2002)
P. fluorescens PL9	Ser-AOHOrn-Ala-Gly-(Ser-Ser-OHAsp-Thr)	*Pseudomonas* sp. A214	Vossen et al. (2000)
P. fluorescens 18.1	Ser-Lys-Gly-FOHOrn-Ser-Ser-Gly-Lys-FOHOrn-Ser	Pfl DSM 50106	Amann et al. (2000)
P. fluorescens A6	Lys-AOHOrn-Gly-aThr-Thr-Gln-Gly-Ser-cOHOrn	*P. costantinii*	Beiderbeck et al. (1999)
P. fluorescens G170	Ser-Lys-Ala-AOHOrn-Thr-Ala-Gly-Gln-Ala-Ser-Ser-cOHOrn	None	Ruangviryachai et al. (2000)
P. fluorescens G173	Ser-Ala-AOHOrn-(Orn-Asp-AOHOrn-Ser)	None	Fernandez et al. (2003a)
P. fluorescens G176	Asp-Lys-OHAsp-Ser-Ala-Ser-cOHOrn	None	Budzikiewicz et al. (1999)
P. marginalis G76	Ser-Ser-FOHOrn-Ser-Ser-(Lys-FOHOrn-Ser)[b]	None	Budzikiewicz (2004)

[a] Abbreviations: usual amino acids, three letter code; aThr, *allo*-Thr; εLys, Lys linked by its ε-NH2, AOHOrn, δ-N-acetyl-δ-N-hydroxy-ornithine; FOHOrn, δ-N-formyl-δ-N-hydroxy-ornithine; cOHOrn, *cyclo*-hydroxy-ornithine (3-amino-1-hydroxy-piperidone-2); OHHis, *threo*-β-hydroxy-histidine; OHAsp, *threo*-β-hydroxy-aspartic acid; Dab, diamino-butanoic acid; Pfl, *Pseudomonas fluorescens*; Pa., *Pseudomonas aeruginosa*; Pp, *Pseudomonas putida*. The D-amino acids are underlined

[b] Incomplete structures (stereochemistry is missing)

Table 2.3. *(continued)* Structures of 25 pyoverdines detected by siderotyping as novel pyoverdines and their respective iron-uptake specificity

Strains	Pyoverdine peptide chain [a]	Crossreactivity with pyoverdine of strain	References
P. putida L1/WCS358	Asp-εLys-OHAsp-Ser-aThr-Ala-Thr-Lys-cOHOrn	None	Fernandez et al. (2003b)
P. putida CFML 90-33	Asp-Lys-Thr-OHAsp-Thr-aThr-cOHOrn	None	Sultana et al. (2001a)
P. putida CFML 90-44	Asp-Lys-AOHOrn-Thr-Ser-Ser-Gly-Ser-Ser-cOHOrn	None	Sultana et al. (2001b)
P. putida CFML 90-51	Asp-Lys-OHAsp-Ser-Gly-aThr-Lys-cOHOrn	None	Sultana et al. (2000b)
P. rhodesiae CFML 92-104	Ser-Lys-FOHOrn-Ser-Ser-Gly-c(Lys-FOHOrn-Ser-Ser)[b]	*P. veronii*	Budzikiewicz (2004)
P. syringae ATCC 19310	εLys-OHAsp-Thr-Thr-Ser-OHAsp-Ser	*P. cichorii*	Jülich et al. (2001)
Pseudomonas sp. D47	Ser-Orn-FOHOrn-(Lys-FOHOrn-Glu-Ser)[b]	None	Budzikiewicz (2004)
Pseudomonas sp. CFML95-275	Ser-Ser-FOHOrn-Ser-Ser-(Lys-FOHOrn-Lys-Ser)	None	Sultana et al. (2000a)
Pseudomonas sp. CFML96-188	Ser-Lys-FOHOrn-(Lys-FOHOrn-Glu-Ser)	*P. rhodesiae, P. veronii*	Weber et al. (2000)
Pseudomonas sp. CFML96-312	Ser-Ser-FOHOrn-(Lys-FOHOrn-Lys-Ser)	None	Schlegel et al. (2001)
Pseudomonas sp. CFML96-318	Ser-Orn-FOHOrn-Ser-Ser-(Lys-FOHOrn-Ser)	*P. rhodesiae, P. veronii*	Schlegel et al. (2001)

[a] Abbreviations: usual amino acids, three letter code; aThr, *allo*-Thr; εLys, Lys linked by its ε-NH2, AOHOrn, δ-N-acetyl-δ-N-hydroxy-ornithine; FO-HOrn, δ-N-formyl-δ-N-hydroxy-ornithine; cOHOrn, *cyclo*-hydroxy-ornithine (3-amino-1-hydroxy-piperidone-2); OHHis, *threo*-β-hydroxy-histidine; OHAsp, *threo*-β-hydroxy-aspartic acid; Dab, diamino-butanoic acid; Pfl, *Pseudomonas fluorescens*; Pa., *Pseudomonas aeruginosa*; Pp, *Pseudomonas putida*. The D-amino acids are underlined

[b] Incomplete structures (stereochemistry is missing)

Table 2.4. Structurally related pyoverdines demonstrating cross-uptake properties

Strains	pHi values	Pyoverdine (peptide part) structure	References
P. fluorescens 1.3	7.5/9.0	Ala-Lys-Gly-Gly-**OHAsp**-**Gln**-**Dab**-Gly-Ser-**cOHOrn**	Georgias et al. (1999)
P. fluorescens ATCC 17400	7.6/9.2	Ala-Lys-Gly-Gly-**OHAsp**-**Gln**-**Dab**-Ser-Ala-**cOHOrn**	Demange et al., (1990)
P. putida Gwose	7.6/9.2	Ser-Thr-Ser-Orn-**OHAsp**-**Gln**-**Dab**-Ser-aThr-**cOHOrn**	Gwose and Taraz (1992)
P. fluorescens PL7	5.0/5.3/7.6	Ser-AOHOrn-Ala-**Gly-aThr**-Ala-**cOHOrn**	Barelmann et al. (2002)
P. fluorescens PL8	7.6/8.9/9.0	Lys-AOHOrn-Ala-**Gly-aThr**-Ser-**cOHOrn**	Barelmann et al. (2002)
P. aeruginosa ATCC 27853	7.4/7.5/8.6/8.9	Ser- FOHOrn-Orn-**Gly-aThr**-Ser-**cOHOrn**	Tappe et al. (1993)
P. chlororaphis D-TR133	4.1/5.4/7.5	**Asp-FoOHOrn-Lys-(Thr-D/LAla-L/DAla-FoOHOrn**-Ala)	Barelmann et al. (2003)
P. fluorescens CHA0	5.1/7.5/8.5	**Asp-FoOHOrn-Lys-(Thr-D/LAla-L/DAla-FoOHOrn**-Lys)	Wong-Lun-Sang et al. (1996)
P. fluorescens PL9	4.4/5.2	**Ser-AOHOrn**-Ala-Gly-(**Ser**-Ser-**OHAsp**- **Thr**)	Vossen et al. (2000)
Pseudomonas sp. A214	4.0/4.5/5.1	**Ser-AOHOrn**-Ala-Gly-(**Ser**-Ala-**OHAsp**-**Thr**)	Fernandez et al. (2003b)
P. fluorescens A6	7.6/8.9	Lys-**AOHOrn**-Gly-aThr-Thr-**Gln**-Gly-**Ser-cOHOrn**	Beiderbeck et al. (1999)
P. costantinii CFBP 5705	4.9/5.2/7.6	Ser-**AOHOrn**-Gly-Thr-aThr-**Gln**-**Gly**-**Ser-cOHOrn**	Fernandez et al. (2001)
P. syringae ATCC 19310	3.9/4.5	**Lys-OHAsp**-Thr-Thr-Ser-**OHAsp**-Ser	Jülich et al. (2001)
P. cichorii	3.9/4.5/7.2	**Lys-OHAsp**-Thr-Thr-Gly-**OHAsp**-Ser	Bultreys et al. (2004)
P. aeruginosa R/Pa6	5.2/7.3	**Ser**-Dab-**FOHOrn**-Gln-**Gln-FOHOrn**-Gly	Gipp et al. (1991)
P. aeruginosa R'	5.2/7.3	**Ser**-Dab-**FOHOrn**-**Gln-FOHOrn**-Gly	Ruangviriyachai et al. (2001)
P. aeruginosa ATCC 15692	7.1/7.2/8.8	**Ser**-Arg-Ser-**FOHOrn**-(Lys-**FOHOrn**-Thr-Thr)	Briskot et al. (1989)
P. fluorescens ATCC 13525	7.1/7.3/8.7	**Ser**-Lys-Gly-**FOHOrn**-(Lys-**FOHOrn**-Ser)	Linget et al. (1992)

2.4
Siderotyping as a Powerful Tool
for the Search of New Siderophores

While the description of new pyoverdines differing by their peptide chain was increasing in number during the 1990s, a rapid way to differentiate efficiently these siderophores one from each other became necessary in order to recognize rapidly those pyoverdines already described at the structure level in the literature. Thus, the two siderotyping methods described above were first developed for such a goal. Applied to a collection of fluorescent *Pseudomonas* of more than 2000 isolates received from many laboratories and harvested from different worldwide geographical and environmental areas, these methods allowed within a few years the recognition of many pyoverdines corresponding to none of the already described compounds. Thanks to a very efficient collaboration with Professor Budzikiewicz at Köln University and his collaborators, numerous new structures were published within a few years, doubling the number of the already known pyoverdines. Table 2.3 describes the bacterial origin and the peptidic structure of such 25 pyoverdines; these structure and siderotyping properties have been published during the period 1997–2004. At present the two methods have allowed one to distinguish, according to their respective IEF pattern and iron uptake specificity, a total of 110 pyoverdines, 60 of them remaining as yet structurally unknown. The efficiency of the two methods applied together for the differentiation of pyoverdines is close to 100%: pyoverdines produced by the two strains *P. aeruginosa* R and *P. aeruginosa* R', differing by a supplementary Gln residue for the first one, were the only ones among the more than 50 structurally known pyoverdines, which were not distinguishable by siderotyping since presenting an identical IEF pattern (Table 2.4) and a 100% cross-reactivity. All the others were seen as differing by at least one character at the level of the IEF pattern (number or respective intensity of PVD-isoform bands) or at the level of iron uptake (percentage of cross-reactivity).

2.5
Siderotyping as a Powerful Tool
for *Pseudomonas* Taxonomy and Phylogeny

No evident correlations between *Pseudomonas* taxonomy and siderophore structure diversity were suspected until a polyphasic approach (Vandamme et al. 1996) was systematically used for bacterial identification. For "old" species, identified thanks to a few strong and specific phenotypic characters like growth temperature for *P. aeruginosa*, mushroom pathogenicity for *P. tolaasii* or plant-pathogenicity for *P. syringae*, siderotyping studies concluded there was a limited variability among pyoverdines produced within each species, with one, two or three different compounds for *P. syringae*, *P. tolaasii* and *P. aeruginosa*, respectively (Meyer et al. 1997, 2002b; Munsch et al. 2000). However, pyoverdine structures were so diverse among strains of the other major *P. fluorescens* and *P. putida* species, that specificity was not recognized at the species level but, rather, at the strain level. Indeed, the definition of many new species among such

Table 2.5. *Pseudomonas* spp. and their respective siderovars

Species and siderovar number		Number of strains analyzed	Siderovar type	Species sharing the same siderovar	References
Fluorescent Pseudomonas					
P. aeruginosa	sv. 1	>200	sv. PAO1		Meyer et al. (1997)
	sv. 2	>100	sv. 27853		Meyer et al. (1997)
	sv. 3	>50	sv. Pa6		Meyer et al. (1997)
P. brassicacearum		10	sv. PL9	*P. lini* sv. 1	Achouak et al. (2000)
P. cichorii		4	sv. cich		Meyer et al. (2004)
P. costantinii		10	sv Ps3a		Munsch et al. (2002)
P. fuscovaginae		16	sv. G17		Meyer et al. (2004)
P. jessenii		8	sv. 9AW		Meyer et al. (2002b)
P. lini	sv. 1	7	sv.PL9	*P. brassicacearum*	Delorme et al. (2002)
	sv. 2	2	sv. B10		Delorme et al. (2002)
P. mandelii		43	sv. SB8.3		Meyer et al. (2002b)
P. monteilii		10	sv. Lille 1		Meyer et al. (2002b)
P. mosselii		12	sv. Lille 17		Dabboussi et al. (2002)
P. palleroniana		10	sv. 13525	*P. veronii*	Gardan et al. (2002)
P. rhodesiae		7	sv. Lille 25		Meyer et al. (2002b)
P. salomonii		15	sv. 96-318		Gardan et al. (2002)
P. syringae and related species[a]		79	sv. syr		Geoffroy et al. (2000)
P. thivervalensis		6	sv. thiv		Achouak et al. (2000)
P. tolaasii		17	sv tol		Munsch et al. (2000)
P. veronii		8	sv. 13525	*P. palleronian*	Meyer et al. (2002b)
Non-fluorescent Pseudomonas					
P. corrugata		13	sv. corr		Meyer et al. (2002b)
P. graminis		19	sv. gram		Meyer et al. (2002b)
P. frederiksbergensis		5	sv. fred		Meyer et al. (2002b)
P. plecoglossicida		6	sv. plec		Meyer et al. (2002b)
P. fragi		4	No siderophore		Champomier et al. (1996)

[a] Related species with number of strains analyzed under brackets are *P. viridiflava* (1), *P. savastanoi* (3), *P. ficuserectae* (1), *P. meliae*(1), *P. amygdali* (1), *P. tremae* (3), *P. avellanae* (1) and *P. cannabina* (1)

strains first recognized as *P. fluorescens* or *P. putida* greatly helped in raising the present correlation concept established between *Pseudomonas* taxons (strains grouped according to polyphasic taxonomy) and siderovars (group of strains producing an identical pyoverdine). As detailed in Table 2.5, a large majority of *Pseudomonas* species (28 of 30) are composed of strains producing within a species a unique and identical siderophore and, therefore, belonging to the same siderovar. Only two species demonstrated a heterogeneity at the pyoverdine level with two siderovars recognized among *Pseudomonas lini* strains and three within the numerous (more than 350) *P. aeruginosa* isolates so far analyzed. Also, it is remarkable that, in most cases, species are each characterized by a structurally different siderophore. Only a few species have been described so far as sharing an identical siderophore, as illustrated in Table 2.5 for *P. lini* sv. 1 and *P. brassicacearum* or for strains of *P. veronii* and *P. palleroniana* producing the same pyoverdine as the one produced by the type-strain *P. fluorescens* ATCC 13525. It is indeed of interest to highlight that these species, closely related at the siderophore level, are also phylogenetically related, as seen through 16S rDNA homology. Thus, siderophores appear as molecules of particular interest not only for iron metabolism but also for bacterial taxonomy and phylogeny.

2.6
Siderotyping and Environmental/Ecological Microbiology

Isoelectrophoresis is the most convenient siderotyping method, especially when numerous strains have to be analyzed. Thus, studies related to taxonomy, epidemiology, and population diversity should benefit from a technique which allows at low cost the analysis of up to 100 isolates within a day. For instance, we recently concluded by analyzing more than 1200 strains isolated from different geographical areas, that one particular siderovar described for a single strain isolated in Antarctica, appeared as specific to cold regions since found in another unique strain originating from Finland (Geoffroy and Meyer 2004). In another domain related to rhizobacteria, it was concluded that strains colonizing a precise plant rhizospheric compartment all belonged to a unique siderovar, suggesting an efficient bacterial adaptation to a specific plant location (Founoune et al. 2002). As developed in a recent review (Meyer and Geoffroy 2005), other goals which could be explored using siderotyping are:
- Tracing the survival of field-released bacteria, or the plant colonization efficiency of inoculated rhizobacteria: many fluorescent *Pseudomonas* are used as biocontrol agents or biofertilizers in various crop culture experimentations. Siderotyping is a convenient tool to discriminate these bacteria from the indigenous bacterial populations, as far as they belong to a particular siderovar, and to specifically trace them in soil or in plant tissues.

- Determining very quickly the possible clonal origin or, on the contrary, the level of biodiversity among a natural population.
- Searching for new isolates very likely belonging to a given species, by selecting for strains able to specifically use the siderophore characteristic to that species,
- Enriching a bacterial population of very poorly represented fluorescent isolates by supplementing the inoculated growth medium with the corresponding siderophore in presence of EDDHA to inhibit the other bacteria.

2.7
Raising a Pyoverdine Bank Based on IEF Patterns

Of the structurally known pyoverdines, most of them have been studied and their respective IEF pattern obtained in our laboratory. Thus, assembling the IEF data of a maximum of such pyoverdines considered as reference compounds, should constitute a powerful tool for comparison purposes when analyzing new collections of strains. It is what has been done in Fig. 2.2 for 45 fluorescent *Pseudomonas* strains each producing a different pyoverdine as established by structural studies or based on specific siderotyping behaviours for a few representatives. For clarity, the IEF patterns were ranked according to the pHi values of the different isoforms, starting with strains presenting the most acidic pyoverdine-IEF patterns. Thus, four sub-groups were designed as delineated in Fig. 2.2. Sub-group I includes seven strains producing pyoverdines with a predominance of acidic amino acids, e.g., *P. syringae* which pyoverdine peptide contains two hydroxy-aspartic acid residues as iron-chelating groups instead of the classical two hydroxy-ornithine residues for most pyoverdines (see also Tables 2.3 and 2.4 for structure comparisons). The major sub-group (sub-group IV, 18 strains), is composed of strains with more alkaline PVD-IEF patterns which means usually one or two bands with pHi values up to pH 8 and one or two bands in the range of pH 7–8. Another important sub-group (sub-group II, 17 strains) is also showing neutral band(s) together with some bands at acidic pH values. Finally, sub-group III is limited to three strains characterized by PVD-isoform bands at acidic, neutral and alkaline pH values.

The IEF patterns designed in Fig. 2.2 represent a reconstructed image of the electrophoretic profiles obtained when analyzing the culture supernatant of strains grown in CAA medium (Casamino acids from Difco n° 0230-01, 5 g/L; PO_4HK_2, 1.18 g/L, $SO_4Mg.7H_2O$, 0,25 g/L), under precise conditions of temperature and aeration. To reach the profiles shown in Fig. 2.1, usually 1 µL of a 20-fold concentrated (through lyophilisation) CAA culture supernatant is applied onto the gel. Then the pHi values of the fluorescent bands appearing under UV light on the electrophoresed gel are determined by comparison

IEF-bank sub-group	Strain	Siderovar type	PVD-isoelectrofocusing pattern (3.5 4.0 4.5 5.0 5.5 6.0 6.5 7.0 7.5 8.0 8.5 9.0)	References
I	*P. putida* strain C	putC		Fuchs et al. 2001
	P. syringae ATCC 19310	Syr		Jülich et al. 2001
	P. putida ATCC 12633	12633		Fuchs et al. 2001
	P. putida CFML 90-33	90.33		Sultana et al. 2001a
	Pseudomonas sp. A214	A214		Fernandez et al. 2003b
	P. brassicacearum	PL9		Achouak et al. 2000
	P. fluorescens G173	G173		Fernandez et al. 2003a
II	*P. fluorescens* 12	Pfl12		Weber et al. 2000
	P. putida CFML 90-44	90-44		Sultana et al. 2001b
	P. cichorii CFBP 2101	CFBP 2101		Munsch et al. 2002
	P. marginalis G76	G76		Budzikiewicz 2004
	Pseudomonas sp. D47	D 47		Budzikiewicz 2004
	P. aeruginosa Pa6 (sv. 3)	Pa 6		Meyer et al. 1997
	P. putida CFML 90-51	90-51		Sultana et al. 2000b
	P. monteilii CFML 90-54	Lille 1		Meyer 2000
	Pseudomonas sp. CFML 96-188	96-188		Weber et al. 2000
	P. putida L1	L1		Fernandez et al. 2003b
	P. lini sv. 2	B10		Delorme et al. 2002
	P. chlororaphis D-TR133	D-TR133		Barelmann et al 2003
	P. costantinii Ps3a	Ps3a		Fernandez et al. 2001
	P. fluorescens PL7	PL7		Barelmann et al. 2002
	P. thivervalensis ML45	ML45		Achouak et al. 2000
	P. fluorescens 51W	51W		Meyer et al. 1998
	P. fluorescens ii	Pflii		Fuchs et al. 2001

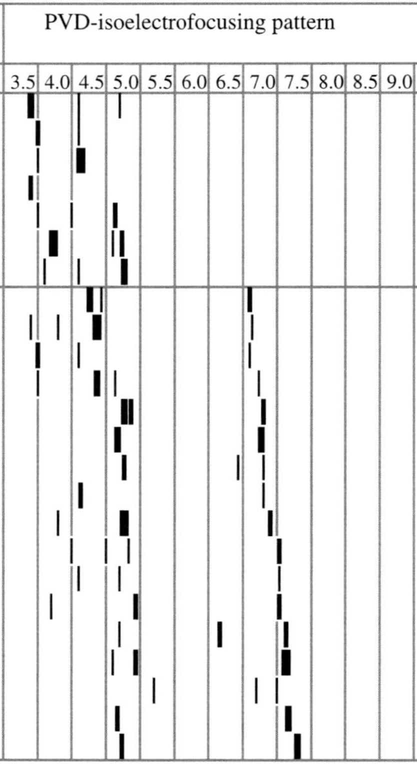

Fig. 2.2. Pyoverdine bank according to IEF patterns

IEF-bank sub-group	Strain	Siderovar type	PVD-isoelectrofocusing pattern (3.5 4.0 4.5 5.0 5.5 6.0 6.5 7.0 7.5 8.0 8.5 9.0)	References
III	*P. fuscovaginae* CFBP 2065[T]	G17		Munsch et al. 2002
	P. mosselii CFML 90-77	Lille17		Dabboussi et al. 2002
	P. fluorescens CHA0	CHAO		Barelmann et al. 2003
IV	*P. rhodesiae* CFML 92-104	Lille 25		Weber et al. 2000
	P. fluorescens DSM 50106	50106		Fuchs et al. 2001
	P. fluorescens 18.1	Pfl18.1		Amann et al. 2000
	P. veronii CFML 92-124	Lille 31		Weber et al. 2000
	P. fluorescens ATCC 13525	Pfl13525		Weber et al. 2000
	P. aeruginosa PAO1 (sv. 1)	PAO1		Meyer et al. 1997
	P. salomonii CFBP 2022[T]	96-318		Gardan et al. 2002
	Pseudomonas sp. CFML 95-275	95-275		Sultana et al. 2000a
	P. tolaasii CFBP 2068[T]	tol		Munsch et al. 2000
	Pseudomonas sp. CFML 96-312	96-312		Schlegel et al. 2001
	P. aeruginosa ATCC 27853 (sv. 2)	Pa 27853		Meyer et al. 1997
	P. mandelii	SB8.3		Meyer et al. 2002b
	P. fluorescens 1.3	Pfl1.3		Georgias et al. 1999
	P. fluorescens PL8	PL8		Barelmann et al. 2002
	P. fluorescens ATCC 17400	Pfl17400		Munsch et al. 2000
	P. fluorescens A6	A6		Beiderbeck et al. 1999
	P. "reactans" Ps6-10	Ps6-10		Munsch et al. 2000
	P. jessenii 9AW	9AW		Budzikiewicz et al. 1997

Fig. 2.2. *(continued)* Pyoverdine bank according to IEF patterns

with an internal standard (Fuchs et al. 2001). The corresponding bands are then represented on a computer scale with a width proportional to the fluorescence intensity demonstrated on the gel. Indeed, in the absence of fluorescence quantification, the width value is arbitrarily chosen.

A small change in growth conditions, including the previous mentioned physico-chemical ones, but also the medium iron-contaminating level (which depends on the chemicals and also and mainly on the vessel cleanness) and, indeed, the quality of the inoculum and the age of the culture, will induce very likely small change(s) in the PVD-IEF pattern of a given strain. Usually, changes occur mainly as a modification in the intensity of fluorescence of the respective bands, less often as a change in number of bands (usually due to the appearance, or disappearance, of a minor band). Such slight modifications in the experimental PVD-IEF pattern obtained for a given strain are in most cases without incidence on the final conclusion which is the grouping of the strain under study within a siderovar characterized by the same pyoverdine-IEF pattern. Indeed, with the increasing number of pyoverdines differing in their structures, it could happen that IEF patterns of two structurally different pyoverdines present very small differences, sometimes no difference at all as already mentioned above for the pyoverdines of strains *P. aeruginosa* R and R'. In such cases, cross-incorporation studies will tell if the strain was grouped in the correct siderovar or not.

Of the 45 strains represented in Fig. 2.2, 17 of them belong to well identified fluorescent *Pseudomonas* species. As already demonstrated in one case by DNA-DNA hybridization (Meyer et al. 2002b), any unidentified isolate presenting the same pyoverdine-IEF pattern as one of these reference strains could be at least expected to belong to the corresponding species. Thus, the identification at the species level of a pseudomonad could be possible in a very short period of time and based on a single phenotypic character, thanks to siderotyping.

2.8
Conclusions

Siderotyping is defined as the characterization of bacterial strains according to the siderophore(s) they produce. The method was first developed to detect novel pyoverdine molecules, the main siderophores of the fluorescent *Pseudomonas*. Moreover, as soon as well classified pseudomonad collections, i.e., collections studied through polyphasic taxonomy, were available, siderotyping proved to be a very efficient tool for a rapid identification, usually at the species level, of fluorescent as well as non-fluorescent *Pseudomonas* spp. In pioneering work based on the analysis of close to 400 strains grouped in 28 taxons, among them 15 well defined species, it was concluded, as a general rule, that strains belonging to a given species were all producing an identical siderophore, whereas each species was characterized by a specific siderophore (Meyer et al. 2002b).

Acknowledgments

The author thanks his collaborators C. Gruffaz and V. Geoffroy as well as the many students who actively participated to the development of siderotyping over the last ten years. The many generous donators of bacterial strains are deeply acknowledged as well as Prof. H. Budzikiewicz and his collaborators for the wonderful structural chemistry work done on many pyoverdines detected during the siderotyping studies.

References

Achouak W, Sutra L, Heulin T, Meyer JM, Fromin N, Degreave S, Christen R, Gardan L (2000) Description of *Pseudomonas brassicacearum* sp. nov. and *Pseudomonas thivervalensis* sp. nov., root-associated bacteria isolated from *Arabidopsis thaliana* and *Brassica napus*. Int J Syst Evol Microbiol 50:9–18

Amann C, Taraz K, Budzikiewicz H, Meyer JM (2000) The siderophores of *Pseudomonas fluorescens* 18.1 and the importance of cyclopeptidic substructures for the recognition at the cell surface. Z Naturforsch 55c:671–680

Anthony U, Christophersen C, Nielsen PH, Gram l, Petersen BO (1995) Pseudomonine, an isoxazolidone with siderophoric activity from *Pseudomonas fluorescens* AH2 isolated from lake Victorian Nile Perch. J Nat Prod 58:1786–1789

Barelmann I, Taraz K, Budzikiewicz H, Geoffroy V, Meyer JM (2002) The structures of the pyoverdins from two *Pseudomonas fluorescens* strains accepted mutually by their respective producers. Z Naturforsch 57:9–16

Barelmann I, Fernandez DU, Budzikiewicz H, Meyer JM (2003) The pyoverdine from *Pseudomonas chlororaphis* D-TR133 showing mutual acceptance with the pyoverdine of *Pseudomonas fluorescens* CHA0. BioMetals 16:263–270

Beiderbeck H, Taraz K, Meyer JM (1999) Revised structures of the pyoverdins from *Pseudomonas putida* CFBP 2461 and from *Pseudomonas fluorescens* CFBP 2392. Biometals 12:331–338

Bossis E, Lemanceau P, Latour X, Gardan L (2000) The taxonomy of *Pseudomonas fluorescens* and *Pseudomonas putida*: current status and need for revision. Agronomie 20:51–63

Briskot G, Taraz K, Budzikiewicz H (1989) Pyoverdine-Type siderophores from *Pseudomonas aeruginosa*. Liebigs Ann Chem 375–384

Budzikiewicz H (2004) Siderophores of the Pseudomonadaceae sensu stricto (fluorescent and non-fluorescent *Pseudomonas* spp.) Prog Chem Org Nat Prod 87:81–235

Budzikiewicz H, Kilz S, Taraz K, Meyer JM (1997) Identical pyoverdines from *Pseudomonas fluorescens* 9AW and from *Pseudomonas putida* 9BW. Z Naturforsch 52c:721–728

Budzikiewicz H, Fernandez DU, Fuchs R, Michalke R, Taraz K, Ruangviriyachai C (1999) Pyoverdines with a Lys ε-amino link in the peptide chain. Z Naturforsch 54c:1021–1026

Bultreys A, Gheysen I (2000) Production and comparison of peptide siderophores from strains of distantly related pathovars of *Pseudomonas syringae* and *Pseudomonas viridiflava* LMG 2352. Appl Environ Microbiol 66:325–331

Bultreys A, Gheysen I, Whatelet B, Maraite H, De Hoffmann E (2003) High-performance liquid chromatography analyses of pyoverdin siderophores differentiate among phytopathogenic fluorescent *Pseudomonas* species. Appl Environ Microbiol 69:1143–1153

Bultreys A, Gheysen I, Wathelet B, Schäfer M, Budzikiewicz H (2004) The pyoverdins of *Pseudomonas syringae* and *Pseudomonas cichorii*. Z Naturforsch 59c:613–618

Buyer JS, de Lorenzo V, Neilands JB (1991) Production of the siderophore aerobactin by a halophilic pseudomonad. Appl Environ Microbiol 57:2246–2250

Chakraborty RN, Patel HN, Desai SB (1990) Isolation and partial characterization of catechol-type siderophore from *Pseudomonas stutzeri* RC 7. Curr Microbiol 20:283–286

Champomier-Vergès MC, Stintzi A, Meyer JM (1996) Acquisition of iron by the non-siderophore producing *Pseudomonas fragi*. Microbiology 142:1191–1199

Cladera AM, Bennasar A, Barcelo M, Lalucat J, Garcia-Valdes E (2004) Comparative genetic diversity of *Pseudomonas stutzeri* genomovars, clonal structure, and phylogeny of the species. J Bacteriol 186:5239–5248

Cornelis P, Hohnadel D, Meyer JM (1989) Evidence for different Pyoverdine-mediated iron uptake systems among *Pseudomonas aeruginosa* strains. Infect Immun 57:3491–3497

Cox CD, Rinehart KL, Moore ML, Cook JC (1981) Pyochelin: novel structure of an iron-chelating growth promoter for *Pseudomonas aeruginosa*. Proc Natl Acad Sci USA 78:4256–4260

Crosa JH, Mey AR, Payne SM (2004) Iron transport in bacteria. ASM Press, Washington DC

Dabboussi F, Hamzé M, Singer E, Geoffroy V, Meyer JM, Izard, D (2002) *Pseudomonas mosselii* sp. nov., a new species isolated from clinical specimens. Int J Syst Evol Microbiol 52:363–376

Delorme S, Lemanceau P, Christen R, Corberand T, Meyer JM, Gardan L (2002) *Pseudomonas lini* sp. nov., a novel species from bulk and rhizospheric soils. Int J Syst Evol Microbiol 52:513–523

Demange P, Bateman A, Mertz C, Dell A, Piémont Y, Abdullah M (1990) Structures of pyoverdins Pt, siderophores of *Pseudomonas tolaasii* NCPPB 2192, and pyoverdins Pf, siderophores of *Pseudomonas fluorescens* CCM 2798. Identification of an unusual natural amino acid. Biochemistry 29:11041–11051

De Vos D, Lim A Jr, Pirnay JP, Duinslaeger L, Revets H, Vanderkelen A, Hamers R, Cornelis P (1997) Analysis of epidemic *Pseudomonas aeruginosa* isolates by isoelectric focusing of pyoverdine and RAPD-PCR: modern tools for an integrated anti nosocomial infection strategy in burn wound centres. Burns 23:379–386

Fernandez DU, Fuchs R, Taraz K, Budzikiewicz H, Munsch P, Meyer JM (2001) The structure of a pyoverdine produced by a *Pseudomonas tolaasii*-like isolate. BioMetals 14:81–84

Fernandez DU, Fuchs R, Schäfer M, Budzikiewicz H, Meyer JM (2003a) The pyoverdin of *Pseudomonas fluorescens* G173, a novel structural type accompanied by unexpected natural derivatives of the corresponding ferribactin. Z Naturforsch 58c:1–10

Fernandez DU, Geoffroy V, Schäfer M, Meyer JM, Budzikiewicz H (2003b) Structure revision of pyoverdines produced by plant-growth promoting and plant-deleterious *Pseudomonas* species. Monatshefte für Chemie 134:1421–1431

Founoune H, Duponnois R, Meyer JM, Thioulouse J, Masse D, Chotte JL, Neyra M (2002) Interactions between ectomycorrhizal symbiosis and fluorescent pseudomonads on *Acacia holosericea*: isolation of Mycorrhiza Helper Bacteria (MHB) from a soudano-sahelian soil. FEMS Microbiol Ecology, 41:37–46

Fuchs R, Schäfer M, Geoffroy V, Meyer JM (2001) Siderotyping – a powerful tool for the characterization of pyoverdines. Curr Topics Med Chem 1:31–57

Gardan L, Shafik H, Belouin S, Grimont F, Grimont PAD (1999) DNA relatedness among the pathovars of *Pseudomonas syringae* and description of *Pseudomonas tremae* sp. nov. and *Pseudomonas cannabina* sp. nov. (ex Sutic and Dowson 1959). Int J Syst Bacteriol 49:469–478

Gardan L, Bella P, Meyer JM, Christen R, Rott P, Achouak W, Samson R (2002) *Pseudomonas salomonii* sp. nov. pathogenic on garlic, and *Pseudomonas palleroniana* sp. nov., isolated from rice. Int J Syst Evol Microbiol 52:2065–2074

Geoffroy VA, Meyer JM (2004) Siderotyping of Antarctic fluorescent *Pseudomonas* strains. Cell Mol Biol 50:585–590

Geoffroy V, Meyer JM, Gardan L (2000) Pathovars of *Pseudomonas syringae* are characterized by a unique pyoverdine-mediated iron uptake system. Abstracts of the 10th International Conference on Plant Pathogenic bacteria, Charlottetown, Prince Edward Island, Canada

Georgias, H, Taraz K, Budzikiewicz H, Geoffroy V, Meyer JM (1999) The structure of the pyoverdin from *Pseudomonas fluorescens* 1.3. Structural and biological relationships of pyoverdins from different strains. Z Naturforsch 54:301–308

Gipp S, Hahn J, Taraz K, Budzikiewicz H (1991) Zwei Pyoverdine aus *Pseudomonas aeruginosa* R. Z Naturforsch 46:534–541

Grimont PAD, Vancanneyt M, Lefèvre M, Vandermeulebroecke K, Vauterin L, Brosch R, Kersters K, Grimont F (1996) Ability of Biolog and Biotype-100 systems to reveal the taxonomic diversity of the pseudomonads. Syst Appl Microbiol 19:510–527

Gwose I, Taraz K (1992) Pyoverdine aus *Pseudomonas putida*, Z Naturforsch 47:487–502

Haas D, Défago G (2005) Biological control of soil-borne pathogens by fluorescent pseudomonads. Nat Rev Microbiol doi:10.1038/nrmicro1129

Hohnadel D, Meyer JM (1988) Specificity of pyoverdine-mediated iron uptake among fluorescent *Pseudomonas* strains. J Bacteriol 170:4865–4873

Jülich M, Taraz K, Budzikiewicz H, Geoffroy V, Meyer JM, Gardan L (2001) The structure of the pyoverdin isolated from various *Pseudomonas syringae* pathovars. Z Naturforsch 56c:687–694

Kilz S, Lenz C, Fuchs R, Budzikiewicz H (1999) A fast screening method for the identification of siderophores from fluorescent *Pseudomonas* spp. by liquid chromatography/electrospray mass spectrometry. J Mass Spectrom 34:281–290

Kloepper JW, Leong J, Teintze M, Schroth MN (1980) *Pseudomonas* siderophores : a mechanism explaining disease-suppressive soils. Curr Microbiol 4:317–320

Koedam N, Wittouck E, Gaballa A, Gillis A, Höfte M, Cornelis P (1994) Detection and differenciation of microbial siderophores by isoelectric focusing and chrome azurol S overlay. BioMetals 7:287–291

Lewis TA, Lynch L, Morales S, Austin PR, Hartwell HJ, Kaplan B, Forker C, Meyer JM (2004) Physiological and molecular genetic evaluation of the dechlorination agent pyridine-2,6-bis(monothiocarboxylic acid) (PDTC) as a secondary siderophore of *Pseudomonas*. Environ Microbiol 6:159–169

Lindberg GD, Larkin JM, Whaley HA (1980) Production of tropolone by a *Pseudomonas*. J Nat Prod 43:592–594

Linget C, Azadi P, MacLeod JK, Dell A, Abdallah, MA (1992) Bacterial siderophores: the structures of the pyoverdins of *Pseudomonas fluorescens* ATCC 13525. Tetrahedron Lett 33:1737–1740

Mercado-Blanco J, van der Drift KMGM, Olsson PE, Thomas-Oates JE, van Loon LC, Bakker PAHM (2001) Analysis of the *pmsCEAB* gene cluster involved in biosynthesis of salicylic acid and the siderophore pseudomonine in the biocontrol strain *Pseudomonas fluorescens* WCS354. J Bacteriol 183:1909–1920

Meyer JM (1998) Siderotyping methods for *Pseudomonas* and related bacteria. In: Pandey A (ed) Advances in biotechnology. Educational Publishers and Distributors, New Delhi, pp 172–182

Meyer JM, Abdallah M (1980) The siderochromes of non-fluorescent pseudomonads: nocardamine production by *Pseudomonas stutzeri*. J Gen Microbiol 118:125–129

Meyer JM, Geoffroy VA (2004) Environmental fluorescent *Pseudomonas* and pyoverdine diversity: how siderophores could help microbiologists in bacterial identification and taxonomy. In: Crosa JH, Mey AR, Payne SM (eds) Iron transport in bacteria. ASM Press, Washington DC, pp 451–468

Meyer JM, Geoffroy VA (2005) Analyzing *Pseudomonas* biodiversity: an easy challenge thanks to siderotyping. In: Satyaranayana T, Johri BN (eds) Microbial diversity: current perspectives and potential applications. IK International Publishing House Private Limited, New Delhi, pp 307–322

Meyer JM, Hohnadel D, Khan HA, Cornelis P (1990) Pyoverdine-facilitated iron uptake in *Pseudomonas aeruginosa*: immunological characterization of the ferripyoverdine receptor. Mol Microbiol 4:1401–1405

Meyer JM, Azelvandre P, Georges C (1992) Iron metabolism in *Pseudomonas*: salicylic acid, siderophore of *Pseudomonas fluorescens* CHA0. Biofactors 4:23–27

Meyer JM, Tran Van V, Stintzi A, Berge O, Winkelmann G (1995) Ornibactin production and transport properties in strains of *Burkholderia vietnamiensis* and *Burkholderia cepacia* (formerly *Pseudomonas cepacia*). BioMetals 8:309–317

Meyer JM, Stintzi A, De Vos D, Cornelis P, Tappe R, Taraz K, Budzikiewicz H (1997) Use of siderophores to type pseudomonads: the three *Pseudomonas aeruginosa* pyoverdine systems. Microbiology 143:35–43

Meyer JM, Stintzi A, Coulanges V, Shivaji S, Voss JA, Taraz K, Budzikiewicz H (1998) Siderotyping of fluorescent pseudomonads : characterization of pyoverdines of *Pseudomonas fluorescens* and *Pseudomonas putida* strains from Antarctica. Microbiology 144:3119–3126

Meyer JM, Geoffroy VA, Baysse C, Cornelis P, Barelmann I, Taraz K, Budzikiewicz H (2002a) Siderophore-mediated iron uptake in fluorescent *Pseudomonas*: characterization of the pyoverdine-receptor binding site of three cross-reacting pyoverdines. Arch Biochem Biophys 397:179–183

Meyer JM, Geoffroy VA, Baida N, Gardan L, Izard D, Lemanceau P, Achouak W, Palleroni N (2002b) Siderophore typing, a powerful tool for the identification of fluorescent and non-fluorescent *Pseudomonas*. Appl Environ Microbiol 68:2745–2753

Meyer JM, Fischer-Le Saux M, Moura L, Geoffroy VA, Gardan L (2004) Taxonomie des *Pseudomonas* phytopathogènes: étude comparative entre sidérotypage et taxonomie numérique. VIème Congrès SFM Bordeaux, France

Mossialos D, Meyer JM, Wolf U, Budzikiewicz H, Koedam N, Baysse C, Anjaiah V, Cornelis P (2000) Quinolobactin, a new siderophore of *Pseudomonas fluorescens* ATCC 17400 whose production is repressed by the cognate pyoverdine. Appl Environ Microbiol 66:487–492

Munsch P, Geoffroy V, Alatossava T, Meyer JM (2000) Application of siderotyping for the characterization of *Pseudomonas tolaasii* and *Pseudomonas reactans* isolates associated with brown blotch disease of cultivated mushrooms. Appl Environ Microbiol 66:4834–4841

Munsch P, Alatossava T, Meyer JM, Marttinen N, Christen R, Gardan L (2002) *Pseudomonas costantinii* sp. nov., another causal agent of brown blotch disease, isolated from cultivated mushroom sporophores in Finland. Int J Syst Evol Microbiol 52:1973–1983

Neilands JB (1981) Microbial iron compounds. Annu Rev Biochem 50:715–731

Nishimori E, Kita-Tsukamoto K, Wakabaiyashi H (2000) *Pseudomonas plecoglossicida* sp. nov., the causative agent of bacterial haemorrhagic ascites of ayu, *Plecoglossus altivelis*. Int J Syst Evol Microbiol 50:83–89

Palleroni NJ (1984) *Pseudomonas*, In: Krieg NR (ed) Bergey's manual of systematic bacteriology, vol 1. Williams and Wilkins, Baltimore, pp 141–199

Risse D, Beiderbeck H, Taraz K, Budzikiewicz H, Gustine D (1998) Corrugatin, a lipopeptide siderophore from *Pseudomonas corrugata*. Z Naturforsch 53c:295–304

Ruangviriyachai C, Barelmann I, Fuchs R, Budzikiewicz H (2000) An exceptionally large pyoverdin from a *Pseudomonas* strain collected in Thailand. Z Naturforsch 55c:323–327

Ruangviriyachai C, Fernandez DU, Fuchs R, Meyer JM, Budzikiewicz H (2001) A new pyoverdin from *Pseudomonas aeruginosa* R'. Z Naturforsch 56c:933–938

Schlegel K, Fuchs R, Schäfer M, Taraz K, Budzikiewicz H, Geoffroy V, Meyer JM (2001) The pyoverdins of *Pseudomonas* sp. 96-312 and 96-318. Z Naturforsch 56c:680–686

Schwyn B, Neilands JB (1987) Universal chemical assay for the detection and determination of siderophores. Anal Biochem160:47–56

Smith EE, Sims EH, Spencer DH, Kaul R, Olson MV (2005) Evidence for diversifying selection at the pyoverdine locus of *Pseudomonas aeruginosa*. J Bacteriol 187:2138–2147

Sultana R, Fuchs R, Schmickler H, Schlegel K, Budzikiewicz H, Siddiqui BS, Geoffroy V, Meyer JM (2000a) A pyoverdin from *Pseudomonas* sp. CFML 95-275. Z Naturforsch 55c:857–865

Sultana R, Siddiqui BS, Taraz K, Budzikiewicz H, Meyer JM (2000b). A pyoverdine from *Pseudomonas putida* CFML 90-51 with a Lys ε-amino link in the peptide chain. BioMetals 13:147–152

Sultana R, Siddiqui BS, Taraz K, Budzikiewicz H, Meyer JM (2001a) An isopyoverdin from *Pseudomonas putida* CFML 90-33. Tetrahedron 57:1019–1023

Sultana R, Siddiqui BS, Taraz K, Budzikiewicz H, Meyer JM (2001b) An isopyoverdin from *Pseudomonas putida* CFML 90-44. Z Naturforsch 56c:303–307

Tappe R, Taraz K, Budzikiewicz H, Meyer JM, Lefèvre JF (1993) Structure elucidation of a pyoverdin produced by *Pseudomonas aeruginosa* ATCC 27853. J Prakt Chem 335:83–87

Vandamme P, Pot B, Gillis M, De Vos P, Kersters K, Swings J (1996) Polyphasic taxonomy a consensus approach to bacterial systematics. Microbiol Rev 60:407–438

Visca P, Ciervo A, Sanfilippo V, Orsi N (1993) Iron-regulated salicylate synthesis by *Pseudomonas* spp. J Gen Microbiol 139:1995–2001

Voss J, Taraz K, Budzikiewicz H (1999) A pyoverdin from the Antarctica strain 51W of *Pseudomonas fluorescens*. Z Naturforsch 54c:156–162

Vossen W, Fuchs R, Taraz K, Budzikiewicz H (2000) Can the peptide chain of a pyoverdin be bound by an ester to the chromophore? –The old problem of pseudobactin 7SR1. Z Naturforsch 55c:153–164

Weber M, Taraz K, Budzikiewicz H, Geoffroy V, Meyer JM (2000) The structure of a pyoverdine from *Pseudomonas* sp. CFML 96.188 and its relation to other pyoverdines with a cyclic C-terminus. BioMetals 13:301–309

Wong-Lun-Sang S, Bernardini JJ, Hennard C, Kyslic P, Dell A, Abdallah MA (1996) Bacterial siderophores: structure elucidation, 2D 1H and 13C NMR assignments of pyoverdins produced by *Pseudomonas fluorescens* CHA0. Tetrahedron Lett 37:3329–3332

Yamamoto S, Harayama S (1998) Phylogenetic relationships of *Pseudomonas putida* strains deduced from the nucleotide sequences of *gyrB*, *rpoD* and 16S rRNA genes. Int J Syst Bacteriol 48:813–819

3 Siderotyping, a Tool to Characterize, Classify and Identify Fluorescent Pseudomonads

Alain Bultreys

3.1
Introduction

The genus *Pseudomonas* sensu stricto belongs to the γ subclass of the Proteobacteria (Kersters et al. 1996); it is limited to species of the previous *Pseudomonas* rRNA group I (Palleroni 1984). The revised genus contains about 100 species (http://www.dsmz.de/bactnom/nam2400.htm) found in all the major natural environments and using a wide range of substrates. The fluorescent species produce pyoverdins, a siderophore fluorescent under UV light. Among the fluorescent pseudomonads, pathogenic strains are harmful to human, plants or mushroom; saprophytic ones can be useful in bioremediation, biocontrol or plant growth promotion. Bacterial siderophores can be important determinants of these processes and they can be a characteristic of a species. Therefore, it is useful to determine the siderophores produced or used by a strain in a characterization, classification or identification process, a practice sometimes called siderotyping (Meyer et al. 1997).

In this chapter we will discuss the specificity of siderophores in bacteria. Methods to detect a specific siderophore will then be described. Finally, the presently published siderophores produced by the fluorescent pseudomonads, their biological importance and techniques of detection will be presented. The siderophores corrugatin, cepabactin and norcadamine produced by non-fluorescent pseudomonads (Budzikiewicz 2004) will not be described.

Département Biotechnologie, Centre Wallon de Recherches Agronomiques,
Chaussée de Charleroi 234, B-5030 Gembloux, Belgium,
email: bultreys@cra.wallonie.be

Soil Biology, Volume 12
Microbial Siderophores
A. Varma, S.B. Chincholkar (Eds.)
© Springer-Verlag Berlin Heidelberg 2007

3.2
Siderophore Specificity

3.2.1
Specificity of Production

Siderophore production is generally specific at the genus level; for example, pyoverdins are produced only by *Pseudomonas* spp., ornibactin by *Burkholderia* spp. and mycobactin by *Mycobacterium* spp. However, there are exceptions: pyochelin and cepabactin are produced by *Pseudomonas* and *Burkholderia* species; enterobactin by *Klebsiella*, *Enterobacter* and *Erwinia* species; and other examples include corynebactin, ferrioxamines and salmochelin (Budzikiewicz 2004; Winkelmann 2004). In the case of yersiniabactin found in the genera *Yersinia*, *Escherichia*, *Citrobacter*, *Klebsiella*, *Salmonella*, *Enterobacter*, *Pseudomonas* and, perhaps, *Photorhabdus*, horizontal gene transfer is responsible for siderophore propagation (Bach et al. 2000; Schubert et al. 2000; Oeschlaeger et al. 2003; Mokracka et al. 2004; Bultreys et al. 2006). Siderophore production can also be specific at the species level. In pseudomonads, corrugatin, pseudomonine and quinolobactin have been found in one species (Budzikiewicz 2004). Also, pyoverdins can be specific at the species level because the peptide part of the molecule varies (Meyer et al. 2002a).

3.2.2
Specificity of Utilization and Heterologous Uptake

The rule is that for each siderophore produced there is a specific receptor to translocate the iron-bound siderophore back to the cell. However, a siderophore can be incorporated by a strain that is unable to produce it. Cross interactions have been observed with pyoverdins of similar structures (Poole and McKay 2003), and the genomes of *P. aeruginosa* PAO1, *P. putida* KT2440, *P. fluorescens* Pf0 and *P. syringae* DC3000 contain 35, 29, 26 and 23 genes, respectively, that encode outer-membrane receptors (Cornelis and Matthijs 2002; Martins dos Santos et al. 2004). *P. aeruginosa* PAO1 produces only pyoverdin and pyochelin, but its pyoverdin receptor binds a structurally different pyoverdin; other receptors are specific for aerobactin, enterobactin, another pyoverdin, cepabactin, deferrioxamines, deferrichrysin, deferrirubin, coprogen, citrate and *myo*-inositol hexakisphosphate (Poole 2004). In these systems, the heterologous siderophore generally activates the production of its receptor. Siderophore uptake experiments using purified siderophore enable the detection of these siderophore-inducible or constitutive iron transport systems (Poole et al. 1990; Champomier-Verges et al. 1996; Ongena et al. 2002). The ability to use heterologous siderophores is important for competitiveness (Champonier-Verges et al. 1996; Martins dos Santos et al. 2004).

3.3
Siderotyping Methods

In fluorescent pseudomonads, the culture media often used to produce sidero-phores are succinate (SMM) (Meyer and Abdhalla 1978), CAA (Meyer et al. 1998) and Na-gluconate (Budzikiewicz 1993). The glucose asparagine GASN medium (Bultreys and Gheysen 2000) has also already been used to produce pyoverdin, dihydropyoverdin, yersiniabactin and pyridine-2,6-bis(monothiocarboxylic acid).

The bacteria are generally grown in shaken liquid medium, but they have also been grown in a still Petri dish containing liquid medium and one block of agar medium. This technique enabled a considerable improving in pyoverdin and yersiniabactin productions by *P. syringae*, and in yersiniabactin production by *Escherichia coli*, compared to the technique in shaken Erlenmeyer flasks (Bul-treys and Gheysen 2000; Bultreys et al. 2006).

3.3.1
Siderophore Uptake Experiments

A positive response in these tests is not always indicative of the ability to pro-duce the siderophore because heterologous uptake can occur (Fuchs et al. 2001). Growth stimulation tests: a solid culture medium containing the strong iron chelator ethylenediaminedihydroxyphenyl-acetic acid (Meyer et al. 1997) or di-pyridyl (Bultreys et al. 2001) is used. Plates are inoculated and a paper disc im-pregnated with a siderophore is placed on the agar. Growth stimulation around the paper disc is indicative of the uptake of the tested siderophore.

Siderophore-mediated ^{59}Fe uptake: a ^{59}Fe-siderophore complex is incubated in the presence of iron-depleted bacteria and the suspension is filtered. After wash-ing, the radioactivity of the cells on the filter is determined (Munsch et al. 2000).

3.3.2
Electrophoretic Methods

Pyoverdin and yersiniabactin receptors and peptide synthetases can be de-tected after *SDS-PAGE*, either by immunoblotting or by radiography after growth with ^{35}S-labeled amino acids (Meyer et al. 1997; Schubert et al. 1998). *Isoelectric Focusing Electrophoresis (IEF)* followed by an overlay with an iron-containing blue chrome azurol S (CAS) agarose gel (Schwyn and Neilands 1987) is a method used to detect siderophores (Koedam et al. 1994). The siderophores are separated according to their isoelectric point (pI). Pyoverdins are detected after IEF under UV light. Other siderophores are detected after overlay; in the presence of a siderophore, the iron is locally removed from the blue CAS gel, which becomes orange. Fe(III)-chelates of the pyoverdin and dihydropyoverdin of *P. syringae* can be detected after IEF by their natural color (Bultreys et al. 2001).

3.3.3
Chromatographic Methods

Pyochelin can be detected by thin layer chromatography (TLC) (Sokol 1984). Siderophores are detected by high performance liquid chromatography (HPLC), either after extraction, as for pyochelin and pseudomonine (Serino et al. 1997; Kilz et al. 1999; Mercado-Blanco et al. 2001), or in the culture medium, as for pyoverdins and yersiniabactin (Bultreys et al. 2003, 2006). They are identified by their retention times and UV spectra analyzed with a photodiode array detector.

3.3.4
Mass Spectrometry (MS)

One method couples HPLC with electrospray ionisation (ESI)-MS; it enables the determination of the molecular ion of pyoverdins (Kilz et al. 1999). Also, free pyoverdin extracts analyzed by ESI-MS and collision activation can provide information on pyoverdin structures (Fuchs and Budzikiewics 2001).

3.3.5
Use of Modified Indicator Strains

In uptake tests, a strain unable to produce a siderophore but able to use it detects this siderophore in the culture supernatant of tested strains (Mokracka et al. 2004). In another test, the up-regulation of *fyuA* in the presence of yersiniabactin, monitored by a *fyuA-gfp* (green fluorescent protein) reporter fusion, indicates the presence of yersiniabactin in the culture supernatant of tested strains (Schubert et al. 2000).

3.3.6
Genetic Tests

DNA hybridization and/or PCR were used for aerobactin, a novel catecholate siderophore, yersiniabactin, pyridine-2,6-bis(monothiocarboxylic acid) and pseudomonine (Johnson et al. 2001; Mercado-Blanco et al. 2001; Sepúlveda-Torres et al. 2002; Bultreys et al. 2006). Repressed siderophores can be detected. However, the possession of a gene does not always correlate with the ability to produce the siderophore. Also, sequence variations can induce false PCR negatives.

3.4
Siderophores of Fluorescent Pseudomonads

3.4.1
Pyochelin and its By-Product Salicylic Acid

3.4.1.1
Description and Biological Importance

Pyochelin is a salicylic acid-derived siderophore with the formula $C_{14}H_{16}N_2O_3S_2$ (molecular mass 324) produced by strains of *P. aeruginosa*, *P. fluorescens*, *Burkholderia cepacia* and *Burkholderia multivorans* (Cox and Graham 1979; Cox et al. 1981; Sokol 1984). Pyochelin exists in nature as two interconvertible stereoisomers: pyochelin I and II (Rinehart et al. 1995). Fe(III)-dipyochelin has a low stability constant of 5×10^5 (Visca et al. 1992). Pyochelin complexes with Zn(II), Cu(II), Co(II), Mo(VI), and Ni(II) might deliver these metal ions to the cell (Visca et al. 1992). Complexes of pyochelin with vanadium have antibacterial effects (Baysse et al. 2000).

Pyochelin is synthesized from chorismate and two moles of cysteine. Salicylic acid and the iron-chelator and antibiotic dihydroaeruginoic acid (Carmi et al. 1994) are by-products (Crosa and Walsh 2002). PchA and PchB transform chorismate into salicylate (Gaille et al. 2002, 2003). Salicylic acid plays a role in plant defense by inducing systemic acquired resistance (SAR) (Durrant and Dong 2004), and bacteria secreting salicylic acid can induce SAR in plants (De Meyer and Höfte 1997; Maurhofer et al. 1998; De Meyer et al. 1999).

Pyochelin contributes to the virulence of *P. aeruginosa* in mice and humans (Cox 1982; Britigan et al. 1997; Takase et al. 2000), possibly because of siderophore activity (Ankenbauer et al. 1985), but ferripyochelin also enhances hydroxyl radical formation and pulmonary epithelial and artery endothelial cell injury in presence of pyocianin (Britigan et al. 1992, 1997).

Pyochelin and especially ferripyochelin have the capacity of degrading toxic organotins found in the environment, like triphenyltin chloride, by a mechanism involving hydroxyl radical formation (Sun et al. 2006; Sun and Zhong 2006).

Pseudomonads producing pyochelin can play a role in biocontrol. Pyochelin and pyocianin induce resistance against *Botrytis cinerea* in tomato, probably resulting from the formation of reactive oxygen species which play a role in plant defense (Audenaert et al. 2002). Also, pyochelin- and pyoverdin-mediated iron competition protects tomato against *Pythium* (Buysens et al. 1996).

3.4.1.2
Detection Methods

The media used to produce pyochelin are CAA, SMMCA and GGP (Cox and Graham 1979; Visca et al. 1992; Serino et al. 1997; Darling et al. 1998; Reimmann et al. 1998; Takase et al. 2000). The use of 1/10-strength nutrient broth-yeast extract amended with glucose or glycerol increases pyochelin and salicylic acid production (Duffy and Défago 1999).

Pyochelin is a light yellow siderophore with a yellowish-green fluorescence, which can be masked by the pyoverdin. In methanol, it forms a wine-red (pH 2.5) to orange (pH 7.0) non-fluorescent complex with iron. Iron free pyochelin displays absorption maxima at 218, 248 and 310 nm and iron-saturated pyochelin at 237, 255, 325, 425 and 520 (pH 2.5) or 488 (pH 7.0) nm (Cox and Graham 1979). The most widely used methods of detection are TLC and HPLC (Sokol 1984; Ankenbauer et al. 1988; Serino et al. 1997; Darling et al. 1998; Reimmann et al. 1998; Duffy and Défago 1999; Takase et al. 2000; Visser et al. 2004). In TLC, salicylic acid and pyochelin are detected in concentrated chloroform or dichloromethane extracts of acidified culture supernatants. In HPLC, pyochelin isomerizes spontaneously to pyochelin I and II (3:1 ratio); ethyl acetate extracts of acidified culture supernatants are concentrated before injection. Salicylic acid, dihydroaeruginoic acid and pyochelin I and II are identified by their retention times and UV spectra. Pyochelin can be detected by IEF and CAS overlay (Meyer and Geoffroy 2004). The PCR primers 5'-AGATGGACAAAGC-GCCCTGC-3' and 5'-GATGGGCGGAGACGAACAGG-3' amplify (Tm 60 °C) 2139 bp of *pchD* of *P. aeruginosa* PAO1 encoding the salicyl-AMP ligase used in pyochelin synthesis (Serino et al. 1997; Takase et al. 2000).

3.4.2
Pseudomonine

Pseudomonine is a salicylic acid-based siderophore with the formula $C_{16}H_{18}N_4O_4$ (molecular mass 330). It is produced with salicylic acid by *P. fluorescens* AH2 isolated from spoiled Nile perch from Lake Victoria (Anthoni et al. 1995). It is also produced with pyoverdin and salicylic acid by the plant growth-promoting *P. fluorescens* WCS374 (Mercado-Blanco et al. 2001). Iron-regulated metabolites produced by WCS374 induce systemic resistance and the suppression of *Fusarium* wilt in radish, but the role of pseudomonine is not clarified.

The genes *pmsCEAB* are involved in pseudomonine synthesis (Mercado-Blanco et al. 2001). PmsB and PmsC show similarities with PchB and PchA of *P. aeruginosa*, involved in salicylate synthesis.

Pseudomonine was produced in asparagine sucrose broth (Anthoni et al. 1995) or in SMM using a pyoverdin-defective mutant (Mercado-Blanco et al. 2001). Pseudomonine emits a blue fluorescence under UV light, detectable when the pyoverdin is repressed. It shows absorbance maxima at 298, 237 and 203 nm in water (Anthoni et al. 1995), and it is detectable by HPLC (Mercado-Blanco et al. 2001). The PCR primers SAL01 5'-GAACCTCAATGACATTCGAG-3' and SAL02 5'-GTAGAGCTTCTCGACGAAAG-3' amplify (Tm 56 °C) 214 bp of *pmsB* of *P. fluorescens* WCS374. Pseudomonine production was detected by RT-PCR (Mercado-Blanco et al. 2001).

3.4.3
Yersiniabactin

3.4.3.1
Description and Biological Importance

Yersiniabactin is salicylic acid based siderophore of formula $C_{21}H_{27}N_3O_4S_3$ (molecular mass 481) produced by *P. syringae* (Bultreys et al. 2006). It was initially characterized in *Yersinia pestis*, the causal agent of bubonic plague (Haag et al. 1993; Drechsel et al. 1995; Chambers et al. 1996). Yersiniabactin is widespread among human and animal pathogenic enterobacteria, such as *Yersinia* spp., *Escherichia coli*, *Citrobacter* spp., *Klebsiella* spp., *Salmonella enterica* and *Enterobacter* spp. (Bach et al. 2000; Schubert et al. 2000; Carniel 2001; Oelschlaeger et al. 2003; Mokracka et al. 2004). The complete yersiniabactin iron uptake system, called the yersiniabactin locus, is located in a genomic high-pathogenicity island transmissible by horizontal gene transfer (Carniel 2001; Perry 2004; Schmidt and Hensel 2004; Antonenka et al. 2005). Genes that are similar to yersiniabactin genes have been detected in the insect pathogen *Photorhabdus luminescens* (Duchaud et al. 2003) and in the plant pathogen *P. syringae* (Buell et al. 2003), which is divided into many pathovars and nine genospecies (Gardan et al. 1999). A yersiniabactin locus was recently detected in three genospecies of *P. syringae*, but most generally not in the other genospecies and in representatives of other pseudomonads (Bultreys et al. 2006). However, the locus organization and gene sequences are different compared to the enterobacteria, and the locus is not located in a high-pathogenicity island as usual; also, the gene organizations of *P. luminescens* and *P. syringae* are closer to each other than to the *Y. pestis* group. Interestingly, a *Pseudomonas* strain produces the Zn-, Cu- or Fe-containing antimycoplasma agent micacocidin A, B and C, which strongly resembles yersiniabactin (Kobayashi et al. 1998).

Yersiniabactin is synthesized from chorismate by the salicylate synthase Irp9/YbtS, the salicyl-AMP ligase Irp5/YbtE, the peptide synthetase high-molecular-weight protein (HMWP) 2, the polyketide synthase/peptide synthetase HMWP1 and the thiazoline reductase Irp3/YbtU (Crosa and Walsh 2002). Only one protein usually converts chorismate in salicylate in yersiniabactin synthesis (Kerbarh et al. 2005), but two genes homologous to *pchA* and *pchB* of *P. aeruginosa* seem involved in this conversion in *P. syringae* (Bultreys et al. 2006).

In *Yersinia* spp., yersiniabactin is a virulence factor: it is indispensable in the early stage of infection of *Y. pestis*, and defective mutants of *Y. pseudotuberculosis* and *Y. enterocolitica* show a loss of virulence (Perry 2004). In *E. coli*, the involvement of the HPI in virulence is not as clear, and two in vivo studies drew different conclusions (Schubert et al. 2002; Lefranc Nègre et al. 2004).

In *P. syringae*, the existence of the genospecies 1, 2, 4 and 6 within which the strains do not produce yersiniabactin indicates that pathogenicity is possible without producing yersiniabactin. Strains defective in pyoverdin production belonging to the pathovars *tomato* and *persicae* produce yersiniabactin, but the advantage of producing both yersiniabactin and pyoverdin is unclear; the very high stability constant (4×10^{36}; Perry et al. 1999) of ferriyersiniabactin (compared with 10^{25} for the pyoverdin) could carry an adaptive advantage (Bultreys et al. 2006).

3.4.3.2
Yersiniabactin and Taxonomy

Yersiniabactin is informative on the evolution within *P. syringae* and in classification. The different GC contents in the yersiniabactin locus and in the chromosome indicate a yersiniabactin locus acquisition by horizontal gene transfer, either by an ancestor of the producing pathovars followed by stabilization in the chromosome, or by an ancestor of *P. syringae* followed by a locus deletion in an ancestor of the non producing pathovars (Bultreys et al. 2006). This is confirmed by a DNA hybridization study (Gardan et al. 1999): only the pathovars of the genospecies 3, 7 and 8 have a yersiniabactin locus, except for two pathovars belonging to the genospecies 2. Only two exceptions in the genospecies 3 (Bultreys et al. 2006) are an error and a possible misidentification: the negative strain *P. syringae* pv. *ribicola* LMG 2276 (CFBP 2348) actually belongs to the genospecies 6 (Gardan et al. 1999); *P. syringae* pv. *maculicola* CFBP 1657 used in the DNA hybridization study has a yersiniabactin locus (Bultreys and Gheysen, unpublished), whereas the genospecies of the negative strain LMG 5295 is unknown. This correlation renders yersiniabactin detection a strong information in early classification of a strain, and in the study of the *P. syringae* evolution.

3.4.3.3
Methods and Yersiniabactin Use in Identification

As evoked in 'Siderotyping methods', yersiniabactin is detected among enterobacteria using growth stimulation tests, SDS-PAGE, HPLC of the culture medium, modified indicator strains and genetic tests.

Among pseudomonads (Bultreys et al. 2006), yersiniabactin was produced in solid-liquid GASN or King B media in still Petri dishes. Yersiniabactin is nearly colorless and orange when chelated to iron. Ferriyersiniabactin was detected in the culture medium by HPLC and identified by its spectral characteristics: absorbance maxima near 227, 255, 305, and 386 nm at pH 5.3 and pH 7.0. The PCR primers PSYE2 5'-GGCACCTGGAACAGG-3' and PSYE2R 5'-GCCA-GATCGTCCATCAT-3' amplify (Tm 64 °C) a fragment of the *irp1* gene (encoding HMWP1) of 943 bp in *P. syringae* and 925 bp in *Escherichia coli*, but they are ineffective for the *P. syringae* pathovars *glycinea* and *phaseolicola* of genospecies 2, which also have a yersiniabactin locus. Dot blot using several washing conditions enabled a general detection of *irp1* in both *P. syringae* and enterobacteria. A PCR test using the primers PT3 and PT3R is specific for *P. syringae* and is proposed to identify all the yersiniabactin producing pathovars on their respective hosts (Bultreys and Gheysen 2006).

3.4.4
Pyridine-2,6-bis(monothiocarboxylic acid) (PDTC)

PDTC (molecular mass 198) was purified from strains of *P. putida* (Ockels et al. 1978) and *P. stutzeri* KC (Lee et al. 1999). It converts the pollutant CCl_4 to CO_2 in iron-limiting conditions (Lee et al. 1999; Lewis et al. 2001). In *P. putida*, PDTC is a siderophore repressed by the pyoverdin (Lewis et al. 2004). It forms 2:1 complexes of comparable stability ($\sim 10^{33}$) with iron, nickel and cobalt (Stolworthy et al. 2001). PDTC forms complexes with 14 metals and can protect bacteria from mercury, cadmium, as well as selenium and tellurium oxyanions; it is involved in an initial line of defense of bacteria against toxicity from various metals and metalloids (Cortese et al. 2002a; Zawadzka et al. 2006). Cu(II) and PDTC render strains able to reduce amorphous Fe(III) oxyhydroxide (Cortese et al. 2002a).

At least five genes are involved in the PDTC system in *P. stutzeri* KC (Lewis et al. 2000; Sepúlveda-Torres et al. 2002). Two of them, *orfF* and *orfI*, were not detected in seven *P. stutzeri*, one *P. balearica*, or a *P. putida* producing PDTC. This suggests that *P. stutzeri* KC may possess a distinct biosynthetic pathway

(Sepúlveda-Torres et al. 2002), which may have been acquired from mycobacteria and cyanobacteria (Cortese et al. 2002b).

PDTC is produced in DRM (Lee et al. 1999) or GASN (Bultreys and Gheysen 2000) shaken liquid media (Zawadzka et al. 2006). It forms a blue Fe(II)-complex and a brown Fe(III)-complex; the absorbance maxima of Fe(III)-(PDTC)$_2$ are 345, 468, 604 and 740 nm and of Fe(II)-(PDTC)$_2$ are 314 and 687 nm (Cortese et al. 2002a). PDTC concentration is usually determined by measuring the absorbance of Fe(II)-(PDTC)$_2$ at 687 nm (Budzikiewicz et al. 1983). CCl$_4$ degradation can indicate PDTC production: cultures in medium D supplemented with CCl$_4$ are incubated under denitrifying conditions and CCl$_4$ is measured by gas chromatography (Tatara et al. 1993). The PCR primers CC109f 5'-GTTA-CAGCCGCCACCTACTGAT-3' and CC110r 5'-GCTAGGCAGAGAAGAGTC-CACG-3' amplify 1112 bp of *orfF* and the primers CC111f 5'-GGCTGCTCAG-TATCGGCAGTAT-3' and CC112r 5'-GGGGCGTTGACAGAGAAGTAAG-3' 1385 bp of *orfI* of *P. stutzeri* KC, and a Southern hybridization method is described (Sepúlveda-Torres et al. 2002).

3.4.5
Quinolobactin

Quinolobactin (8-hydroxy-4-methoxy-2-quinoline carboxylic acid) is a secondary siderophore with a low affinity constant for Fe(III) produced by *P. fluorescens* (Neuenhaus et al. 1980; Mossialos et al. 2000). Quinolobactin is produced in the first 16 h of iron stress before it is suppressed by the pyoverdin; this could be the first way of producing strains dealing with iron limitation (Mossialos et al. 2000; Cornelis and Matthijs 2002). A pathway for quinolobactin synthesis from xanthurenic acid has been proposed (Matthijs et al. 2004).

A quinolobactin spot was detected by IEF and CAS overlay of CAA culture supernatants for a pyoverdin-defective mutant; the spot was detected in the wild-type preparation only after concentration (Mossialos et al. 2000).

3.4.6
Pyoverdin (Pseudobactin)

3.4.6.1
Description and Biological Importance

A pyoverdin is made up of (i) a quinoline chromophore, (ii) a peptide chain of 6 to 12 amino acids containing about half *d*-amino acids and (iii) an acid (amide) side chain consisting of a dicarboxylic acid (amide) (Budzikiewicz

1993, 1997, 2004). The peptide chain is strain specific, except in one case (Barelmann et al. 2003), and is variable among strains and species. About 50 peptide chains are known, but 106 are predicted (Meyer and Geoffroy 2004). The catechol of the chromophore and two amino acids are involved in iron chelation; the amino acids are either β-hydroxy aspartic acid (in one case, β-hydroxy histidine) or hydroxamic acids derived from ornithine. Several pyoverdins varying according to the presence of a cycle in the peptide chain and by the side chain can be found in the culture medium; some are degradation products of the secreted forms (Schäfer et al. 1991; Bultreys et al. 2004). Pyoverdin precursors such as ferribactin and dihydropyoverdin, as well as isopyoverdins, vary in the nature of the chromophore; with rare exceptions (Jacques et al. 1995; Bultreys et al. 2001) they are produced at a much lower concentration than the pyoverdin. Azotobactin produced by pseudomonads and by *Azotobacter vinelandii* differs in the structure of the chromophore and the absence of the acid side chain (Demange et al. 1988).

The uptake of the Fe(III)-chelated pyoverdin is carried out by a specific receptor located in the outer membrane, which recognizes the peptide chain of its cognate pyoverdin. Pyoverdins with small differences (Ruangviriyachai et al. 2001; Barelmann et al. 2002, 2003; Fernández et al. 2003; Bultreys et al. 2004) or a common motif (Meyer et al. 1999, 2002b; Weber et al. 2000; Schlegel et al. 2001) in their peptide chain are incorporated at a reduced or high rate by a same receptor. Heterologous pyoverdins can stimulate the production of specific additional receptors and be incorporated (Morris et al. 1992; Koster et al. 1993, 1995; Leoni et al. 2000).

Pyoverdins are the principal siderophore of the fluorescent pseudomonads. The pyoverdin of *P. aeruginosa* is involved in virulence in animal models (Meyer et al. 1996; Handfield et al. 2000; Takase et al. 2000). It is able to acquire iron from transferrin and lactoferrin (Xiao and Kisaalita 1997) and it regulates the production of three virulence factors: exotoxine A, an endoprotease and pyoverdin itself (Lamont et al. 2002; Beare et al. 2003). On the other hand, the pyoverdin of *P. syringae* is not involved in virulence in cherry fruits (Cody and Gross 1987), but its production is stimulated in conditions found on plant surface when *P. syringae* has to use amino acids as carbon sources (Bultreys and Gheysen 2000).

Because of their high affinity constants for iron, between 10^{24} and 10^{27} (Budzikiewicz 2004), pyoverdins can be involved in competitiveness, growth promotion and biocontrol (Kloepper et al. 1980; Loper and Buyer 1991; Lemanceau et al. 1992; O'Sullivan and O'Gara 1992; Raaijmakers et al. 1995a; Buysens et al. 1996; Ambrosi et al. 2000; Mirleau et al. 2001). Strains producing a specific pyoverdin and using heterologous siderophores are favored (Buyer and Leong 1986; Jurkevitch et al. 1992; Raaijmakers et al. 1995b; Mirleau et al. 2000; Martins dos Santos et al. 2004). Pyoverdins can also play a role in bioremediation by degrading triphenyltin, an aquatic pollutant armful to plankton, gastropods and fish (Inoue et al. 2000, 2003). Pyoverdins bind and oxidize Fe(II) (Xiao and Kisaalita 1998).

3.4.6.2
The Peptide Chain of Pyoverdins and its Evolution

In *P. aeruginosa*, the pyoverdin genes are located in the *pvd* locus, or in a distant place for several chromophore-related genes (Poole 2004). The peptide synthetases PvdD, PvdI and PvdJ are responsible for the non-ribosomal synthesis of the peptide chain (Merriman et al. 1995; Lehoux et al. 2000). They contain as many enzymatic modules as there are amino acids in the pyoverdin; each module is specific for one amino acid (Kleinkauf and von Döhren 1996; von Döhren et al. 1999). The synthesis occurs by transfer of the intermediate from one module to the next without releasing it into the cytoplasm. One domain in each module is selective for one amino acid. A modification in this domain can induce the replacement of one amino acid by another. This is probably one way that pyoverdins evolve, as noted between the pyoverdins of *P. syringae* and *P. cichorii* differing by the replacement of one serine by glycine (Bultreys et al. 2004). Also, a deletion in a peptide synthetase can give a shorter peptide, as noted for the rare fourth type pyoverdin of *P. aeruginosa* which differs from the third type in that there is a missing glutamine (Smith et al. 2005).

The central part of the *pvd* locus is the most divergent locus between strains of the three principal pyoverdin types in the *P. aeruginosa* genome; a high variation is localized in the genes encoding the membrane receptor FpvA, the ABC transporter PvdE and the peptide synthetases PvdD, PvdJ and PvdI (Spencer et al. 2003; Smith et al. 2005). Horizontal gene transfers probably explain these differences because there are unusual codon and tetranucleotide usages. The pyoverdin and its receptor co-evolve and the changes in the receptor, resulting from horizontal gene transfers probably from other pseudomonads, *Agrobacterium tumefaciens* and *Azotobacter vinelandii*, appear to lead to further changes in the pyoverdin (Smith et al. 2005).

3.4.6.3
Pyoverdins and Phylogeny

In phytopathogenic fluorescent pseudomonads, an evolution is apparent in the peptide chains of pyoverdins (Bultreys et al. 2003, 2004): *P. syringae*, *P. viridiflava* and *P. ficuserectae* produce the same pyoverdin; the related species *P. cichorii* produces a pyoverdin differing in the replacement of one serine by glycine; and the distant species *P. fuscovaginae* and *P. asplenii* produce a clearly different, but related, pyoverdin. All these pyoverdins contain two Asp-based iron ligands and, interestingly, the producing species, apart from *P. fuscovaginae* and *P. asplenii*, are arginine dihydrolase-negative. In the arginine dihydrolase-positive species, the 4 pyoverdins of *P. aeruginosa* contain

2 Orn-based ligands, 19 pyoverdins of *P. fluorescens* contain either 2 Orn- or 1 Orn- and 1 Asp-based ligands, and 13 pyoverdins of *P. putida* always contain 1 Orn- and 1 Asp-based ligands; the rest of the peptide chains of these pyoverdins, however, are highly variable. Then, it appears that the amino acids involved in iron chelation evolve slowly because of their necessity for the pyoverdin activity and they could therefore be useful markers in phylogeny (Bultreys et al. 2003). This is confirmed because *pvdA* and *pvdF* necessary for the synthesis of the iron ligand formyl-hydroxyornithine from Orn (Visca et al. 1994; Wilson et al. 2001), and the Orn-based ligands, are conserved in the 4 *P. aeruginosa* pyoverdin types (Smith et al. 2005), although the conserved *pvdF* is positioned just beside the highly variable region of the *pvd* locus.

It is difficult to find filiations between pyoverdins of the arginine dihydrolase-positive species, except when a pyoverdin is produced by different species (Fuchs and Budzikiewicz 2001; Meyer and Geoffroy 2004). The diversifying selection observed in the *pvd* locus in *P. aeruginosa* (Smith et al. 2005) indicates that the heterogeneity apparent in pyoverdins can be higher than the general heterogeneity in a species. Horizontal gene transfers could explain the numerous pyoverdin structures found in *P. fluorescens* and *P. putida* and restrict the phylogenic information available from pyoverdins. It is also a sign of a high selection pressure for new specific iron-chelating systems in the rhizosphere (Smith et al. 2005).

3.4.6.4
Pyoverdins and Taxonomy

Siderotyping of pyoverdins by IEF and siderophore uptake is used to revise the genus *Pseudomonas*, alongside polyphasic taxonomic approach, and the general rules are defined: (i) all strains belonging to a given species produce an identical pyoverdin; and (ii) each species is characterized by an original pyoverdin. Indeed, complex taxonomic studies were elegantly and rapidly confirmed by the description of one corresponding siderotype for at least 11 species: *P. mandelii, P. monteilii, P. rhodesiae, P. tolaasii, P. costantinii, P. brassicacearum, P. thivervalensis, P. salomonii, P. mosselii, P. libanensis* and *P. kilonensis* (Meyer et al. 2002a; Meyer and Geoffroy 2004). However, the same pyoverdin can be produced by related species: *P. syringae, P. viridiflava* and *P. ficuserectae; P. asplenii* and *P. fuscovaginae; P. fluorescens, P. cedrella, P. orientalis, P. palleroniana* and *P. veronii; P. brenneri* and *P. gessardii;* and *P. jessenii* and *P. migulae* (Bultreys et al. 2003; Meyer and Geoffroy 2004). Also, several pyoverdins can be produced in one species: *P. aeruginosa, P. fluorescens, P. putida, P. grimontii* and *P. lini* (Fuchs and Budzikiewicz 2001; Meyer and Geoffroy 2004). The recent observation that the heterogeneity in pyoverdins can be higher than in the species (Smith et al. 2005) should be noted in future work.

3.4.6.5
Pyoverdins and Identification

The strains producing the same pyoverdin are grouped into siderovars. Once the siderovars are defined in a species, pyoverdins are accurate tools of identification; this can be achieved by comparison with a reference (Bultreys et al 2001, 2003), or a general database regrouping the characteristics of all the pyoverdins can be consulted (Meyer et al. 2002a; Meyer and Geoffroy 2004).

Identification is often required for phytopathogenic species. The presence of two Asp-based iron ligands in the peptide chain of the pyoverdins of *P. syringae*, *P. cichorii* and *P. fuscovaginae* influences the color and spectral characteristics of the Fe(III)-chelates between pH 3 and 7. This is easily detected using visual and spectrophotometric tests differentiating phytopathogenic and saprophytic species; the pathogens are identified by HPLC (Bultreys et al. 2001, 2003).

3.4.6.6
Methods

Yellowish-green pyoverdins are detected in King B medium (King et al. 1954) under UV light (365 nm) by their bluish-green fluorescence. Absorbance maxima of free pyoverdins in the visible are 365 and 380 nm (pH<5), 402 nm (pH 7) and 410 nm (pH 10) (Meyer and Abdallah 1978). The absorbance maximum of Fe(III)-chelated typical pyoverdins is near 400 nm (pH 3–8), with broad charge transfer bands at 470 and 550 nm (Budzikiewicz 1993). The Fe(III)-chelated atypical pyoverdins containing two Asp-based iron ligands of *P. syringae* and *P. cichorii* behave as a typical pyoverdin at pH<3.5 (brown), but the maximum shifts at 408 nm at pH>5.5, without charge transfer bands (orange); the pyoverdin of *P. fuscovaginae* behaves identically at neutral pH but the maximum shifts to only 402.5 nm at pH 3, without marked charge transfer bands (dark orange). This can be observed visually and by spectrophotometry in GASN medium, which enables the detection of phytopathogenic species (Bultreys et al. 2001, 2003).

Three principal methods are used to analyse pyoverdin diversity. *MS*-related methods are the most powerful but they are expensive (Kilz et al. 1999; Fuchs et al. 2001). *IEF* of iron-free pyoverdins coupled with ^{59}Fe uptake experiments have become the methods of choice (Meyer et al. 2002a). The pI of pyoverdins detected under UV light from concentrated CAA pyoverdin extracts (generally 2 or 3 isoforms differing in the side chain) are determined. Each strain is defined by its IEF pattern. This method enables a database to be constituted. The problems encountered are the migration of the pyoverdins at the cathode or the anode because the limits of analysis are between pH 4 and 9 (Achouak et al. 2000; Fuchs et al. 2001), the observation of the same pI for pyoverdins varying by a neutral amino acid (Bultreys et al. 2003; Fernández et al. 2003) and the observation of different patterns for strains producing the same pyoverdin (Fuchs et al. 2001; Meyer and Geoffroy 2004). Free or Fe(III)-chelated pyoverdins of *P. syringae* and *P. cichorii* are visually detected but not differentiated by IEF (Bultreys et al. 2003). *HPLC* analysis of GASN culture medium is used to identify phytopathogenic fluorescent pseudomonads (Bultreys et al. 2003). Atypical pyoverdins of *P. syringae*, *P. cichorii* and *P. fuscovaginae* containing two Asp-based iron ligands are differentiated and identified by their retention time and their absorbance maximum being near 408 nm at pH 5.3. The technique is easy to use and more accurate than IEF, but less suited to developing a general database.

3.5
Conclusions

The diversifying evolution detected in pyoverdin genes in *P. aeruginosa* and the abundance of outer membrane receptors in pseudomonads indicate that iron competition is an important selection pressure in the rhizosphere. The existence of secondary siderophores and the production of outer membrane receptors in the presence of heterologous siderophores imply sophisticated regulatory mechanisms in iron-deficient environments. Therefore, the importance of siderophores in fitness and competitiveness seems clear and it can explain the observed involvement of siderophores in virulence, plant growth promotion and biocontrol. Also, siderophores can have useful or toxic secondary effects, and the information from siderophores produced by a pseudomonad is important in understanding, controlling and using pseudomonad behavior. Recent findings on the horizontal gene transfers of complete or partial siderophore systems indicate that gene exchange has been a creative force during evolution, alongside clonal divergence and periodic selection. This probably restricts the phylogenic information available from siderophores, although this information is sometimes clear, and can occasionally be linked to the transfer event itself. Some siderophores currently appear to be specific to certain species or to a group of closely related species and they are therefore of interest in taxonomy and identification. Existing methods enable researchers to determine rapidly the siderophores produced by a pseudomonad, and they are of interest for medicine, agronomy, environmental science and systematic bacteriology.

References

Achouak W, Sutra L, Heulin T, Meyer J-M, Fromin N, Degraeve S, Christen R, Gardan L (2000) *Pseudomonas brassicacearum* sp. nov. and *Pseudomonas thivervalensis* sp. nov., two root-associated bacteria isolated from *Brassica napus* and *Arabidopsis thaliana*. Int J Syst Evol Microbiol 50:9–18

Ambrosi C, Leoni L, Putignani L, Orsi N, Visca P (2000) Pseudobactin biogenesis in the plant growth-promoting rhizobacterium *Pseudomonas* strain B10: identification and functional analysis of the L-Ornithine N^5-oxygenase (*psbA*) gene. J Bacteriol 182:6233–6238

Ankenbauer R, Sriyosachati S, Cox CD (1985) Effects of siderophores on the growth of *Pseudomonas aeruginosa* in human serum and transferrin. Infect Immun 49:132–140

Ankenbauer RG, Toyokuni T, Staley A, Rinehart KL, Cox CD (1988) Synthesis and biological activity of pyochelin, a siderophore of *Pseudomonas aeruginosa*. J Bacteriol 170:5344–5351

Anthoni U, Christophersen C, Nielsen PH, Gram L, Petersen BO (1995) Pseudomonine, an isoxazolidone with siderophoric activity from *Pseudomonas fluorescens* AH2 isolated from Lake Victoria Nile perch. J Nat Prod 58:1786–1789

Antonenka U, Nölting C, Heesemann J, Rakin A (2005) Horizontal transfer of *Yersinia* high-pathogenicity island by the conjugative RP4 *attB* target-presenting shuttle plasmid. Mol Microbiol 57:727–734

Audenaert K, Pattery T, Cornelis P, Höfte M (2002) Induction of systemic resistance to *Botrytis cinerea* in tomato by *Pseudomonas aeruginosa* 7NSK2: role of salicylic acid, pyochelin, and pyocyanin. Mol Plant Microbe Interact 15:1147–1156

Bach S, de Almeida A, Carniel E (2000) The *Yersinia* high-pathogenicity island is present in different members of the family *Enterobacteriaceae*. FEMS Microbiol Lett 183:289–294

Barelmann I, Taraz K, Budzikiewicz H, Geoffroy V, Meyer J-M (2002) The structures of the pyoverdins from two *Pseudomonas fluorescens* strains accepted mutually by their respective producers. Z Naturforsch 57c:9–16

Barelmann I, Fernandez DU, Budzikiewicz H, Meyer J-M (2003) The pyoverdine from *Pseudomonas chlororaphis* D-TR133 showing mutual acceptance with the pyoverdine of *Pseudomonas fluorescens* CHAO. BioMetals 16:263–270

Baysse C, De Vos D, Naudet Y, Vandermonde A, Ochsner U, Meyer JM, Budzikiewicz H, Schäfer M, Fuchs R, Cornelis P (2000) Vanadium interferes with siderophore-mediated iron uptake in *Pseudomonas aeruginosa*. Microbiology 146:2425–2434

Beare PA, For RJ, Martin LW, Lamont IL (2003) Siderophore-mediated cell signalling in *Pseudomonas aeruginosa*: divergent pathways regulate virulence factor production and siderophore receptor synthesis. Mol Micobiol 47:195–207

Britigan BE, Roeder TL, Rasmussen GT, Shasby DM, McCormick ML, Cox CD (1992) Interaction of the *Pseudomonas aeruginosa* secretory products pyocyanin and pyochelin generates hydroxyl radical and causes synergistic damage to endothelial cells: implications for pseudomonas-associated tissue injury. J Clin Invest 90:2187–2196

Britigan BE, Rasmussen GT, Cox CD (1997) Augmentation of oxidant injury to human pulmonary epithelial cells by the *Pseudomonas aeruginosa* siderophore pyochelin. Infect Immun 65:1071–1076

Budzikiewicz H (1993) Secondary metabolites from fluorescent pseudomonads. FEMS Microbiol Rev 104:209–228

Budzikiewicz H (1997) Siderophores of fluorescent pseudomonads. Z Naturforsch 52c:713–720

Budzikiewicz H (2004) Siderophores of the Pseudomonadaceae *sensu sticto* (Fluorescent and non fluorescent *Pseudomonas* spp.). In: Herz W, Falk H, Kirby GW (eds) Progress in the chemistry of organic natural products, vol. 87. Springer, Berlin Heidelberg NewYork, pp 81–237

Budzikiewicz H, Hildebrand U, Ockels W, Reiche M, Taraz K (1983) Weitere aus dem Kulturmedium von *Pseudomonas putida* isolierte Pyridinderivate - Genuine Metaboliten oder Artefackte? Z Naturforsch 38b:516–520

Buell CR, Joardar V, Lindeberg M, Selengut J, Paulsen IT, Gwinn ML, Dodson RJ, Deboy RT, Durkin AS, Kolonay JF, Madupu R, Daugherty S, Brinkac L, Beanan MJ, Haft DH, Nelson WC, Davidsen T, Zafar N, Zhou L, Liu J, Yuan Q, Khouri H, Fedorova N, Tran B,

Russell D, Berry K, Utterback T, Van Aken SE, Feldblyum TV, D'Ascenzo M, Deng W-L, Ramos AR, Alfano JR, Cartinhour S, Chatterjee AK, Delaney TP, Lazarowitz SG, Martin GB, Schneider DJ, Tang X, Bender CL, White O, Fraser CM, Collmer A (2003) The complete genome sequence of the *Arabidopsis* and tomato pathogen *Pseudomonas syringae* pv. *tomato* DC3000. Proc Nat Acad Sci USA 100:10181–10186

Bultreys A, Gheysen I (2000) Production and comparison of peptide siderophores from strains of distantly related pathovars of *Pseudomonas syringae* and *Pseudomonas viridiflava* LMG 2352. Appl Environ Microbiol 66:325–331

Bultreys A, Gheysen I (2006) Yersiniabactin and pyoverdin siderophore uses in *Pseudomonas syringae* identification. In: Elphinstone J, Weller S, Thwaites R, Parkinson N, Stead D, Saddler G (eds) Proceedings of the 11th International Conference on Plant Pathogenic Bacteria, July 10–14, Edinburgh, United Kingdom, p 121

Bultreys A, Gheysen I, Maraite H, de Hoffmann E (2001) Characterization of fluorescent and nonfluorescent peptide siderophores produced by *Pseudomonas syringae* strains and their potential use in strain identification. Appl Environ Microbiol 67:1718–1727

Bultreys A, Gheysen I, Wathelet B, Maraite H, de Hoffmann E (2003) High-performance liquid chromatography analyses of pyoverdin siderophores differentiate among phytopathogenic fluorescent *Pseudomonas* species. Appl Environ Microbiol 69:1143–1153

Bultreys A, Gheysen I, Wathelet B, Schäfer M, Budzikiewicz H (2004) The pyoverdins of *Pseudomonas syringae* and *Pseudomonas cichorii*. Z Naturforsch 59c:613–618

Bultreys A, Gheysen I, de Hoffmann E (2006) Yersiniabactin production by *Pseudomonas syringae* and *Escherichia coli*, and description of a second yersiniabactin locus evolutionary group. Appl Environ Microbiol 72:3814–3825

Buyer JS, Leong J (1986) Iron transport-mediated antagonism between plant growth-promoting and plant-deleterious *Pseudomonas* strains. J Biol Chem 261:791–794

Buysens S, Heungens K, Poppe J, Höfte M (1996) Involvement of pyochelin and pyoverdin in suppression of *Pythium*-induced damping-off of tomato by *Pseudomonas aeruginosa* 7NSK2. Appl Environ Microbiol 62:865–871

Carmi R, Carmeli S, Levy E, Gough FJ (1994) (+)-(*S*)-dihydroaeruginoic acid, an inhibitor of *Septoria tritici* and other phytopathogenic fungi and bacteria, produced by *Pseudomonas fluorescens*. J Nat Prod 57:1200–1205

Carniel E (2001) The *Yersinia* high-pathogenicity island: an iron-uptake island. Microbes Infect 3:561–569

Chambers CE, McIntyre DD, Mouck M, Sokol PA (1996) Physical and structural characterization of yersiniophore, a siderophore produced by clinical isolates of *Yersinia enterocolitica*. BioMetals 9:157–167

Champomier-Verges MC, Stintzi A, Meyer J-M (1996) Acquisition of iron by the non-siderophore-producing *Pseudomonas fragi*. Microbiology 142:1191–1199

Cody YS, Gross DC (1987) Outer membrane protein mediating iron uptake via pyoverdin$_{pss}$, the fluorescent siderophore produced by *Pseudomonas syringae* pv. *syringae*. J Bacteriol 169:2207–2214

Cornelis P, Matthijs S (2002) Diversity of siderophore-mediated iron uptake systems in fluorescent pseudomonads: not only pyoverdines. Environ Microbiol 4:787–798

Cortese MS, Paszczynski A, Lewis TA, Sebat JL, Borek V, Crawford RL (2002a) Metal chelating properties of pyridine-2,6-bis (thiocarboxylic acid) produced by *Pseudomonas* spp. and the biological activities of the formed complexes. Biometals 15:103–120

Cortese MS, Caplan AB, Crawford RL (2002b) Structural, functional, and evolutionary analysis of *moeZ*, a gene encoding an enzyme required for the synthesis of the *Pseudomonas* metabolite, pyridine-2,6-bis (thiocarboxylic acid). BMC Evol Biol 2:1–11

Cox CD (1982) Effect of pyochelin on the virulence of *Pseudomonas aeruginosa*. Infect Immun 36:17–23

Cox CD, Graham R (1979) Isolation of an iron-binding compound from *Pseudomonas aeruginosa*. J Bacteriol 137:357–364

Cox CD, Rinehart KL, Moore ML, Cook JC (1981) Pyochelin: novel structure of an iron-chelating growth promoter for *Pseudomonas aeruginosa*. Proc Natl Acad Sci USA 78:4256–4260

Crosa JH, Walsh CT (2002) Genetics and assembly line enzymology of siderophore biosynthesis in bacteria. Microbiol Mol Biol Rev 66:223–249

Darling P, Chan M, Cox AD, Sokol PA (1998) Siderophore production by cystic fibrosis isoltes of *Burkholderia cepacia*. Infect Immun 66:874–877

De Meyer G, Höfte M (1997) Salicylic acid produced by the rhizobacterium *Pseudomonas aeruginosa* 7NSK2 induces resistance to leaf infection by *Botrytis cinerea* on bean. Phytopathology 87:588–593

De Meyer G, Capieau K, Audenaert K, Buchala A, Métraux JP, Höfte M (1999) Nanogram amounts of salicylic acid produced by the rhizobacterium *Pseudomonas aeruginosa* 7NSK2 activate the systemic acquired resistance pathway in bean. Mol Plant Microbe Interact 12:450–458

Demange P, Bateman A, Dell A, Abdallah MA (1988) Structure of azotobactin D, a siderophore of *Azotobacter vinelandii* strain D (CCM 289). Biochemistry 27:2745–2752

Drechsel H, Stephan H, Lotz H, Haag H, Zähner H, Hantke K, Jung G (1995) Structure elucidation of yersiniabactin, a siderophore from highly virulent *Yersinia* strains. Liebigs Ann 1995:1727–1733

Duchaud E, Rusniok C, Frangeul L, Buchrieser C, Givaudan A, Taourit S, Bocs S, Boursaux-Eude C, Chandler M, Charles J-F, Dassa E, Derose R, Derzelle S, Freyssinet G, Gaudriault S, Médigue C, Lanois A, Powell K, Siguier P, Vincent R, Wingate V, Zouine M, Glaser P, Boemare N, Danchin A, Kunst F (2003) The genome sequence of the entomopathogenic bacterium *Photorhabdus luminescens*. Nat Biotechnol 21:1307–1313

Duffy BK, Défago G (1999) Environmental factors modulating antibiotic and siderophore biosynthesis by *Pseudomonas fluorescens* biocontrol strains. Appl Environ Microbiol 65:2429–2438

Durrant WE, Dong X (2004) Systemic acquired resistance. Annu Rev Phytopathol 42:185–209

Fernández DU, Geoffroy V, Schäffer M, Meyer J-M, Budzikiewicz H (2003) Structure revision of several pyoverdins produced by plant-growth promoting and plant-deleterious *Pseudomonas* species. Monatsh Chem 134:1421–1431

Fuchs R, Budzikiewicz H (2001) Structural studies of pyoverdins by mass spectrometry. Curr Org Chem 5:265–288

Fuchs R, Schäfer M, Geoffroy V, Meyer J-M (2001) Siderotyping-a powerful tool for the characterization of pyoverdines. Curr Top Med Chem 1:31–57

Gaille C, Kast P, Haas D (2002) Salicylate biosynthesis in *Pseudomonas aeruginosa*. J Biol Chem 277:21768–21775

Gaille C, Reimmann C, Haas D (2003) Isochorismate synthase (PchA), the first and rate-limiting enzyme in salicylate biosynthesis of *Pseudomonas aeruginosa*. J Biol Chem 278:16893–16898

Gardan L, Shafik H, Belouin S, Broch R, Grimont F, Grimont PAD (1999) DNA relatedness among the pathovars of *Pseudomonas syringae* and description of *Pseudomonas tremae* sp. nov. and *Pseudomonas cannabina* sp. nov. (ex Sutic and Dowson 1959). Int J Syst Bacteriol 49:469–478

Haag H, Hankte K, Drechsel H, Stojiljkovic I, Jung G, Zähner H (1993) Purification of yersiniabactin: a siderophore and possible virulence factor of *Yersinia enterocolitica*. J Gen Microbiol 139:2159–2165

Handfield M, Lehoux DE, Sanschagrin F, Mahan MJ, Woods DE, Levesque RC (2000) In vivo-induced genes in *Pseudomonas aeruginosa*. Infect Immun 68:2359–2362

Inoue H, Takimura O, Fuse H, Murakami K, Kamimura K, Yamaoka Y (2000) Degradation of triphenyltin by a fluorescent pseudomonad. Appl Environ Microbiol 66:3492–3498

Inoue H, Takimura O, Kawaguchi K, Nitoda T, Fuse H, Murakami K, Yamaoka Y (2003) Tin-carbon cleavage of organotin compounds by pyoverdine from *Pseudomonas chlororaphis*. Appl Environ Microbiol 69:878–883

Jacques Ph, Ongena M, Gwose I, Seinsche D, Schröder H, Delfosse Ph, Thonart Ph, Taraz K, Budzikiewicz H (1995) Structure and characterization of isopyoverdin from *Pseudomonas putida* BTP1 and its relation to the biogenetic pathway leading to pyoverdins. Z Naturforsch 50c:622–629

Johnson JR, Delavari P, Kuskowski M, Stall AL (2001) Phylogenetic distribution of extraintestinal virulence-associated traits in *Escherichia coli*. J Infect Dis 183:78–88

Jurkevitch E, Hadar Y, Chen Y (1992) Differential siderophore utilization and iron uptake by soil and rhizospere bacteria. Appl Environ Microbiol 58:119–124

Kerbarh O, Ciulli A, Howard NI, Abell C (2005) Salicylate biosynthesis: overexpression, purification, and characterization of Irp9, a bifunctional salicylate synthase from *Yersinia enterocolitica*. J Bacteriol 187:5061–5066

Kersters K, Ludwig W, Vancanneyt M, De Vos P, Gillis M, Schleifer K-H (1996) Recent changes in the classification of the Pseudomonads: an overview. System Appl Microbiol 19:465–477

Kilz S, Lenz Ch, Fuchs R, Budzikiewicz H (1999) A fast screening method for the identification of siderophores from fluorescent *Pseudomonas* spp. by liquid chromatography/electrospray mass spectrometry. J Mass Spectrom 34:281–290

King EO, Ward MK, Raney DE (1954) Two simple media for the demonstration of pyocyanin and fluorescein. J Lab Clin Med 44:301–307

Kleinkauf H, von Döhren H (1996) A nonribosomal system of peptide biosynthesis. Eur J Biochem 236:335–351

Kloepper JW, Leong J, Teintze M, Schroth MN (1980) Enhanced plant growth by siderophores produced by plant growth promoting rhizobacteria. Nature 286:885–886

Kobayashi S, Nakai H, Ikenishi Y, Sun WY, Ozaki M, Hayase Y, Takeda R (1998) Micacocidin A, B and C, novel antimycoplasma agents from *Pseudomonas* sp. II. Structure elucidation. J Antibiot 51:328–332

Koedam N, Wittouck E, Gaballa A, Gillis A, Höfte M, Cornelis P (1994) Detection and differentiation of microbial siderophores by isoelectric focusing and chrome azurol S overlay. BioMetals 7:287–291

Koster M, van de Vossenberg J, Leong J, Weisbeek PJ (1993) Identification and characterization of the *pupB* gene encoding an inducible ferric-pseudobactin receptor of *Pseudomonas putida* WCS358. Mol Microbiol 8:591–601

Koster M, Ovaa W, Bitter W, Weisbeek P (1995) Multiple outer membrane receptors for uptake of ferric pseudobactins in *Pseudomonas putida* WCS358. Mol gen Genet 248:735–743

Lamont IL, Beare PA, Ochsner U, Vasil AI, Vasil ML (2002) Siderophore-mediated signaling regulates virulence factor production in *Pseudomonas aeruginosa*. Proc Natl Acad Sci USA 99:7072–7077

Lee C-H, Lewis TA, Paszczynski A, Crawford RL (1999) Identification of an extracellular catalyst of carbon tetrachloride dehalogenation from *Pseudomonas stutzeri* strain KC as pyridine-2,6-bis (thiocarboxylate). Biochem Biophys Res Commun 261:562–566

Lefranc Nègre V, Bonacorsi S, Schubert S, Bidet P, Nassif X, Bingen E (2004) The siderophore receptor iroN, but not the high-pathogenicity island or the hemin receptor chuA, contributes to the bacteremic step of *Escherichia coli* neonatal meningitis. Infect Immun 72:1216–1220

Lehoux DE, Sanschagrin F, Levesque RC (2000) Genomics of the 35-kb *pvd* locus and analysis of novel *pvdIJK* genes implicated in pyoverdine biosynthesis in *Pseudomonas aeruginosa*. FEMS Microbiol Lett 190:141–146

Lemanceau P, Bakker PA, De Kogel WJ, Alabouvette C, Schippers B (1992) Effect of pseudobactin 358 production by *Pseudomonas putida* WCS358 on suppression of *Fusarium* wilt of carnations by nonpathogenic *Fusarium oxysporum* Fo47. Appl Environ Microbiol 58:2978–2982

Leoni L, Orsi N, de Lorenzo V, Visca P (2000) Functional analysis of PvdS, an iron starvation sigma factor of *Pseudomonas aeruginosa*. J Bacteriol 182:1481–1491

Lewis TA, Cortese MS, Sebat JL, Green TL, Lee C-H, Crawford RL (2000) A *Pseudomonas stutzeri* gene cluster encoding the biosynthesis of the CCl_4-dechlorination agent pyridine-2,6-bis (thiocarboxylic acid) Environ Microbiol 2:407–416

Lewis TA, Paszczynski A, Gordon-Wylie SW, Jeedigunta S, Lee C-H, Crawford RL (2001) Carbon tetrachloride dechlorination by the bacterial transition metal chelator pyridine-2,6-bis (thiocarboxylic acid). Environ Sci Technol 35:552–559

Lewis TA, Leach L, Morales S, Austin PR, Hartwell HJ, Kaplan B, Forker C, Meyer JM (2004) Physiological and molecular genetic evaluation of the dechlorination agent, pyridine-2,6-bis(monothiocarboxylic acid) (PDTC) as a secondary siderophore of *Pseudomonas*. Environ Microbiol 6:159–169

Loper JE, Buyer JS (1991) Siderophores in microbial interactions on plant surfaces. Mol Plant Microbe Interact. 4:5–13

Martins dos Santos VAP, Heim S, Moore ERB, Strätz M, Timmis KN (2004) Insights into the genomic basis of niche specificity of *Pseudomonas putida* KT2440. Environ Microbiol 6:1264–1286

Matthijs S, Baysse C, Koedam N, Tehrani KA, Verheyden L, Budzikiewicz H, Schäfer M, Hoorelbeke B, Meyer J-M, De Greve H, Cornelis P (2004) The *Pseudomonas* siderophore quinolobactin is synthesized from xanthurenic acid, an intermediate of the kynurenine pathway. Mol Microbiol 52:371–384

Maurhofer M, Reimmann C, Schmidli-Sacherer P, Heeb S, Haas D, Défago G (1998) Salicylic acid biosynthetic genes expressed in *Pseudomonas fluorescens* strain P3 improve the induction of systemic resistance in tobacco against tobacco necrosis virus. Phytopathology 88:678–684

Mercado-Blanco J, van der Drift KMGM, Olsson PE, Thomas-Oates JE, van Loon LC, Bakker PAHM (2001) Analysis of the *pmsCEAB* gene cluster involved in biosynthesis of salicylic acid and the siderophore pseudomonine in the biocontrol strain *Pseudomonas fluorescens* WCS374. J Bacteriol 183:1909–1920

Merriman TR, Merriman ME, Lamont IL (1995) Nucleotide sequence of *pvdD*, a pyoverdine biosynthetic gene from *Pseudomonas aeruginosa*: PvdD has similarity to peptide synthetases. J Bacteriol 177:252–258

Meyer J-M, Abdallah MA (1978) The fluorescent pigment of *Pseudomonas fluorescens*: byosynthesis, purification and physicochemical properties. J Gen Microbiol 107:319–328

Meyer J-M, Geoffroy VA (2004) Environmental fluorescent *Pseudomonas* and pyoverdine diversity: how siderophores could help microbiologists in bacterial identification and taxonomy. In: Crosa JH, Mey AR, Payne SM (eds) Iron transport in bacteria. ASM Press, Washington pp 451–468

Meyer J-M, Neely A, Stintzi A, Georges C, Holder IA (1996) Pyoverdin is essential for virulence of *Pseudomonas aeruginosa*. Infect Immun 64:518–523

Meyer J-M, Stintzi A, De Vos D, Cornelis P, Tappe R, Taraz K, Budzikiewicz H (1997) Use of siderophores to type pseudomonads, the three *Pseudomonas aeruginosa* pyoverdin systems. Microbiology 143:35–43

Meyer J-M, Stintzi A, Coulanges V, Shivaji S, Voss JA, Taraz K, Budzikiewicz H (1998) Siderotyping of fluorescent pseudomonads: characterization of pyoverdines of *Pseudomonas fluorescens* and *Pseudomonas putida* strains from Antarctica. Microbiology 144:3119–3126

Meyer J-M, Stintzi A, Poole K (1999) The ferripyoverdine receptor FpvA of *Pseudomonas aeruginosa* PAO1 recognizes the ferripyoverdines of *P. aeruginosa* PAO1 and *P. fluorescens* ATCC 13525. FEMS Microbiol Lett 170:145–150

Meyer J-M, Geoffroy VA, Baida N, Gardan L, Izard D, Lemanceau P, Achouak W, Palleroni NJ (2002a) Siderophore typing, a powerful tool for the identification of fluorescent and nonfluorescent pseudomonads. Appl Environ Microbiol 68:2745–2753

Meyer J-M, Geoffroy VA, Baysse C, Cornelis P, Barelmann I, Taraz K, Budzikiewicz H (2002b) Siderophore-mediated iron uptake in fluorescent *Pseudomonas*: characterization of the pyoverdine-receptor binding site of three cross-reacting pyoverdines. Arch Biochem Biophys 397:179–183

Mirleau P, Delorme S, Philippot L, Meyer JM, Lmazurier S, Lemanceau P (2000) Fitness in soil and rhizosphere of *Pseudomonas fluorescens* C7R12 compared with a C7R12 mutant affected in pyoverdine synthesis and uptake. FEMS Microbiol Ecol 34:35–44

Mirleau P, Philippot L, Corberand T, Lemanceau P (2001) Involvement of nitrate reductase and pyoverdine in competitiveness of *Pseudomonas fluorescens* strain C7R12 in soil. Appl Environ Microbiol 67:2627–2635

Mokracka J, Koczura R, Kaznowski A (2004) Yersiniabactin and other siderophores produced by clinical isolates of *Enterobacter* spp. and *Citrobacter* spp. FEMS Immunol Med Micobiol 40:51–55

Morris J, O'Sullivan DJ, Koster M, Leong J, Weisbeek PJ, O'Gara FO (1992) Characterization of fluorescent siderophore-mediated iron-uptake in *Pseudomonas* sp. strain M114: evidence for the existence of an additional ferric siderophore receptor. Appl Environ Microbiol 58:630–635

Mossialos D, Meyer J-M, Budzikiewicz H, Wolff U, Koedam N, Baysse C, Anjaiah V, Cornelis P (2000) Quinolobactin, a new siderophore of *Pseudomonas* fluorescens ATCC 17400, the production of which is repressed by the cognate pyoverdine. Appl Environ Microbiol 66:487–492

Munsch P, Geoffroy VA, Alatossava T, Meyer J-M (2000) Application of siderotyping for characterization of *Pseudomonas tolaasii* and "*Pseudomonas reactans*"isolates associated with brown blotch disease of cultivated mushrooms. Appl Environ Microbiol 66:4834–4841

Neuenhaus W, Budzikiewicz H, Korth H, Pulverer G (1980) 8-Hydroxy-4-methoxy-monothiochinaldinsäure – eine weitere Thiosäure aus *Pseudomonas*. Z Naturforsch 35b:1569–1571

O'Sullivan D, O'Gara F (1992) Traits of fluorescent *Pseudomonas* spp. involved in suppression of plant root pathogens. Microbiol Rev 56:662–676

Ockels W, Römer A, Budzikiewicz H (1978) An Fe(III) complex of pyridine-2,6-di-(monothio-carboxylic acid) - a novel bacterial metabolic product. Tetrahedron Lett 36:3341–3342

Oelschlaeger TA, Zhang D, Schubert S, Carniel E, Rabsch W, Karch H, Hacker J (2003) The high-pathogenicity island is absent in human pathogens of *Salmonella enterica* subspecies I but present in isolates of subspecies III and VI. J Bacteriol 185:1107–1111

Ongena M, Jacques P, Delfosse P, Thonart P (2002) Unusual traits of the pyoverdin-mediated iron acquisition system in *Pseudomonas putida* strain BTP1. BioMetals 15:1–13

Palleroni NJ (1984) Family I. *Pseudomonadaceae*. In: Krieg NR, Holt JG (eds.) Bergey's manual of Systematic Bacteriology. William &Wilkins Co, Baltimore, pp 141–199

Perry RD (2004) *Yersinia*. In: Crosa JH, Mey AR, Payne SM (eds) Iron transport in bacteria. ASM Press, Washington pp 219–240

Perry RD, Balbo PB, Jones HA, Fetherston JD, DeMoll E (1999) Yersiniabactin from *Yersinia pestis*: biochemical characterization of the siderophore and its role in iron transport and regulation. Microbiology 145:1181–1190

Poole K (2004) *Pseudomonas*. In: Crosa JH, Mey AR, Payne SM (eds) Iron transport in bacteria. ASM Press, Washington pp 293–310

Poole K, McKay GA (2003) Iron acquisition and its control in *Pseudomonas aeruginosa*: many roads lead to Rome. Front Biosci 8:661–686

Poole K, Young L, Neshat S (1990) Enterobactin-mediated iron transport in *Pseudomonas aeruginosa*. J Bacteriol 172:6991–6996

Raaijmakers JM, van der Sluis L, Koster M, Bakker PAHM, Weisbeek PJ, Schippers B (1995a) Utilization of heterologous siderophores and rhizosphere competence of fluorescent *Pseudomonas* spp. Can J Microbiol 41:126–135

Raaijmakers JM, Leeman M, van Oorschot MMP, van der Sluis I, Schippers B, Bakker PAHM (1995b) Dose-response relationships in biological control of *Fusarium* wilt of radish by *Pseudomonas* spp. Phytopathology 85:1075–1081

Reimmann C, Serino L, Beyeler M, Haas D (1998) Dihydroaeruginoic acid synthetase and pyochelin synthetase, products of the *pchEF* genes, are induced by extracellular pyochelin in *Pseudomonas aeruginosa*. Microbiology 144:3135–3148

Rinehart KL, Staley AL, Wilson SR, Ankenbauer RG, Cox CD (1995) Stereochemical assignment of the pyochelins. J Org Chem 60:2786–2791

Ruangviriyachai C, Uría Fernández D, Fuchs R, Meyer J-M, Budzikiewicz H (2001) A new pyoverdin from *Pseudomonas aeruginosa* R'. Z Naturforsch 56:933–938

Schäfer H, Taraz K, Budzikiewicz H (1991) Zur Genese der Amidisch an den Chromophor von Pyoverdinen gebundenen Dicarbonsäuren. Z Naturforsch 46c:398–406

Schlegel K, Fuchs R, Schäfer M, Taraz K, Budzikiewicz H, Geoffroy V, Meyer J-M (2001) The pyoverdins of *Pseudomonas* sp. 96-312 and 96-318. Z Naturforsch 56:680–686

Schmidt H, Hensel M (2004) Pathogenicity islands in bacterial pathogenesis. Clin Microbiol Rev 17:14–56

Schubert S, Rakin A, Karch H, Carniel E, Heeseman J (1998) Prevalence of the "high-pathogenicity island" of *yersinia* species among *Escherichia coli* strains that are pathogenic to humans. Infect Immun 66:480–485

Schubert S, Cuenca S, Fischer D, Heesemann J (2000) High-pathogenicity island of *Yersinia pestis* in enterobacteriaceae isolated from blood cultures and urine samples: prevalence and functional expression. J Infect Dis 182:1268–1271

Schubert S, Picard B, Gouriou S, Heesemann J, Denamur E (2002) *Yersinia* high-pathogenicity island contributes to virulence in *Escherichia coli* causing extraintestinal infections. Infect Immun 70:5335–5337

Schwyn B, Neilands JB (1987) Universal chemical assay for the detection and determination of siderophores. Anal Biochem 160:47–56

Sepúlveda-Torres L, Huang A, Kim H, Criddle CS (2002) Analysis of regulatory elements and genes required for carbon tetrachloride degradation in *Pseudomonas stutzeri* strain KC. J Mol Microbiol Biotechnol 4:151–161

Serino L, Reimmann C, Visca P, Beyeler M, Chiesa VD, Haas D (1997) Biosynthesis of pyochelin and dihydroaeruginoic acid requires the iron-regulated *pchDCBA* operon in *Pseudomonas aeruginosa*. J Bacteriol 179:248–257

Smith EE, Sims EH, Spencer DH, Kaul R, Olson MV (2005) Evidence for diversifying selection at the pyoverdine locus of *Pseudomonas aeruginosa*. J Bacteriol 187:2138–2147

Sokol PA (1984) Production of the ferripyochelin outer membrane receptor by *Pseudomonas* species. FEMS Microbiol Lett 23:313–317

Spencer DH, Kas A, Smith EE, Raymond CK, Sims EH, Hastings M, Burns JL, Kaul R, Olson MV (2003) Whole-genome sequence variation among multiple isolates of *Pseudomonas aeruginosa*. J Bacteriol 185:1316–1325

Stolworthy JC, Paszczynski A, Korus R, Crawford RL (2001) Metal binding by pyridine-2,6-bis(monothiocarboxylic acid), a biochelator produced by *Pseudomonas stutzeri* and *Pseudomonas putida*. Biodegradation 12:411–418

Sun G-X, Zhong J-J (2006) Mechanism on augmentation of organotin decomposition by ferripyochelin: formation of hydroxyl radical and organotin/pyochelin/iron ternary complex. Appl Environ Microbiol 72:7264–7269

Sun G-X, Zhou W-Q, Zhong J-J (2006) Organotin decomposition by pyochelin, secreted by *Pseudomonas aeruginosa* even in an iron-sufficient environment. Appl Environ Microbiol 72:6411–6413

Takase H, Nitanai H, Hoshino K, Otani T (2000) Impact of siderophore production on *Pseudomonas aeruginosa* infections in immunosuppressed mice. Infect Immun 68:1834–1839

Tatara GM, Dybas MJ, Criddle CS (1993) Effects of medium and trace elements on kinetics of carbon tetrachloride transformation by *Pseudomonas* sp. strain KC. Appl Environ Microbiol 59:2126–2131

Visca P, Colotti G, Serino L, Verzili D, Orsi N, Chiancone E (1992) Metal regulation of siderophore synthesis in *Pseudomonas aeruginosa* and functional effects of siderophore-metal complexes. Appl Environ Microbiol 58:2886–2893

Visca P, Ciervo A, Orsi N (1994) Cloning and nucleotide sequence of the *pvdA* gene encoding the pyoverdin biosynthetic enzyme L-ornithine N5-oxygenase in *Pseudomonas aeruginosa*. J Bacteriol 176:1128–1140

Visser MB, Majumbar S, Hani E, Sokol PA (2004) Importance of the ornibactin and pyochelin siderophore transport systems in *Burkholderia cenocepacia* lung infections. Infect Immun 72:2850–2857

von Döhren H, Dieckmann R, Pavela-Vrancic M (1999) The nonribosomal code. Chem Biol 6:R273–R279

Weber M, Taraz K, Budzikiewicz H, Geoffroy V, Meyer J-M (2000) The structure of a pyoverdine from *Pseudomonas* sp. CFML 96.188 and its relation to other pyoverdins with a cyclic C-terminus. BioMetals 13:301–309

Wilson MJ, McMorran BJ, Lamont IL (2001) Analysis of promoters recognized by PvdS, an extracytoplasmic-function sigma factor protein from *Pseudomonas aeruginosa*. J Bacteriol 183:2151–2155

Winkelmann G (2004) Ecology of siderophores. In: Crosa JH, Mey AR, Payne SM (eds) Iron transport in bacteria. ASM Press, Washington, pp 437–450

Xiao R, Kisaalita WS (1997) Iron acquisition from transferrin and lactoferrin by *Pseudomonas aeruginosa* pyoverdin. Microbiology 143:2509–2515

Xiao R, Kisaalita WS (1998) Fluorescent pseudomonad pyoverdines bind and oxidize ferrous ion Appl Envir Microbiol 64:1472–1476

Zawadzka AM, Crawford RL, Paszczynski AJ (2006) Pyridine-2,6-bis(thiocarboxylic acid) produced by *Pseudomonas stutzeri* KC reduces and precipitates selenium and tellurium oxyanions. Appl Environ Microbiol 72:3119–3129

4 Siderophores of Symbiotic Fungi

Kurt Haselwandter and Günther Winkelmann

4.1
Introduction

A total of approximately 500 different siderophore structures have been described so far which mainly consist of one or more of the basic ligand structures shown in Fig. 4.1. While catecholate ligands are common in bacterial siderophores, they have so far not been found in fungal siderophores, although transporters for enterobactin in fungi have been found to occur (Heymann et. al. 2000; Haas et al. 2003; Winkelmann 2004). Thus hydroxamates predominate in the asco- and basidiomycota of which the ferrichromes, fusarinines and coprogens represent the major siderophore classes (Fig. 4.2). Within the Mucorales (Zygomycota) hydroxamate ligands seem to be unknown and instead carboxylate ligands like rhizoferrin are used as iron transporting agents (Drechsel et al. 1992; Winkelmann 1992). For a more comprehensive treatise of fungal siderophores and transport systems the reader is referred to a chapter in a book on microbial transport systems (Winkelmann 2001).

Compilation of data on the taxonomic distribution of siderophores amongst the fungi will pay particular reference to mutualistic fungal symbionts of plants. In this chapter special emphasis is laid upon mycorrhizal fungi as typical fungal root associates. Another type of mutualistic association is represented by the lichen symbiosis. The main aim of this chapter is to summarize what is currently known with regard to the siderophores released by symbiotic fungi, but at the same time to highlight some major gaps in our knowledge. Furthermore, we will discuss the potential role siderophores may play with regard to the establishment and function of the respective types of symbioses.

Kurt Haselwandter: Department of Microbiology, University of Innsbruck, Technikerstr. 25, A-6020 Innsbruck, Austria, email: Kurt.Haselwandter@uibk.ac.at

Günther Winkelmann: Department of Microbiology and Biotechnology, University of Tübingen, Auf der Morgenstelle 28, D-72076, Germany

Soil Biology, Volume 12
Microbial Siderophores
A. Varma, S.B. Chincholkar (Eds.)
© Springer-Verlag Berlin Heidelberg 2007

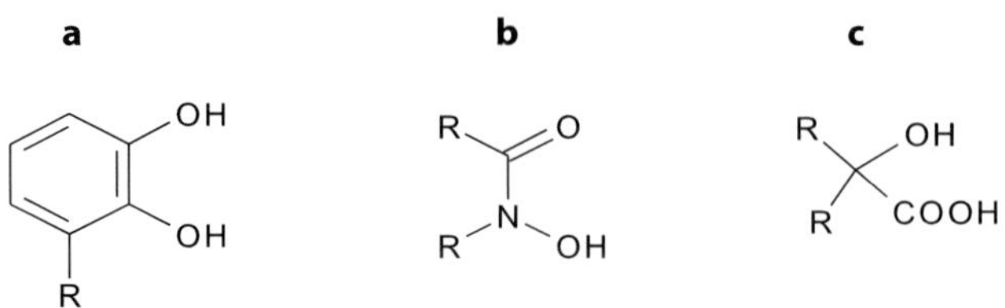

Fig. 4.1a–c. Structures of the main iron-binding ligands of siderophores: **a** catecholate as in bacterial enterobactin; **b** hydroxamate as in fungal ferricrocin; **c** citrate as in fungal rhizoferrin

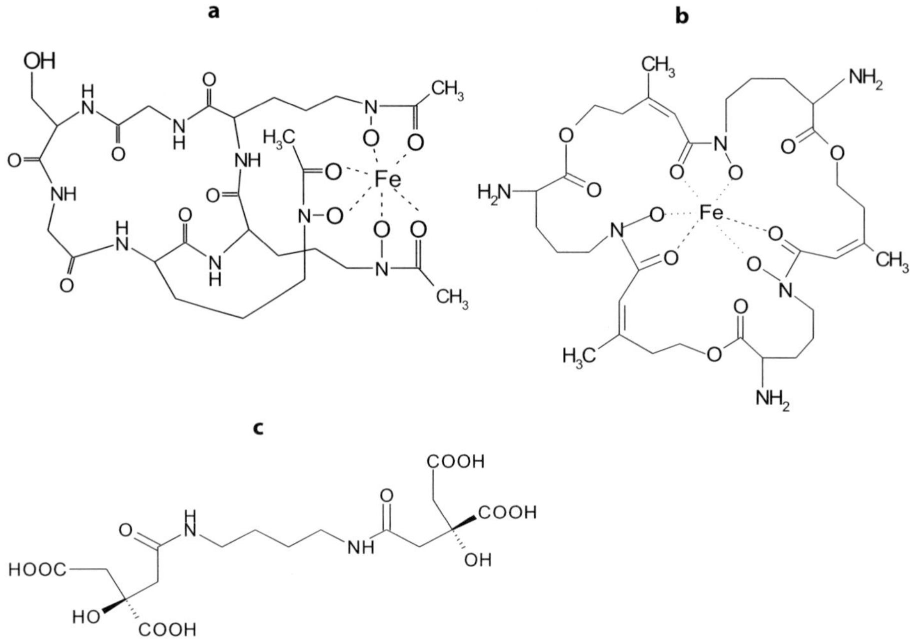

Fig. 4.2a–c. Characteristic fungal siderophores: **a** ferricrocin, a hexapeptide siderophore of the ferrichrome class; **b** fusigen – both hydroxamates have been found in a variety of fungi including ectomycorrhizal, ectendomycorrhizal, ericoid mycorrhizal fungi and dark septate endophytes; **c** rhizoferrin is a member of the carboxylate siderophores which is produced by Mucorales (Zygomycota)

4.2
Fungi Forming Mutualistic Symbioses with Plants

The associations formed by symbiotic fungi with plants range from bryophytes and pteridophytes to vascular plants in case of the mycorrhizal symbiosis, whereas algae or cyanobacteria represent the photobiont of lichens.

Mycorrhizal fungi associate with the majority of terrestrial plants, and provide protection from various environmental stresses as well as essential nutrients (Haselwandter and Bowen 1996; Smith and Read 1997; Peterson et al. 2004).

They contribute to the micronutrient uptake by plants (George et al. 1994). The potential to produce siderophores was demonstrated some time ago for a range of mycorrhizal fungi (cf. Haselwandter 1995).

4.2.1
Ectomycorrhizal Fungi

Based on a bioassay with *Microbacterium flavescens* JG-9, Szaniszlo et al. (1981) have shown that a number of ectomycorrhizal fungi produce hydroxamate siderophores when grown in pure culture under iron-limiting conditions. One of the strains tested in this study was the ascomyceteous fungus *Cenococcum geophilum* which is known for its global distribution and broad host range, thus covering an enormous ecological amplitude. It was demonstrated that *C. geophilum* releases in vitro ferricrocin as main siderophore when this fungus grows in a deferrated nutrient medium (Haselwandter and Winkelmann 2002). This was the first report on the precise structure elucidated for the main siderophore of an ectomycorrhizal fungus. Undoubtedly, more research on the chemical structure of main siderophores excreted by a wider range of ectomycorrhizal fungi including the Basidiomycota is needed.

Information of this kind is of paramount importance with regard to an evaluation of the potential role ectomycorrhizal fungi may play in mediating iron uptake from soil. In addition, siderophores may act as ligand-based weathering agents leading to the dissolution of rock material (Hoffland et al. 2004). The solubilisation of iron from goethite (ferric oxyhydroxide) by *Suillus granulatus* was attributed to siderophores released by this ectomycorrhizal basidiomycete (Watteau and Berthelin 1994). Hydroxamate siderophores can be detected in aqueous extracts of a variety of soils (Powell et al. 1980). In soil solutions of podzolic forest profiles Essen et al. (2006) have detected concentrations up to 2 and 12 nM of ferrichrome and ferricrocin, respectively. Van Hees et al. (2006) have claimed that the extramatrical mycelium of *Hebeloma crustuliniforme*, forming ectomycorrhizae with *Pinus sylvestris* seedlings, may release ferricrocin in addition to higher amounts of oxalate, thus altering the chemical conditions of soil microsites and affecting mineral dissolution therein.

4.2.2
Arbuscular Mycorrhizal Fungi

Mycorrhizal *Hilaria jamesii* grass, which showed greater iron uptake than non-mycorrhizal controls, tested positively when bioassayed for hydroxamate siderophores (Cress et al. 1986). However, it has to be stressed that these tests were carried out under non-axenic conditions; hence these results cannot be taken as proof that *Glomus* species may produce hydroxamate siderophores.

It may well be that the hydroxamate siderophores detected were produced by microorganisms other than arbuscular mycorrhizal fungi. Sound evidence for the nature of siderophores released by arbuscular mycorrhizal fungi can only be obtained when the fungi together with suitable host plants are grown under pure culture conditions. Results of such experiments are not available yet.

It could well be that the main siderophores produced by arbuscular mycorrhizal fungi do not belong to the hydroxamates. A phylogenetic analysis of small subunit (SSU) rRNA gene sequences has provided clear evidence that the arbuscular mycorrhizal fungi are best placed into a new monophyletic phylum, the Glomeromycota (Schüssler et al. 2001). The Glomeromycota represent a taxon clearly separated from the Zygomycota and at an equivalent level to the phyla of the Ascomycota and Basidiomycota. According to Schüssler et al. (2001) the Glomeromycota may share a common ancestor with the Ascomycota and Basidiomycota. Therefore, also from a phylogenetic point of view it will be of great interest to know the chemical structure of the main siderophores released by arbuscular mycorrhizal fungi.

The arbuscular mycorrhizal fungi are of ecological and economical importance. Rai (1988) claimed that on high pH calcareous soil, mycorrhizal inoculation with *Glomus albidum* not only enhanced the nodulation and N_2 fixation rates, but also the iron uptake by two genotypes of chick pea (*Cicer arietinum*). When broccoli (*Brassica oleracea* var. *italica*) was inoculated with *Glomus macrocarpum* and *G. fasciculatum*, levels of Fe in the leaf tissue and total uptake of iron was higher than in the absence of arbuscular mycorrhizal fungi (Purakayastha et al. 1998). Hyphae of the arbuscular mycorrhizal fungus *G. mosseae* can mobilise and/or take up Fe from soil and translocate it to the host plant (Caris et al. 1998). At a low level of micronutrients, colonization of maize (*Zea mays*) with *G. intraradices* increased the total shoot iron content (Liu et al. 2000). In the case of a maize cultivar known for its sensitivity to Fe deficiency chlorosis arbuscular mycorrhizal fungi enhanced Fe acquisition in plants grown on alkaline soil (Clark and Zeto 1996). Hence arbuscular mycorrhizal fungi may play an important role in remediation of iron deficiency in agricultural crops. This is an aspect which ought to be taken into account in addition to a range of widely accepted ecological considerations for successful application of arbuscular mycorrhizal fungi as inoculum (Vosatka and Dodd 2002).

4.2.3
Ericoid Mycorrhizal Fungi

The first evidence that ericoid mycorrhizal fungi release hydroxamate siderophores was provided in investigations which used a *Microbacterium flavescens* JG-9 based bioassay (Schuler and Haselwandter 1988). Using the same detection system as described by Haselwandter et al. (1988), it was shown that the supplementation of the nutrient medium with l-ornithine favoured the siderophore biosynthesis by the ericoid mycorrhizal fungus *Rhizoscyphus* (previously

named *Hymenoscyphus) ericae* (Federspiel et al. 1991). In subsequent studies
the main siderophore released by ericoid mycorrhizal fungi like *Rhizoscyphus
ericae* and *Oidiodendron griseum,* both typical for acid soils, was identified as
ferricrocin, whereas a fungal endophyte isolated from a calcicolous ericaceous
plant produced fusigen as principal siderophore (Haselwandter et al. 1992).

Interestingly, and possibly of great ecological importance, the different ericoid
mycorrhizal fungi differ in their response to ferric iron with regard to siderophore
biosynthesis. In the case of *R. ericae*, within six days of incubation the siderophore
concentration in the nutrient solution fell to about 50% of the control (0 ng Fe^{3+}
ml^{-1}) when the nutrient medium was supplemented with 80 ng Fe^{3+} ml^{-1}. Under
the same experimental conditions the reduction was in the range of 70% in the
case of *O. griseum*, but only around 10% in the case of the endophyte of the calcicolous ericaceous plant (Dobernigg and Haselwandter 1992). It is assumed that
the potential of ericoid mycorrhizal fungi to produce siderophores has implications for the iron nutrition of the associated host plants. Shaw et al. (1990) have
demonstrated that mycorrhizal roots of *Calluna vulgaris* absorbed significantly
more iron per unit time than those which were uninfected.

4.2.4
Orchidaceous Mycorrhizal Fungi

A fungal endophyte isolated from the alpine orchid *Nigritella nigra* was shown
to release ferrichrome as principal siderophore in addition to smaller amounts
of a range of other hydroxamate siderophores like neocoprogen I and II, fer-
ricorocin, ferrirubin, coprogen and ferrirhodin. On the other hand, some orchidaceous mycorrhizal isolates covering various *Rhizoctonia* and *Ceratobasidium* species were found to produce a hydroxamate siderophore with an as
yet undescribed structure (Passler 1998; Reiter 1999). Subsequently, a thor-
ough investigation was undertaken in order to elucidate the precise nature of
this compound. This study revealed a novel structure named basidiochrome
which represents a linear trishydroxamate siderophore containing a tripeptide
sequence of L-N^5-hydroxyornithine each of which linked to a *cis*-methylgluta-
conic acid residue (Haselwandter et al. 2006). Thus this finding corroborates an
earlier hypothesis that mycorrhizal fungi may be a good source for detection of
novel siderophores with a hitherto unknown structure (Haselwandter 1995).

4.2.5
Ectendomycorrhizal Fungi

Ectendomycorrhizae are characterized by the production of a fungal sheath and
Hartig net, both of which can give rise to intracellular penetration of healthy
root cells by hyphae. The ascomyceteous genus *Wilcoxina* forms ectendomycor-

rhizae with young individuals of conifers and deciduous trees (Wilcox 1991). Under iron-limiting conditions, three different isolates of *Wilcoxina* species were shown to release ferricrocin as the only siderophore into the nutrient medium (Prabhu et al. 1996).

4.2.6
Dark Septate Endophytes

The hyphomycete *Phialocephala fortinii* is the most prominent dark septate endophyte frequently forming mycorrhiza-like associations with healthy looking plant roots in a great variety of ecosystems, especially in alpine and arctic habitats (Haselwandter and Read 1980; Piercey et al. 2004). Dark septate endophytes (DSE) may represent the main fungal colonizers of plant roots, or colonize plant roots simultaneously, e.g. together with ectomycorrhizal, arbuscular or ericoid mycorrhizal fungi (Jumpponen and Trappe 1998).

In a comparative study five different isolates of *P. fortinii including* the type strain were tested for their potential to synthesize siderophores (Bartholdy et al. 2001). Ferricrocin was produced by all the isolates investigated in significant quantities, while only two of the isolates released ferrirubin in substantial amounts as well. Ferrichrome C was also detected in the culture filtrate of each isolate, albeit at low concentration in three of the five isolates; only in the case of two isolates did ferrichrome C come second after ferricrocin as principal siderophore. In a growth experiment with two alpine *Carex* species, dry weight of roots and shoots and shoot phosphorus content increased upon inoculation with, e.g. *P. fortinii* isolate C2 (=strain CC3 of Bartholdy et al. 2001) at least in case of one of the *Carex* species tested (Haselwandter and Read 1982). It might well be that the iron nutrition of the host plant may benefit from colonization by DSE in a similar way as is known for the mycorrhizal symbiosis (see above).

4.2.7
Mycobionts of Lichens

Lichens are typically formed by the association of a mycobiont with a photobiont. Many lichens contain more than one photosynthetic partner and lichenicolous fungi may also be involved. The dynamic interactions between the partners range from mutualism through commensalism to antagonism (Richardson 1999). In the case of the most complex structures, the association leads to the formation of long-lived, morphologically distinct entities in which both organisms derive mutual benefit from the integration. A mass of fungal hyphae composes the majority of any lichen. Such thalli are the product of a hyphal polymorphism with multiple switches between polar and apolar growth and hydrophilic or hydrophobic cell wall surfaces (Honegger 1998).

Interestingly, lichen-forming fungi frequently have nuclear small subunit ribosomal DNA (SSU rDNA) longer than the 1800 nucleotides typical for eukaryotes, resulting from insertions in the SSU rDNA (Gargas et al. 1995). Culture experiments and DNA sequence data confirmed that a single mycobiont can simultaneously lichenize two unrelated photobionts, a eukaryote and a prokaryote (Stenroos et al. 2003).

In a project to establish cultures of a broad spectrum of lichen-forming and lichenicolous fungi, close to 500 species were isolated (Crittenden et al. 1995). The mycobionts belonging to the Ostropales were most readily isolated (70% of species attempted were isolated), followed by the Dothideales (63%), Pertusariales (53%) and Lecanorales (45%). Within the Dothideales the Leptosphaeriaceae and Pleosporaceae were already reported to contain siderophore-producing species (Renshaw et al. 2002), while for the other orders listed above no such reports are available yet. Within the Pleosporaceae, fungal genera like *Alternaria*, *Curvularia*, *Leptosphaerulina*, and *Stemphylium* were shown to produce the ferrichrome ferricrocin and/or various coprogens including coprogen, coprogen B, neocoprogen I, neocoprogen II and dimerum acid (Jalal et al. 1988; Jalal and van der Helm 1989; Manulis et al. 1987).

The ability of fungi to interact with a great variety of different substrates including minerals, metals and organic compounds in biophysical and biochemical processes makes them ideally suited as biological weathering agents of rock and minerals (Burford et al. 2003). Because processes in soils are inherently complex, a model based on the lichen-mineral system was applied to identify mineral dissolution processes occurring in the rhizosphere (Banfield et al. 1999). It would be of great interest to know whether lichen-forming fungi of the Dothideales release an array of siderophores similar to that described already for this fungal order. In addition, mycobionts of lichens belonging to the other orders of fungi which can be readily isolated should be investigated both, qualitatively and quantitatively, for their potential to synthesize siderophores. Information of this kind could contribute to a better understanding of the mechanisms underlying the important role lichens play in rock weathering and soil development (Chen et al. 2000). In particular, such knowledge would facilitate in the assessment as to whether and to what extent siderophores released by lichen-forming fungi are involved in these processes. For some time lichen-forming fungi have been considered to represent potential sources of novel metabolites (Crittenden and Porter 1991).

4.3
Outlook

Data on the spectrum of siderophores released by symbiotic fungi are compiled in Table 4.1 with regard to their taxonomic affiliation. It is clear that so far the hydroxamates of the ferricrocin type seem to predominate. At least to some ex-

Table 4.1. Siderophores of symbiotic fungi in relation to taxonomic affiliation

Type of symbiosis	Fungal species analysed	Taxonomic affiliation	Main siderophore	References
Ectomycorrhiza				
	Cenococcum geophilum	Ascomycota	Ferricrocin	Haselwandter and Winkelmann 2002
		Other Ascomycota	Unknown	-
		Basidiomycota	Unknown	-
		Zygomycota	Unknown	-
Arbuscular mycorrhiza				
	No species analysed yet	Glomeromycota	Unknown	-
Ericoid mycorrhiza				
	Rhizoscyphus ericae	Ascomycota	Ferricrocin	Haselwandter et al. 1992
	Oidiodendron griseum	Ascomycota (anamorph)	Ferricrocin	Haselwandter et al. 1992
	Unidentified species	Unknown	Fusigen	Haselwandter et al. 1992
Orchid mycorrhiza				
	Unidentified species	Unknown	Ferrichrome	Passler 1998
	Rhizoctonia spp.	Basidiomycota (anamorph)	Basidiochrome	Haselwandter et al. 2006
	Ceratobasidium spp.	Basidiomycota	Basidiochrome	Haselwandter et al. 2006

Table 4.1. (*continued*) Siderophores of symbiotic fungi in relation to taxonomic affiliation

Type of symbiosis	Fungal species analysed	Taxonomic affiliation	Main siderophore	References
Ectendomycorrhiza				
	Wilcoxina spp.	Ascomycota	Ferricrocin	Prabhu et al. 1996
Dark septate endophytes				
	Phialocephala fortinii	Ascomycota	Ferricrocin (4 strains)/ ferrirubin (1 strain)	Bartholdy et al. 2001
Lichens				
	No species analysed yet	Ascomycota	Unknown	-

tent this picture may be misleading and possibly mainly due to the fact that many of the symbiotic species analysed up to now are ascomycetous fungi many of which are known to release ferrichromes, quite frequently of the ferricrocin type. The high number of 'unknowns' in Table 4.1 also underlines the need to have a much wider range of fungi screened with regard both to types of symbioses and to taxonomic affiliation for their potential to produce siderophores.

For a great variety of fungal species, siderophore mediated iron uptake appears essential not only for survival as free-living organisms but also for the establishment and function of symbiotic relationships in the broadest sense. In case of antagonistic relationships, the deletion of the gene SidA encoding for the enzyme L-ornithine-N^5-monoxygenase catalyzing one important step in the siderophore biosynthesis in *Aspergillus fumigatus* has rendered this fungal pathogen of humans not pathogenic any more (Schrettl et al. 2004). From the guts of the woodlice species *Armadillidium vulgare* a yeast symbiont (*Debaromyces mycophilus*) was isolated the growth of which is obligatorily dependent on the availability of a siderophore like ferrichrome in its environment which in nature may derive from fungal biomass digested in the guts of these animals (Thanh et al. 2002). Thus this source of iron appears to be a prerequisite for the establishment of the symbiosis between the yeast *Debaromyces mycophilus* and the woodlice species *Armadillidium vulgare*.

In case of fungi as symbionts of plants the role of the fungal capacity to synthesize siderophores is not clear with regard to the establishment of the symbiosis. On the other hand, it is evident that such a fungal capacity may affect the iron nutrition of associated host plants as outlined above. A similar approach involving a siderophore deficient mutant of a mycorrhizal fungus which is already known to produce hydroxamate siderophores could be applied for an evaluation of the importance of siderophore biosynthesis for the establishment of the symbiosis with a host plant.

References

Banfield JF, Barker WW, Welch SA, Taunton A (1999) Biological impact on mineral dissolution: application of the lichen model to understanding mineral weathering in the rhizosphere. Proc Natl Acad Sci USA 96:3404–3411

Bartholdy BA, Berreck M, Haselwandter K (2001) Hydroxamate siderophore synthesis by *Phialocephala fortinii*, a typical dark septate fungal root endophyte. BioMetals 14:33–42

Burford EP, Fomina M, Gadd GM (2003) Fungal involvement in bioweathering and biotransformation of rocks and minerals. Mineral Mag 67:1127–1155

Caris C, Hördt W, Hawkins H-J, Römheld V, George E (1998) Studies of iron transport by arbuscular mycorrhizal hyphae from soil to peanut and sorghum plants. Mycorrhiza 8:35–39

Chen J, Blume H-P, Beyer L (2000) Weathering of rocks induced by lichen colonization – a review. Catena 39:121–146

Clark RB, Zeto SK (1996) Mineral acquisition by mycorrhizal maize grown on acid and alkaline soil. Soil Biol Biochem 28:1495–1503

Cress WA, Johnson GV, Barton LL (1986) The role of endomycorrhizal fungi in iron uptake by *Hilaria jamesii*. J Plant Nutr 9:547–556

Crittenden PD, Porter N (1991) Lichen-forming fungi: potential sources of novel metabolites. Trends Biotechnol 9:409–414

Crittenden PD, David JC, Hawksworth DL, Campbell FS (1995) Attempted isolation and success in the culturing of a broad spectrum of lichen-forming and lichenicolous fungi. New Phytol 130:267–297

Dobernigg B, Haselwandter K (1992) Effect of ferric iron on the release of siderophores by ericoid mycorrhizal fungi. In: Read DJ, Lewis DH, Fitter AH, Alexander IJ (eds) Mycorrhizas in ecosystems. University Press, Cambridge, UK, pp 252–257

Drechsel H, Jung G, Winkelmann G (1992) Stereochemical characterization of rhizoferrin and identification of its dehydration products. BioMetals 5:141–148

Essen SA, Bylund D, Holmström SJM, Moberg M, Lundström US (2006) Quantification of hydroxamate siderophores in soil solutions of podzolic soil profiles in Sweden. BioMetals 19:269–282

Federspiel A, Schuler R, Haselwandter K (1991) Effect of pH, L-ornithine and L-proline on the hydroxamate siderophore production by *Hymenoscyphus ericae*, a typical ericoid mycorrhizal fungus. Plant Soil 130:259–261

Gargas A, DePriest PT, Taylor JW (1995) Positions of multiple insertions in SSU rDNA of lichen-forming fungi. Mol Biol Evol 12:208–218

George E, Römheld V, Marschner H (1994) Contribution of mycorrhizal fungi to micronutrient uptake by plants. In: Manthey JA, Crowley DE, Luster DG (eds) Biochemistry of metal micronutrients in the rhizosphere. Lewis Publishers, Boca Raton FL, pp 93–109

Haas H, Schoeser M, Lesuisse E, Ernst JF, Parson W, Abt B, Winkelmann G, Oberegger H (2003) Characterization of *Aspergillus nidulans* transporters for siderophores enterobactin and triacetylfusarinine C. Biochem J 371:505–513

Haselwandter K (1995) Mycorrhizal fungi: siderophore production. Crit Rev Biotechnol 15:287–291

Haselwandter K, Bowen GD (1996) Mycorrhizal relations in trees for agroforestry and land rehabilitation. For Ecol Manage 81:1–17

Haselwandter K, Read DJ (1980) Fungal associations of roots of dominant and sub-dominant plants in high-alpine vegetation systems with special reference to mycorrhiza. Oecologia (Berlin) 45:57–62

Haselwandter K, Read DJ (1982) The significance of a root-fungus association in two *Carex* species of high-alpine plant communities. Oecologia (Berlin) 53:352–354

Haselwandter K, Winkelmann G (2002) Ferricrocin – an ectomycorrhizal siderophore of *Cenococcum geophilum*. BioMetals 15:73–77

Haselwandter K, Krismer R, Holzmann HP, Reid CPP (1988) Hydroxamate siderophore content of organic fertilizers. J Plant Nutr 11:959–967

Haselwandter K, Dobernigg B, Beck W, Jung G, Cansier A, Winkelmann G (1992) Isolation and identification of hydroxamate siderophores of ericoid mycorrhizal fungi. Biometals 5:51–56

Haselwandter K, Passler V, Reiter S, Schmid DG, Nicholson G, Hentschel P, Albert K, Winkelmann G (2006) Basidiochrome – a novel siderophore of the orchidaceous mycorrhizal fungi *Ceratobasidium* and *Rhizoctonia* spp. BioMetals 19:335–343

Heymann P, Ernst JF, Winkelmann G (2000) A gene of the major facilitator superfamily encodes a transporter for enterobactin (Enb1p) in *Saccharomyces cerevisiae*. BioMetals 13:65–72

Hoffland E, Kuyper TW, Wallander H, Plassard C, Gorbushina AA, Haselwandter K, Holmström S, Landeweert R, Lundström US, Rosling A, Sen R, Smits MM, van Hees PAW, van Breemen N (2004) The role of fungi in weathering. Front Ecol Environ 2:258–264

Honegger R (1998) The lichen symbiosis: what is so special about it? Lichenologist 30:193–212

Jalal MAF, van der Helm D (1989) Siderophores of highly phytopathogenic *Alternaria longipes*. Biol Metals 2:11–17

Jalal MAF, Love SK, van der Helm D (1988) N-alpha-dimethylcoprogens. Three novel trihydroxamate siderophores from pathogenic fungi. Biol Metals 1:4–8

Jumpponen AM, Trappe JM (1998) Dark septate endophytes: a review of facultative biotrophic root-colonizing fungi. New Phytol 140:295–310

Liu A, Hamel C, Hamilton RI, Ma BL, Smith DL (2000) Acquisition of Cu, Zn, Mn and Fe by mycorrhizal maize (*Zea mays* L.) grown in soil at different P and micronutrient levels. Mycorrhiza 9:331–336

Manulis S, Kashman Y, Barash I (1987) Identification of siderophores and siderophore-mediated uptake of iron in *Stemphylium botryosum*. Phytochemistry 26:1317–1320

Passler V (1998) Siderophorenproduktion von drei Orchideenmykorrhizapilzen im Vergleich zu *Rhizoctonia endophytica* var. *endophytica* in Abhängigkeit von pH und Eisenverfügbarkeit. MSc Thesis, University of Innsbruck, Austria

Peterson RL, Massicotte HB, Melville LH (2004) Mycorrhizas: anatomy and cell biology. NRC Research Press, Ottawa

Piercey MM, Graham SW, Currah RS (2004) Patterns of genetic variation in *Phialocephala fortinii* across a broad latitudinal transect in Canada. Mycol Res 108:955–964

Powell PE, Cline GR, Reid CPP, Szaniszlo PJ (1980) Occurrence of hydroxamate siderophore iron chelators in soils. Nature (London) 287:833–834

Prabhu V, Biolchini PF, Boyer GL (1996) Detection and identification of ferricrocin produced by ectendomycorrhizal fungi in the genus *Wilcoxina*. BioMetals 9:229–234

Purakayastha TJ, Singh CS, Chonkar PK (1998) Growth and iron nutrition of broccoli (*Brassica oleracea* L. var. *italica* Plenck), grown in a Typic Ustochrept, as influenced by vesicular-arbuscular mycorrhizal fungi in the presence of pyrite and farmyard manure. Biol Fertil Soils 27:35–38

Rai R (1988) Interaction response of *Glomus albidus* and *Cicer Rhizobium* strains on iron uptake and symbiotic N_2-fixation in calcareous soil. J Plant Nutr 11:863–869

Reiter S (1999) Einfluss von pH-Wert und Eisenverfügbarkeit auf die Siderophorenbildung durch vier typische Mykorrhizapilzarten der Orchidaceae. MScThesis, University of Innsbruck, Austria

Renshaw JC, Robson GD, Trinci APJ, Wiebe MG, Livens FR, Collison D, Taylor RJ (2002) Fungal siderophores: structures, functions and applications. Mycol Res 106:1123–1142

Richardson DHS (1999) War in the world of lichens: parasitism and symbiosis as exemplified by lichens and lichenicolous fungi. Mycol Res 103:641–650

Schrettl M, Bignell E, Kragl C, Joechl C, Rogers T, Arst HN Jr, Haynes K, Haas H (2004) Siderophore biosynthesis but not reductive iron assimilation is essential for *Aspergillus fumigatus* virulence. J Exp Med 200:1213–1219

Schüssler A, Schwarzott D, Walker C (2001) A new fungal phylum, the *Glomeromycota*: phylogeny and evolution. Mycol Res 105:1413–1421

Schuler R, Haselwandter K (1988) Hydroxamate siderophore production by ericoid mycorrhizal fungi. J Plant Nutr 11:907–913

Shaw G, Leake JR, Baker AJM, Read DJ (1990) The biology of mycorrhiza in the Ericaceae. XVII. The role of mycorrhizal infection in the regulation of iron uptake by ericaceous plants. New Phytol 115:251–258

Smith SE, Read DJ (1997) Mycorrhizal symbiosis, 2nd edn. Academic Press, London

Stenroos S, Stocker-Woergoetter E, Yoshimura I, Myllys L, Thell A, Hyvonen J (2003) Culture experiments and DNA sequence data confirm the identity of *Lobaria* photomorphs. Can J Bot 81:232–247

Szaniszlo PJ, Powell PE, Reid CPP, Cline GR (1981) Production of hydroxamate siderophore iron chelators by ectomycorrhizal fungi. Mycologia 73:1158–1174

Thanh VN, van Dyk MS, Wingfield MJ (2002) *Debaromyces mycophilus* sp.nov., a siderophore-dependent yeast isolated from woodlice. FEMS Yeast Res 2:415–427

van Hees PAW, Rosling A, Essen S, Godbold DL, Jones DL, Finlay RD (2006) Oxalate and ferricrocin exudation by the extramatrical mycelium of an ectomycorrhizal fungus in symbiosis with *Pinus sylvestris*. New Phytol 169:367–377

Vosatka M, Dodd JC (2002) Ecological considerations for successful application of arbuscular mycorrhizal fungi inoculum. In: Gianinazzi S, Schüepp H, Barea JM, Haselwandter K (eds) Mycorrhizal technology in agriculture: from genes to bioproducts. Birkhäuser, Basel, pp 235–247

Watteau F, Berthelin J (1994) Mineral dissolution of iron and aluminium from soil minerals: efficiency and specificity of hydroxamate siderophores compared to aliphatic acids. Eur J Soil Biol 30:1–9

Wilcox HE (1991) Mycorrhizae. In: Waisel Y, Eshel A, Kafkafi U (eds) Plant roots: the hidden half. Dekker, New York, pp 731–765

Winkelmann G (1992) Structures and functions of fungal siderophores containing hydroxamate and complexone type iron binding ligands. Mycol Res 96:529–534

Winkelmann G (2001) Siderophore transport in fungi. In: Winkelmann G (ed) Microbial transport systems. Wiley-VCH, Weinheim, pp 463–480

Winkelmann G (2004) Ecology of siderophores. In: Crosa JH, Mey AR, Payne SM (eds) Iron transport in bacteria. ASM Press, Washington DC, pp 437–450

5 Protein-mediated Siderophore Uptake in Gram-negative Bacteria: A Structural Perspective

José D. Faraldo-Gómez

5.1
Introduction

Iron is amongst the most important nutrients of bacteria and also one of the most abundant chemical elements composing the Earth's crust. Under normal conditions, however, free iron is scarce due to its rapid oxidation and the subsequent formation of insoluble hydroxides. Thus, bacteria have had to evolve highly efficient mechanisms of iron uptake in order to ensure a sufficient supply.

These mechanisms generally rely on receptor proteins residing in the various compartments of the bacterial cell envelope that specifically bind and/or translocate a wide range of iron-containing molecules (Fig. 5.1). For many pathogens, these ligands can be the iron-binding proteins that the host itself utilizes to keep iron in solution, or other small proteins that capture heme. Alternatively, bacteria and other microorganisms are able to synthesize, excrete and finally reabsorb low-molecular-weight compounds that chelate environmental iron with outstanding affinity, known as siderophores (Wandersman and Delepelaire 2004).

This chapter will focus on the mechanisms of protein-mediated siderophore uptake in Gram-negative bacteria, from the perspective of structural biology. Consistent with iron's scarcity and importance, these mechanisms involve all the challenging and often poorly understood complexities characteristic of biology at the atomic level, such as allosteric regulation, signaling mechanisms, protein-protein recognition, energy transduction, etc.

The chapter's organization follows the uptake pathway, starting from the outermost surface and then moving towards the cytoplasm. It has been my intention to provide an overview of this process that is accessible, and hopefully not too tedious, to non-specialists in the field of structural protein biology. As usual, I refer those seeking a more detailed understanding to the research papers and reviews in the bibliography list, and to the references therein.

Center for Integrative Sciences, University of Chicago, Chicago, USA,
email: faraldo@uchicago.edu

Soil Biology, Volume 12
Microbial Siderophores
A. Varma, S.B. Chincholkar (Eds.)
© Springer-Verlag Berlin Heidelberg 2007

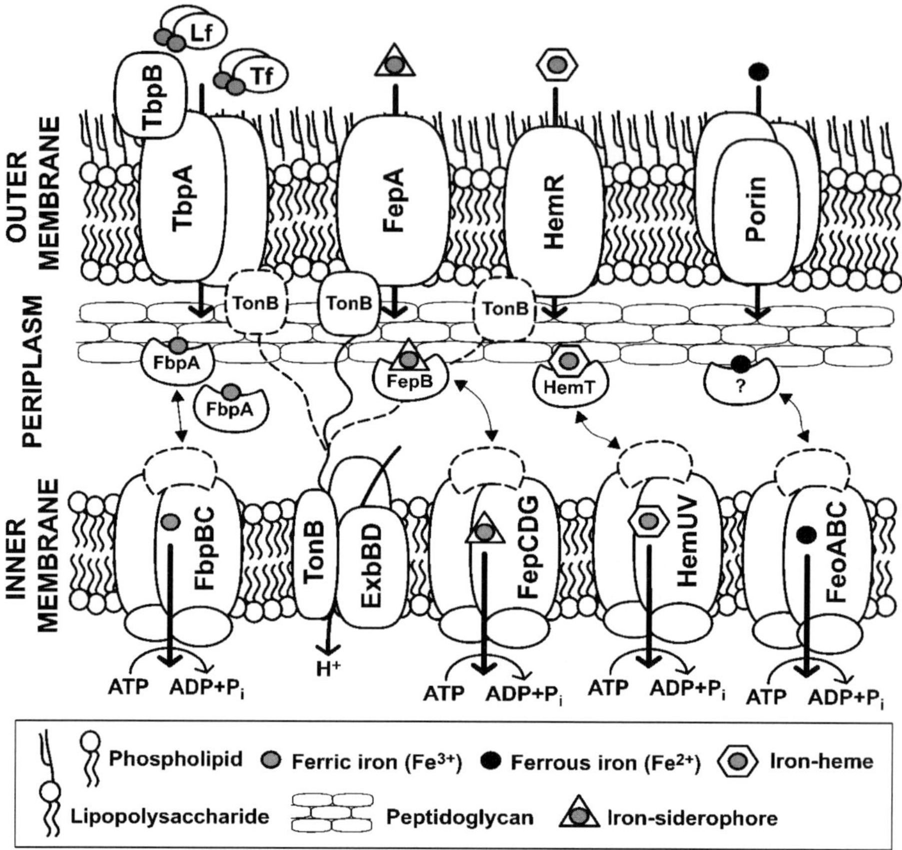

Fig. 5.1. Representative systems for uptake of iron in Gram-negative bacteria (Braun and Killmann 1999; Ratledge and Dover 2000; Crichton 2001; Wandersman and Delepelaire 2004). Under aerobic conditions, iron uptake is mediated by various high-affinity receptor proteins that bind Fe^{3+}-containing compounds on the cell's surface and that subsequently facilitate their translocation into the periplasm. This process is thought to require an input of mechanical energy that is provided by the TonB-ExbBD complex, or homologues thereof, anchored in the inner membrane. Periplasmic binding proteins and ATP-driven transporters in the cytoplasmic membrane ensure further transport into the cell. For example, *Escherichia coli* reabsorbs the endogenous siderophore enterobactin via the FepA receptor and the FepBCDG system. Pathogenic bacteria can also use iron sources from the host; for instance, in *Neisseria* species ferric iron is removed from transferrin (Tf) and lactoferrin (Lf), and transported into the cell by the TbpAB-FbpABC system; in *Yersinia enterocolitica* heme can be taken up directly through the HemRTUV system. Under anaerobic conditions, soluble Fe^{2+} can diffuse across outer membrane porins, and is subsequently imported by energy-dependent systems such as the FeoABC

5.2
Transport Across the Outer Membrane

5.2.1
The Outer-Membrane Receptor and Transporter Proteins

In contrast to Gram-positive bacteria, the outermost surface of Gram-negative bacteria is a double-layered lipid membrane, which encloses both the *murein sacculus* and the inner or cytoplasmic membrane. The external layer of this outer membrane is primarily composed of lipopolysaccharide (LPS) lipids. The strong interactions between LPS lipids, mediated by Mg^{2+} and Ca^{2+}, as well as their reduced intramolecular mobility, make the outer membrane highly impermeable. The outer membrane thus serves as a resistance barrier against environmental hazards such as toxic agents, host-defense proteins and digestive enzymes (Nikaido 1996).

A large number of pore-forming proteins exist in the outer membrane in order to facilitate the uptake of nutrients across the LPS barrier. Ions and other small molecules such as sugars permeate the membrane across water-filled protein channels known as porins, driven by self-diffusion (Koebnik et al. 2000; Nikaido 2003). For larger solutes, or for molecules that are too scarce to rely on diffusion-driven transport, more sophisticated protein architectures are employed. These may include specific binding sites, as well as gating mechanisms whereby a permeation pore is opened or closed depending on whether the appropriate substrate is present, and, in some cases, also on whether an external source of mechanical energy is available.

Regardless of their function, all of the integral outer membrane proteins known to date are structurally distinct in that they fold in the so-called β-barrel architecture, instead of forming bundles of transmembrane α-helices like other membrane proteins do (Schulz 2000; Wimley 2003). The β-barrel architecture results from the lower-than-usual content of hydrophobic amino acids of these proteins, which in turn may be required for the successful translocation of newly synthesized proteins from the cell interior to the outer membrane (Koebnik et al. 2000; Postle 2002).

From the structural and functional viewpoints, the β-barrels involved in siderophore uptake are distinct from other outer membrane proteins in several aspects. First, they display a nanomolar-range affinity for their substrates, in contrast to, e.g., fatty-acid transporters (μM) (van den Berg et al. 2004) or the various porins (mM) (Nikaido 2003). Second, they form very large β-barrels (ca. 25 Å in diameter) that contain an additional protein domain in their interior, which provides the recognition site for the substrates to be transported; this so-called "plug" domain effectively blocks the pore formed by the barrel, and thus constitutes the permeation gate. And third, they require an external input of mechanical energy in order for the permeation gate to open. Incidentally, uptake of vitamin B_{12} across the outer membrane is analogous to that of sidero-phores (Kadner 1990).

To date, four members of the family of outer-membrane siderophore trans-porters have been characterized structurally at the atomic level, by means of X-ray crystallography; these are FhuA (Ferguson et al. 1998; Locher et al. 1998), FepA (Buchanan et al. 1999) and FecA (Ferguson et al. 2002; Yue et al. 2003) from *Escherichia coli*, and FpvA from *Pseudomonas aeruginosa* (Cobessi et al. 2005). These membrane proteins mediate the uptake of the siderophores known as ferrichrome, enterobactin, ferric citrate and pyoverdin, respectively. In addi-tion to these, the structure of the receptor for vitamin B_{12} is also known (Chi-mento et al. 2003a).

Schematic representations of some of these structures are shown in Fig. 5.2. In all cases, the β-barrel domain is made up of 22 strands; adjacent β-strands are generally connected by short stretches of amino acids on the periplasmic face of the proteins, and with longer stretches in the extracellular side. Interestingly, the conformation of some of these extracellular loops is strongly dependent on the presence of the iron-loaded substrate; specifically, these loops appear to change their structure upon ligand binding, precisely so as to obstruct any possible re-lease of the ligand back into the extracellular solution.

In the interior of the β-barrel, the plug domain is held in place by numer-ous hydrogen bonds and polar contacts, and effectively prevents any significant permeation (Faraldo-Gómez et al. 2002; Chimento et al. 2005). Thus, transloca-tion of siderophores across these proteins appears to involve a significant struc-tural rearrangement of the plug domain, though the nature of these structural changes is unknown. It is possible that the plug domain comes out of the bar-rel in its entirety, dragging the bound siderophore with it; alternatively, partial unfolding of its structure while still residing within the barrel domain might be sufficient to open up a permeation pore (Faraldo-Gómez and Sansom 2003; Chimento et al. 2005).

Whatever the case may be, it has been long established that this step in the transport process necessitates the interaction of the outer membrane receptors with a protein anchored in the cytoplasmic membrane, known as TonB (Postle 1990), which is believed to provide the mechanical energy required to alter the conformation of the plug domain (see next section).

That the energy required for transport across the outer membrane is trans-duced from the cytoplasmic membrane is not striking, since no energy is avail-able in the outer membrane in the form of electrochemical gradients, due to the numerous porins through which ionic species can diffuse freely. However, it is less obvious how the plug domain provides a functional advantage at the outer membrane level, given the energetic cost that it imposes on the uptake process.

One possible rationale is that, in terms of self-protection against environ-mental hazards, bacteria cannot afford to employ porins of size large enough to accommodate relatively large solutes such as siderophores, vitamin B_{12} or long-chain fatty acids, since these would also allow for the non-specific permeation of other toxic compounds (Nikaido 2003). Alternatively, or in addition, com-

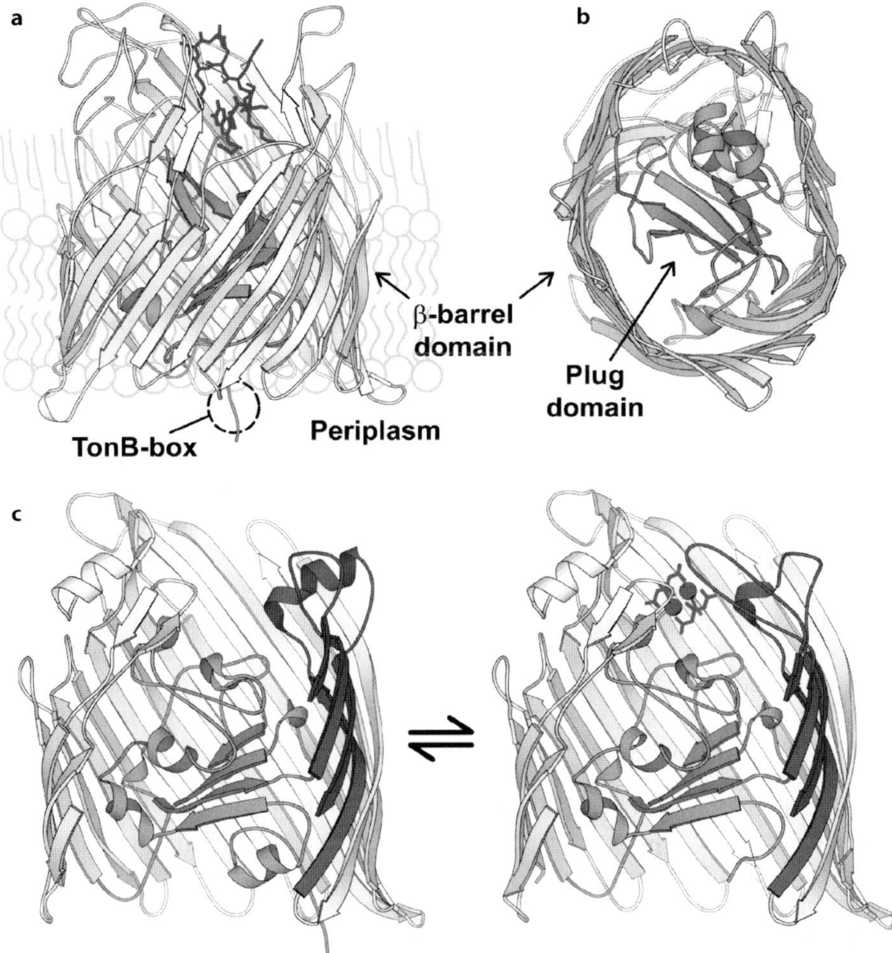

Fig. 5.2. a Schematic representation of the three-dimensional atomic structure of FpvA, the outer membrane receptor and transporter of pyoverdin from *Pseudomonas aeruginosa* (Cobessi et al. 2005). *Ribbons* follow the trace of the protein's backbone; *coil-shaped ribbons* represent α-helices and *arrow-shaped forms* represent β-strands; for clarity, amino-acid side chains are omitted. Pyoverdin (*dark grey sticks*) is shown in its binding site at the extracellular side of the plug domain. **b** Structure of FhuA, the outer membrane TonB-dependent transporter for ferrichrome from *E. coli*, viewed from the periplasm (Ferguson et al. 1998; Locher et al. 1998). **c** Ligand-free (*left*) and ligand-bound (*right*) states of FecA, the ferric citrate receptor in *E. coli* (Ferguson et al. 2002). Note the change in the conformation of some of the extracellular loops upon ligand binding

parison of the transport kinetics in porins and TonB-dependent transporters suggests that the presence of the plug domain and the reliance on an energy input might address the challenge of ensuring a sufficient uptake of nutrients such as iron under conditions of starvation.

Transport rates across porins are, within a wide range of solute concentrations, largely correlated with the electrochemical gradients across the membrane. Some specific porins, such as LamB, enhance their efficiency by possessing low-affinity binding sites, but at very low concentrations, as for iron-complexes under physiological conditions, porin-mediated transport systems are inefficient. By contrast, TonB-dependent receptors such as FepA are rather insensitive to the transmembrane concentration gradient, and reach saturation levels at relatively low concentrations and slow rates, which make them generically worse transporters than porins (Klebba and Newton 1998). However, it is their strong dependence on an external input of energy, rather than on concentration gradients, that, together with the high-affinity binding site on the extracellular face of the plug domain, seem to enable these receptors to operate more efficiently than porins at very low solute concentrations, and thus sustain cell growth.

5.2.2
Energy Transduction and TonB

Although it is widely accepted that TonB plays an essential role in the activation of siderophore uptake across the outer membrane, the actual mechanism of energy transduction remains sketchy. TonB is known to be anchored in the inner membrane by a single N-terminal transmembrane helix, and to project into the periplasm by virtue of a long polypeptide stretch rich in proline amino acids (Postle 1993). At the C-terminal end, the protein folds into a globular domain, which interacts with the outer membrane receptors, namely via a specific sequence of amino acids at the periplasmic side of their plug domain, termed the TonB-box (Fig. 5.2) (Pawelek et al. 2006; Shultis et al. 2006). To date, the only available structural information of TonB at the atomic level corresponds to this C-terminal domain, based on X-ray crystallography and NMR spectroscopy analyses (Ködding et al. 2005; Peacock et al. 2005).

The TonB-box has been shown to adopt alternative configurations dependent on whether the outer membrane receptors are loaded with siderophores, and it is believed that only in the ligand-bound form can TonB interact constructively with the plug domain (Ferguson et al. 1998; Locher et al. 1998; Merianos et al. 2000; Coggshall et al. 2001; Ferguson et al. 2002; Chimento et al. 2003b). Although a detailed structural understanding of this allosteric effect would require further investigations (Fanucci et al. 2003a, b), it would certainly be consistent with the notion that energy must not be transduced to ligand-free receptors in order to ensure efficient transport (Moeck et al. 1997; Braun 1998).

Another open question pertains to how TonB becomes energized, that is, competent to activate transport through the outer membrane receptors. At the present time, the prevailing model hypothesizes that TonB is able to alternate between several conformations (Holroyd and Bradbeer 1984; Larsen et al. 1999), by virtue of its association with the ExbBD protein complex, which also resides in the inner membrane (Postle 1993; Braun 1995). The atomic structure of this macromolecular assembly is unknown, but biochemical analyses yielded a stoichiometry of one TonB per two ExbD per seven ExbB (Held and Postle 2002; Higgs et al. 2002). Sequence analyses also indicate that ExbD is a periplasmic protein anchored to the cytoplasmic membrane by an amino-terminal α-helix (Kampfenkel and Braun 1992), whereas the largest part of ExbB is thought to reside in the cytoplasm, with three membrane-spanning helices anchoring the protein to the cytoplasmic membrane (Kampfenkel and Braun 1993).

More importantly, however, the ExbBD complex is believed to couple the proton motive force sustained across the cytoplasmic membrane to conformational changes in TonB (Larsen et al. 1999; Held and Postle 2002). In other words, the translocation of protons from the periplasm into the cytoplasm, driven by their electrochemical gradient and mediated by ExbBD, would lead to an energized conformation of TonB in which the association of its carboxyl-terminal with the TonB-box of ligand-bound outer membrane receptors would be possible. The subsequent relaxation of TonB to its resting state would induce the rearrangements of the plug domain required for siderophore translocation.

5.3
Transport Across the Periplasm and Cytoplasmic Membrane

5.3.1
Periplasmic Binding Proteins

Once the outer membrane has been permeated, siderophores must traverse the periplasm, that is, the region that separates the outer and inner membrane. The periplasm is believed to be a gel-like environment, 15–25 nm wide, characterized by molecular diffusion rates that are 1000-fold smaller relative to the extracellular solution, and 100-fold smaller relative to the cytoplasm. This extreme viscosity appears to be due to its high protein content, as well as to the presence of the so-called *murein sacculus*, i.e. the complex multilayered network of covalently-linked peptides and sugars that define the shape and volume of the cell (Oliver 1996; Park 1996).

In order to translocate solutes across the periplasm efficiently, bacteria employ a wide range of high-affinity binding proteins that are specific for their respective ligands, which include sugars, amino acids, oligopeptides, ions and other compounds. These periplasmic binding proteins, or PBPs, subsequently deliver their cargo to specific ATP-dependent transporters or chemotaxis receptors residing in the inner membrane.

From the structural viewpoint, most PBPs are alike in that they contain two domains, arranged so as to form a suitable binding site in their interfacial region, although the precise fold and relative mobility of these domains vary across the PBP subfamilies (Dwyer and Hellinga 2004). Although the three-dimensional atomic structure of over 100 PBPs has been resolved by X-ray crystallography, only one member of the siderophore subfamily has been structurally characterized to date, namely the ferrichrome-binding protein FhuD (Clarke et al. 2000). However, the structure of the PBP for vitamin B_{12}, termed BtuF, is also known (Borths et al. 2002; Karpovich et al. 2003), and happens to be similar in its fold to that of FhuD. Since the uptake pathways for B_{12} and siderophores are analogous (see above and below), the similarity of these two PBPs indicates that all members of the siderophore subfamily might share a common architecture and binding mechanism (Köster 2001).

A schematic representation of the atomic structure of FhuD in complex with a gallium-loaded ferrichrome is shown in Fig. 5.3. As for other PBPs, the binding site is located in a cleft in the protein surface that is formed between the so-called N- and C-domains, each of which contains a central sheet of β-strands flanked by several α-helices. The amino acids lining the binding cleft determine the specificity and affinity of FhuD for its ligands, which also include other hydroxamate-type siderophores and derivatives such as coprogen, Desferal and

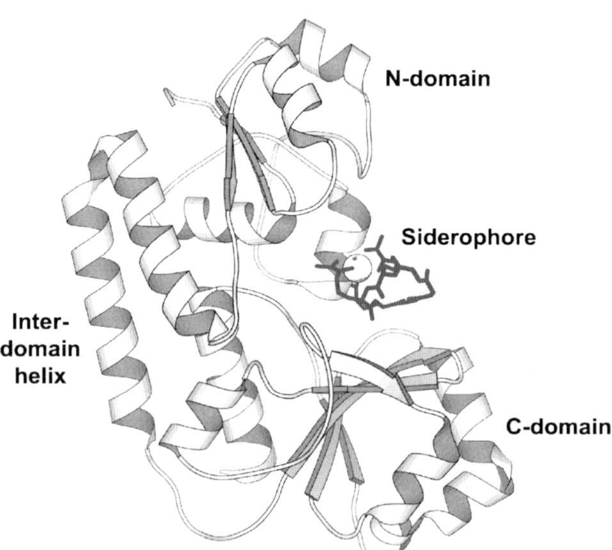

Fig. 5.3. Schematic representation of the three-dimensional atomic structure of the ferrichrome binding protein FhuD, in complex with a gallium-loaded siderophore (Clarke et al. 2000). As before, ribbons follow the trace of the protein's backbone and side chains are omitted for clarity. The gallium ion, which replaces ferric iron in the protein crystal, is shown as a sphere, and the interatomic bonds within ferrichrome are shown as *grey sticks*. The interdomain α-helix that traverses the protein is thought to confer PBPs of the siderophore subfamily with an increased mechanical rigidity

the naturally-occurring antibiotic albomycin (Clarke et al. 2002). Specifically, an arginine, a tyrosine and several tryptophan side chains provide both hydrogen-bonding interactions and a hydrophobic environment that match the chemical nature of the iron-chelating groups in these siderophores. Interestingly, the amino-acid composition of the binding pocket in FhuD is reminiscent of that of the outer membrane receptor FhuA, which illustrates how bacteria evolved very different protein architectures that can support a similar biological function in diverse environments.

Although the atomic structure of a ligand-free FhuD has not been reported as yet, comparison with the structures of ligand-free and ligand-bound BtuF (Fig. 5.4a), as well as with other PBPs in the same structural subfamily, indicates that substrate binding is not likely to result in dramatic structural changes

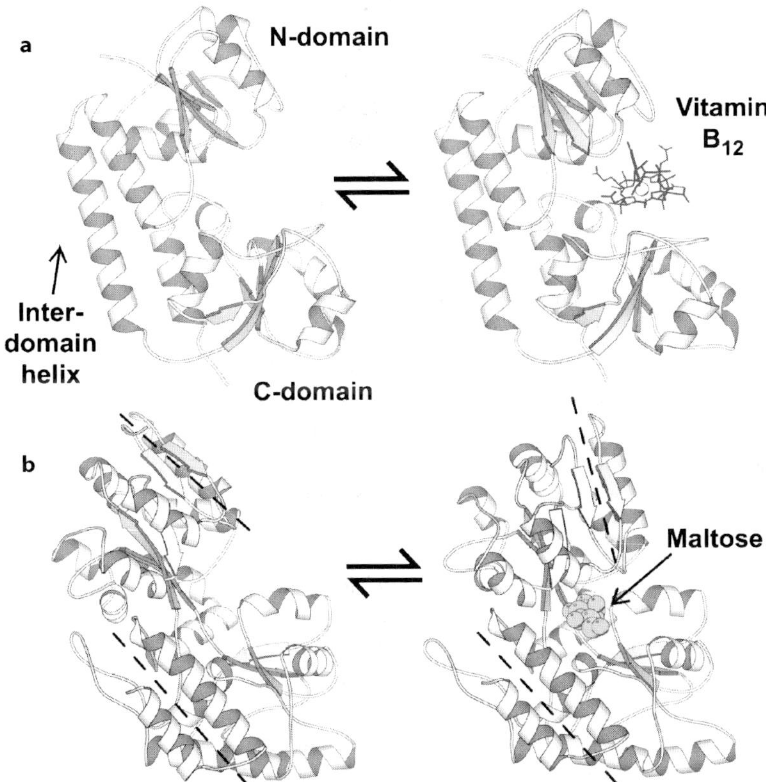

Fig.5.4. a Schematic representation of the three-dimensional atomic structure of BtuF, the PBP for vitamin B$_{12}$, in the ligand-free (*left*) and ligand-loaded (*right*) states (Karpovich et al. 2003). Interatomic bonds within vitamin B$_{12}$ are shown as *sticks*, and the cobalt ion is shown as a *sphere*. **b** Structures of the periplasmic maltose-binding protein MBP, in the ligand-free (*left*) and ligand-bound (*right*) states (Spurlino et al. 1991; Sharff et al. 1992). Maltose atoms are shown as *grey spheres*. Note the dramatic change in the relative orientation and position of the amino- and carboxyl-terminal domains upon ligand binding

beyond the locality of the binding cleft, though a reduction in the overall flexibility in the protein structure is expected. This is in contrast with most of the two-domain PBPs of known structure, which do undergo substantial conformational changes upon ligand binding (Quiocho and Ledvina 1996; Felder et al. 1999; Dwyer and Hellinga 2004); in particular, their binding site, initially exposed to the solvent environment, becomes buried within the protein as a result of the spatial rearrangement of the two flanking domains (Fig. 5.4b). Since PBPs typically deliver their cargo to transport proteins residing in the inner membrane, the different mechanical properties of siderophore PBPs may well correlate with a distinct mechanism of ligand release and transport into the cytoplasm, compared with other members of the family. However, to date no structural information is available that elucidates the nature of the interaction between PBPs and their inner membrane counterparts, so these mechanistic aspects remain a matter of debate.

5.3.2
The Cytoplasmic-membrane ABC Transporters

In contrast to the outer membrane, which primarily serves as a permeability barrier that protects the cell, the inner or cytoplasmic membrane of Gram-negative bacteria is involved in a very wide range of important cellular functions, such as the regulated transport of nutrients and metabolic products, the propagation of signals upon external stimuli, and the generation and conservation of energy. Consistent with this function, a much greater diversity of proteins can be found to reside in this membrane, where they also are in large numbers. In fact, detailed analyses of the inner membrane composition indicate that only three or four layers of phospholipid molecules might separate proteins from each other (Kadner 1996).

In order to deal with this logistical complexity in such a crowded environment, inner-membrane transporters associate specifically with loaded PBPs and pump their ligands into the cytoplasm (Higgins 2001; Davidson and Chen 2004). In their resting state, these membrane transporters, which are in fact aggregates of two identical membrane proteins (i.e. homodimers), are not competent for permeation, but they do provide a suitable docking site for their corresponding PBPs on the membrane surface. Subsequent to this association, large structural changes are thought to take place in the complex that result in (a) the exchange of the substrate between the PBP and transporter, (b) the opening of a translocation pathway across the membrane between the homodimers, (c) the release of the substrate into the cytoplasm, and (d) the disassociation of the PBP from the transporter, which then goes back to the resting state.

Although much remains to be elucidated concerning the actual mechanism of transport, it is well established that this process requires, at one or more stages, the energy derived from the hydrolysis of ATP into ADP. (ATP and ADP stand for adenosine tri- and di-phosphate respectively; the conversion of ATP into

ADP by hydrolysis, i.e. the water-mediated release of the terminal phosphate group from ATP, is a spontaneous reaction provided an appropriate chemical environment. Typically, this reaction results in structural changes in the locality of the ATP/ADP binding sites, which then propagate and become amplified elsewhere in the protein's structure.) Thus, in addition to the two-protein transmembrane domain that provides the permeation pathway, ABC transporters include two extra-membranous nucleotide-binding domains (NBD), which provide suitable sites for ATP binding and hydrolysis. These domains, each of which is non-covalently bound to one of the transmembrane proteins, are also known as ATP binding cassettes – hence the term ABC transporter.

The existing structural information regarding ABC transporters is very limited, and sometimes contradictory (Davidson and Chen 2004). In particular, no three-dimensional atomic structures are available of any of the transporters involved in siderophore uptake across the inner membrane. Fortunately, one of few proteins in this family whose structure has been resolved at atomic resolution is that of the vitamin B_{12} transporter, known as BtuCD (Locher et al. 2002). It is expected that, as for the outer membrane receptors and the PBPs, the structure of siderophore inner-membrane transporters will be similar to that of BtuCD (Köster 2001).

As shown in Fig. 5.5, the transmembrane domain of BtuCD, that is, the BtuC dimer, differs from the outer membrane receptor BtuB in that the protein is folded in the conventional α-helical architecture. In particular, each of monomers contains 10 α-helices, which aggregate to form a so-called helical bundle. As for the outer membrane β-barrel proteins, loops of variable length connect pairs of α-helices, although these are not always adjacent. Each of the NBDs, in this case termed BtuD, binds at the cytoplasmic end of each BtuC monomer, forming an interface that appears to be energetically stabilized primarily by the close contact of hydrophobic amino acids.

From the functional viewpoint, each BtuC bundle and its corresponding BtuD domain is believed to operate as mechanical unit that changes its orientation with respect to the membrane plane during each transport event, opening and closing a permeation pathway that runs along the dimer interface. In particular, hydrolysis of ATP has been proposed to result in a rotational motion of the BtuD domains, which in turn would lead to an increased separation between the BtuC bundles at the cytoplasmic side of the membrane, and a closure at the periplasmic side. Provided that vitamin B_{12} is present in the periplasmic side of BtuC, the proposed structural rearrangements would effectively induce the release of BtuF back into the periplasm and the translocation of B_{12} into the cytoplasm, after which BtuCD would return to its initial state (Locher et al. 2002; Locher and Borths 2004).

Although this mechanism is plausible, it is at odds with previous models proposed for other ABC transporters (Chen et al. 2003), underlying the difficulties in assessing which state in the transport cycle is captured by the crystallographic analysis. In addition, crucial aspects of the transport process are unclear, such as the signalling mechanism that triggers ATP hydrolysis in the cytoplasm upon association of a ligand-loaded PBP at the periplasmic side of the transporter.

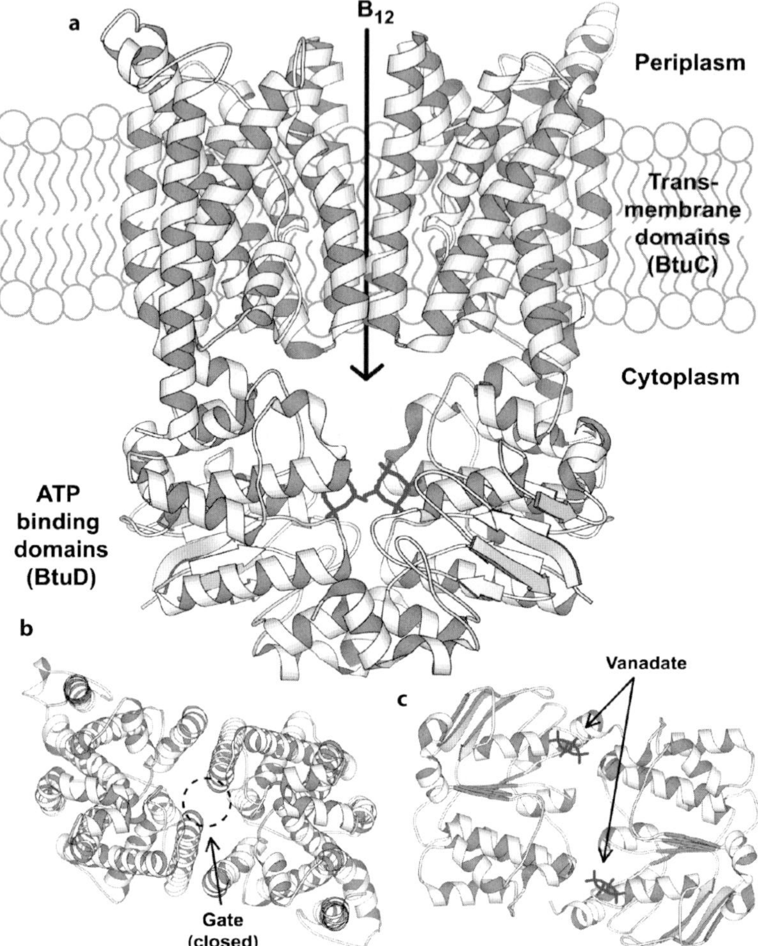

Fig.5.5. **a** Schematic representation of the structure of BtuCD, the inner membrane transporter of vitamin B_{12} (Locher et al. 2002), which is thought to be similar to those of the siderophore ABC transporters. A cyclotetravanadate molecule (shown in *sticks*) occupies the ATP binding site in each of the NBDs. **b** View of BtuC from the periplasm. The cytoplasmic end of the putative permeation pathway is effectively closed due to the close contact of several side chains in the gate region (not shown for clarity). **c** View of the BtuD nucleotide-binding domains from the cytoplasmic side of BtuC. Note that the ATP binding sites are located in the interface between the two BtuD domains, consistent with the hypothesis that rearrangement of this interface upon ATP hydrolysis may lead to structural changes in BtuC that open the gate and allow permeation of B_{12} into the cytoplasm

5.4
Conclusions

As appropriately put by Klebba (2003), siderophore uptake is "a high-affinity, multi-specific, multi-component, energy-dependent reaction that is a paradigm of ligand-gated transport" across biological membranes. In recent years, the determination of the atomic-resolution structure of several proteins involved in this fascinating process, alongside ongoing biochemical, genetic and spectroscopic investigations, has advanced very significantly our understanding thereof. Nonetheless, fundamental questions remain open with regard to signalling and energy transduction mechanisms, large and small-scale conformational changes, and protein-protein and protein-ligand recognition. As it happens, these questions are also the paradigm of twenty-first-century structural biology and biophysics, which suggest that the elucidation of siderophore uptake systems will continue to be at the forefront of research in molecular biology.

References

Borths EL, Locher KP, Lee AT, Rees DC (2002) The structure of *Escherichia coli* BtuF and binding to its cognate ATP binding cassette transporter. Proc Natl Acad Sci USA 99:16642–16647

Braun V (1995) Energy-coupled transport and signal transduction through the Gram-negative outer membrane via TonB-ExbB-ExbD-dependent receptor proteins. FEMS Microbiol Rev 16:295–307

Braun V (1998) Pumping iron through cell membranes. Science 282:2202–2203

Braun V, Killmann H (1999) Bacterial solutions to the iron-supply problem. Trends Biochem Sci 24:104–109

Buchanan SK, Smith BS, Venkatramani L, Xia D, Essar L, Palnitkar M, Chakraborty R, van der Helm D, Deisenhofer J (1999) Crystal structure of the outer membrane active transporter FepA from *Escherichia coli*. Nat Struct Biol 6:56–63

Chen J, Lu G, Lin J, Davidson AL, Quiocho FA (2003) A tweezers-like motion of the ATP-binding cassette dimer in an ABC transport cycle. Mol Cell 12:651–661

Chimento DP, Kadner RJ, Wiener MC (2003a) The *Escherichia coli* outer membrane cobalamin transporter BtuB: structural analysis of calcium and substrate binding, and identification of orthologous transporters by sequence/structure conservation. J Mol Biol 332:999-1014

Chimento DP, Mohanty AK, Kadner RJ, Wiener MC (2003b) Substrate-induced transmembrane signaling in the cobalamin transporter BtuB. Nat Struct Biol 10:394–401

Chimento DP, Kadner RJ, Wiener MC (2005) Comparative structural analysis of TonB-dependent outer membrane transporters: implications for the transport cycle. Proteins 59:240–251

Clarke TE, Ku SY, Dougan DR, Vogel HJ, Tari LW (2000) The structure of the ferric siderophore binding protein FhuD complexed with gallichrome. Nat Struct Biol 7:287–291

Clarke TE, Braun V, Winkelmann G, Tari LW, Vogel HJ (2002) X-ray crystallographic structures of the *Escherichia coli* periplasmic binding protein FhuD bound to hydroxamate-type siderophores and the antibiotic albomycin. J Biol Chem 277:13966–13972

Cobessi D, Celia H, Folschweiller N, Schalk IJ, Abdallah MA, Pattus F (2005) The crystal structure of the pyoverdine outer membrane receptor FpvA from *Pseudomonas aeruginosa* at 3.6 Å resolution. J Mol Biol 347:121–134

Coggshall KA, Cadieux N, Kadner RJ, Cafiso DS (2001) Site directed spin labeling reveals the first step in the transport pathway of a TonB-dependent transporter. Biophys J 80:213

Crichton RR (2001) Inorganic biochemistry of iron metabolism: from molecular mechanisms to clinical consequences. Wiley, New York

Davidson AL, Chen J (2004) ATP-binding cassette transporters in bacteria. Annu Rev Biochem 73:241–268

Dwyer MA, Hellinga HW (2004) Periplasmic binding proteins: a versatile superfamily for protein engineering. Curr Opin Struct Biol 14:495–504

Fanucci GE, Lee JY, Cafiso DS (2003a) Membrane mimetic environments alter the conformation of the outer membrane protein BtuB. J Am Chem Soc 125:13932–13933

Fanucci GE, Lee JY, Cafiso DS (2003b) Spectroscopic evidence that osmolytes used in crystallization buffers inhibit a conformation change in a membrane protein. Biochemistry 42:13106–13112

Faraldo-Gómez JD, Sansom MSP (2003) Acquisition of iron-siderophores in Gram-negative bacteria. Nat Rev Mol Cell Biol 4:105–116

Faraldo-Gómez JD, Smith GR, Sansom MSP (2002) Molecular dynamics simulations of the bacterial outer membrane protein FhuA: a comparative study of the ferrichrome-free and bound states. Biophys J 85:1406–1420

Felder CB, Graul RC, Lee AY, Merkle HP, Sadee W (1999) The Venus flytrap of periplasmic binding proteins: an ancient protein module present in multidrug receptors. AAPS PharmSci 1:A2

Ferguson AD, Hofmann E, Coulton JW, Diederichs K, Welte W (1998) Siderophore-mediated iron transport: crystal structure of FhuA with bound lipopolysaccharide. Science 282:2215–2220

Ferguson AD, Chakraborty R, Smith BS, Esser L, van der Helm D, Deisenhofer J (2002) Structural basis of gating by the outer membrane transporter FecA. Science 295:1715–1719

Held KG, Postle K (2002) ExbB and ExbD do not function independently in TonB-dependent energy transduction. J Bacteriol 184:5170–5173

Higgins CF (2001) ABC transporters: physiology, structure and mechanism – an overview. Res Microbiol 152:205–210

Higgs PI, Larsen RA, Postle K (2002) Quantification of known components of the *Escherichia coli* TonB energy transduction system: TonB, ExbB, ExbD and FepA. Mol Microbiol 44:271–281

Holroyd CD, Bradbeer C (1984) Cobalamin transport in *Escherichia coli*. In: Schlessinger D (ed) Microbiology. American Society for Microbiology, Washington DC, pp 21–23

Kadner RJ (1990) Vitamin B_{12} transport in *Escherichia coli*: energy coupling between membranes. Mol Microbiol 4:2027–2033

Kadner RJ (1996) Cytoplasmic membrane. In: Neidhardt F (ed) *Escherichia coli* and *Salmonella typhimurium*: cellular and molecular biology. American Society for Microbiology, Washington DC, pp 58–87

Kampfenkel K, Braun V (1992) Membrane topology of the *Escherichia coli* ExbD protein. J Bacteriol 174:5485–5487

Kampfenkel K, Braun V (1993) Topology of the ExbB protein in the cytoplasmic membrane of *Escherichia coli*. J Biol Chem 268:6050–6057

Karpovich NK, Huang HH, Smith PC, Hunt JF (2003) Crystal structures of the BtuF periplasmic binding protein for vitamin B12 suggest a functional important reduction in protein mobility upon ligand binding. J Biol Chem 278:8429–8434

Klebba PE (2003) Three paradoxes of ferric enterobactin uptake. Front Biosci 8:1422–1436

Klebba PE, Newton SMC (1998) Mechanisms of solute transport through the outer membrane proteins: burning down the house. Curr Opin Microbiol 1:238–248

Koebnik R, Locher KP, van Gelder P (2000) Structure and function of bacterial outer membrane proteins: barrels in a nutshell. Mol Microbiol 37:239-253

Ködding J, Killig F, Polzer P, Howard SP, Diederichs K, Welte W (2005) Crystal structure of a 92-residue C-terminal fragment of TonB from *Escherichia coli* reveals significant conformational changes compared to structures of smaller TonB fragments. J Biol Chem 280:3022–3028

Köster W (2001) ABC transporter-mediated uptake of iron, siderophores, heme and vitamin B_{12}. Res Microbiol 152:291–301

Larsen RA, Thomas MG, Postle K (1999) Proton motive force, ExbB and ligand-FepA drive conformational changes in TonB. Mol Microbiol 31:1809–1824

Locher KP, Borths EL (2004) ABC transporter architecture and mechanism: implications from the crystal structures of BtuCD and BtuF. FEBS Lett 564:264–268

Locher KP, Rees B, Koebnik R, Mitschler A, Moulinier L, Rosenbusch JP, Moras D (1998) Transmembrane signalling across the ligand-gated FhuA receptor: crystal structures of free and ferrichrome-bound states reveal allosteric changes. Cell 95:771–778

Locher KP, Lee AT, Rees DC (2002) The *E. coli* BtuCD structure: a framework for ABC transporter architecture and mechanism. Science 296:1091–1098

Merianos HJ, Cadieux N, Lin CH, Kadner RJ, Cafiso DS (2000) Substrate-induced exposure of an energy-coupling motif of a membrane transporter. Nat Struct Biol 7:205–209

Moeck G, Coulton JW, Postle K (1997) Cell envelope signalling in *Escherichia coli*: ligand binding to the ferrichrome-iron receptor FhuA promotes interaction with the energy-transducing protein TonB. J Biol Chem 272:28391–28397

Nikaido H (1996) Outer membrane. In: Neidhardt F (ed) *Escherichia coli* and *Salmonella typhimurium*: cellular and molecular biology. American Society for Microbiology, Washington DC. pp 29–47

Nikaido H (2003) Molecular basis of bacterial outer membrane permeability revisited. Microbiol Mol Biol Rev 67:593–656

Oliver DB (1996) Periplasm. In: Neidhardt F (ed) *Escherichia coli* and *Salmonella typhimurium*: cellular and molecular biology. American Society for Microbiology, Washington DC, pp 88–103

Park JT (1996) The *murein sacculus*. In: Neidhardt F (ed) *Escherichia coli* and *Salmonella typhimurium*: cellular and molecular biology. American Society for Microbiology, Washington DC, pp 48–57

Pawelek PD, Croteau N, Ng-Thow-Hing C, Khursigara CM, Moiseeva N, Allaire M, Coulton JW (2006) Structure of TonB in complex with FhuA, *E. coli* outer membrane receptor. Science 312:1399–1402

Peacock RS, Weljie AM, Howard SP, Price FD, Vogel HJ (2005) The solution structure of the C-terminal domain of TonB and interaction studies with TonB-box peptides. J Mol Biol 345:1185–1197

Postle K (1990) TonB and the Gram-negative dilemma. Mol Microbiol 4:2019–2025

Postle K (1993) TonB protein and energy transduction between membranes. J Bioenerg Biomembr 25:591–601

Postle K (2002) Close before opening. Science 295:1658–1659

Quiocho FA, Ledvina PS (1996) Atomic structure and specificity of bacterial periplasmic receptors for active transport and chemotaxis: variation on common themes. Mol Microbiol 20:17–25

Ratledge C, Dover LG (2000) Iron metabolism in pathogenic bacteria. Annu Rev Microbiol 54:881–941

Schulz GE (2000) β-Barrel membrane proteins. Curr Opin Struct Biol 10:443–447

Sharff AJ, Rodseth LE, Spurlino JC, Quiocho FA (1992) Crystallographic evidence of a large ligand-induced hinge-twist motion between two domains of the maltodextrin binding protein involved in active transport and chemotaxis. Biochemistry 31:10657–10663

Shultis DD, Purdy MD, Banchs CN, Wiener MC (2006) Outer membrane active transport: structure of the BtuB:TonB complex. Science 312:1396–1399

Spurlino JC, Lu GY, Quiocho FA (1991) The 2.3-Å resolution structure of the maltose or maltodextrin binding protein, a primary receptor of bacterial active transport and chemotaxis. J Biol Chem 266:5202–5219

van den Berg B, Black PN, Clemons WM Jr, Rapoport TA (2004) Crystal structure of the long-chain fatty acid transporter FadL. Science 304:1506–1509

Wandersman C, Delepelaire P (2004) Bacterial iron sources: from siderophores to hemophores. Annu Rev Microbiol 58:611–647

Wimley WC (2003) The versatile β-barrel membrane protein. Curr Opin Struct Biol 13:404–411

Yue WW, Grizot S, Buchanan SK (2003) Structural evidence for iron-free citrate and ferric citrate binding to the TonB-dependent outer membrane transporter FecA. J Mol Biol 332:353–368

6 Competition for Iron and Induced Systemic Resistance by Siderophores of Plant Growth Promoting Rhizobacteria

Monica Höfte and Peter A.H.M. Bakker

6.1
Introduction

Most aerobic and facultative anaerobic microorganisms produce low molecular weight Fe^{3+} specific ligands, so-called siderophores, under conditions of low iron availability. The siderophores sequester ferric ions in the environment and the ferric siderophores are taken up in the microbial cells after specific recognition by membrane proteins (Höfte 1993). The production of siderophores is an important trait of so-called plant growth promoting rhizobacteria (PGPR) in their ability to suppress soil-borne plant pathogens (Kloepper et al. 1980). Competition for ferric iron between the PGPR and the plant deleterious microorganisms is considered the mode of action of these siderophores (Buysens et al. 1996; Loper and Buyer 1991; Raaijmakers et al. 1995; Schippers et al. 1987). However, it has been reported that disease suppression also occurs when the PGPR and the pathogen are inoculated and remain spatially separated, thus avoiding direct interactions. In this case the protective effect has to be plant mediated and this phenomenon was named induced systemic resistance (Van Loon et al. 1998). For instance, when *Pseudomonas fluorescens* strain WCS417 remained confined to the carnation root system and *Fusarium oxysporum* f.sp. *dianthi* was slash-inoculated into the stem, it was found that the bacteria were still protective (Van Peer et al. 1991). On cucumber similar observations were made for PGPR strains applied to the roots, and subsequent challenge inoculation of the leaves with the anthracnose fungus *Colletotrichum orbiculare* (Wei et al. 1991). The inducing rhizobacteria trigger a reaction in the plant roots leading to a signal that spreads systemically throughout the plant, finally resulting in enhanced

Monica Höfte: Laboratory of Phytopathology, Faculty of Bioscience Engineering,
Ghent University, Coupure Links 653, B-9000 Gent, Belgium, email: monica.hofte@ugent.be

Peter A.H.M. Bakker: Institute of Biology, Section of Phytopathology, Utrecht University,
PO Box 80084, 3508 TB Utrecht, The Netherlands

Soil Biology, Volume 12
Microbial Siderophores
A. Varma, S.B. Chincholkar (Eds.)

defensive capacity to subsequent pathogen infections. The protective action of PGPR against soil-borne pathogens in the rhizosphere is thus extended to a defence-stimulating effect in aboveground tissues against foliar pathogens. This enhanced defensive capacity was expressed in roots as well as in leaves, adding the mechanism of ISR to the list of mechanisms of PGPR effective against soil-borne pathogens (Leeman et al. 1995b). In view of the role of iron-regulated metabolites in suppression of soil borne diseases by PGPR, their possible involvement in ISR has been subject of numerous studies.

6.2
Role of Siderophores and Iron-Regulated Compounds in ISR

Bacterial determinants of ISR that have been identified so far are lipopolysaccharides (LPS) (Leeman et al. 1995b; Van Peer and Schippers 1992), flagella (Meziane et al. 2005), the antibiotics 2,4-diacetylphloroglucinol (Iavicoli et al. 2003; Weller et al. 2004) and pyocyanin (Audenaert et al. 2002), the volatile 2,3-butanediol (Ryu et al. 2004), N-alkylated benzylamine (Ongena et al. 2005) and iron-regulated compounds (Bakker et al. 2003). In this review, we will focus on the compounds produced upon iron-limitation (Table 6.1).

6.2.1
Pseudomonas aeruginosa 7NSK2

Pseudomonas aeruginosa 7NSK2 is a PGPR isolated from the roots of barley (Iswandi et al. 1987). Under iron-limiting conditions, this strain produces three siderophores, pyoverdine, pyochelin and its precursor salicylic acid (SA) and can induce resistance to plant diseases caused by *Botrytis cinerea* on bean and tomato (De Meyer and Höfte 1997; De Meyer et al. 1999b), *Colletotrichum lindemuthianum* on bean (Bigirimana and Höfte 2002) and Tobacco Mosaic Virus on tobacco (De Meyer et al. 1999a). Interestingly, exogenously applied SA induces a systemic resistance in many plant species (Sticher et al. 1997), and therefore this metabolite may be of importance in ISR triggered by strain 7NSK2.

Under iron-limitation, SA-deficient mutants of this strain were not able to induce resistance to the pathogens mentioned above in a pyoverdine-negative or pyoverdine-positive background indicating that SA or pyochelin is essential for ISR in bean, tomato and tobacco. In tomato and tobacco, it was shown that 7NSK2 induces resistance via the SA-dependent signal transduction pathway, since 7NSK2 no longer induced resistance in *NahG* tobacco (De Meyer et al. 1999a) and *NahG* tomato plants (Audenaert et al. 2002). Plants carrying the bac-

Table 6.1. Examples of bacterial strains for which iron-chelating or iron-regulated compounds are involved in induced systemic resistance

Bacterial strain	Plant – pathogen	Determinant(s) involved in ISR	Reference
Pseudomonas aeruginosa 7NSK2	Bean – *Colletotrichum lindemuthianum*	Salicylic acid	Bigirimana and Höfte (2002)
	Bean – *Botrytis cinerea*	Salicylic acid	De Meyer and Höfte (1997)
	Tobacco – Tobacco Mosaic Virus	Salicylic acid	De Meyer et al. (1999a)
	Tomato – *Botrytis cinerea*	Salicylic acid, pyochelin, pyocyanin	Audenaert et al. (2002)
	Rice – *Pyricularia grisea*	Pyocyanin	De Vleesschauwer et al. (2006)
	Rice – *Rhizoctonia solani*	Salicylic acid	De Vleesschauwer et al., unpublished
Pseudomonas fluorescens CHA0	Tobacco – Tobacco mosaic virus	Pyoverdine	Maurhofer et al. (1994)
	Arabidopsis – *Peronospora parasitica*	2,4-diacetylphloroglucinol	Iavicola et al. (2003)
Pseudomonas fluorescens WCS374	Radish – fusarium wilt	Pseudobactin, LPS	Leeman et al. (1995b, 1996)
	Eucalyptus – *Ralstonia solanacearum*	Pseudobactin, unknown determinant(s)	Ran et al. (2005)
Pseudomonas fluorescens WCS417	Carnation – fusarium wilt	LPS	Van Peer and Schippers (1992)
	Radish – fusarium wilt	LPS, unknown iron-regulated determinant(s)	Leeman et al. (1996)
	Arabidosis – *Pseudomonas syringae* pv. *tomato*	LPS	Van Wees et al. (1997)
Pseudomonas putida WCS358	Arabidopsis – *Pseudomonas syringae* pv. *tomato*	Pseudobactin, flagella, LPS	Bakker et al. (2003); Meziane et al. (2005)
	Tomato – *Botrytis cinerea*	Pseudobactin	Meziane et al. (2005)
	Bean – *Botrytis cinerea*	Pseudobactin, LPS	Meziane et al. (2005)
	Bean – *Colletotrichum lindemuthianum*	Pseudobactin, LPS	Meziane et al. (2005)
	Eucalyptus – *Ralstonia solanacearum*	Pseudobactin, LPS	Ran et al. (2005)
Pseudomonas putida BTP1	Bean – *Botrytis cinerea*	N-alkylated benzylamine	Ongena et al. (2005)
Serratia marcescens 90-166	Cucumber – *Colletotrichum orbiculare*	Catechol-type siderophore	Press et al. (2001)

terial *NahG* gene, encoding the enzyme salicylate hydroxylase, which converts SA into the non-inducing product catechol, no longer express SA induced resistance (Gaffney et al. 1993). For *P. aeruginosa* KMPCH, a pyochelin-negative and SA-positive mutant of 7NSK2, it was illustrated that bacterial SA induced phenylalanine ammonia lyase (PAL) activity in bean roots. Moreover, SA levels increased in bean leaves upon root colonization with KMPCH (De Meyer and Höfte 1997). On tomato roots, KMPCH produced SA and induced PAL activity, but surprisingly, this was not the case for the wild type strain 7NSK2 (Audenaert et al. 2002). *P. aeruginosa* 7NSK2 is also able to induce resistance to *Pseudomonas syringae* pv. *syringae* in *Arabidopsis thaliana*. Mutants of 7NSK2 deficient in pyoverdine, pyocheline or SA production were as effective as the wild-type strain in inducing resistance indicating that in *Arabidopsis* these compounds are not necessary for the induction of ISR. Interestingly, strain 7NSK2 still induced resistance to *P. syringae* pv. *syringae* in *NahG Arabidopsis* plants (Ran 2005).

Besides siderophores, *P. aeruginosa* 7NSK2 also produces pyocyanin (5-methyl-1-hydroxyphenazinium betaine), a blue phenazine compound that is considered to be a virulence factor in clinical isolates of *P. aeruginosa* (Britigan et al. 1992, 1997). Abeysinge (1999) has shown that high concentrations of purified pyocyanin (0.1 mM) can induce resistance to *B. cinerea* in bean. Mutant PHZ1 is not able to produce pyocyanin due to an insertion in the *phzM* gene that encodes an *O*-methyltransferase. In infection experiments, PHZ1 did not induce resistance in tomato to *B. cinerea*. In addition, a pyochelin and SA negative mutant 7NSK2-562 did not induce resistance either, although it overproduced pyocyanin. Induced resistance was restored, when both mutants were co-inoculated on tomato roots or when PHZ1 was complemented for pyocyanin production. These results indicate that in strain 7NSK2 pyocyanin and pyochelin (or salicylic acid) act synergistically in induced resistance, probably by generating the very reactive OH-radical on plant roots (Audenaert et al. 2002). We also constructed the *phzM* mutation in mutant KMCPH. Surprisingly, mutant KMPCH-*phzM* lost its ability to induce resistance to *B. cinerea* in bean and tomato (Audenaert et al., unpublished) indicating that also in strain KMPCH, SA and pyocyanin, rather than SA alone are the determinants for induced resistance in bean and tomato.

Studies about bacterial determinants involved in ISR have mainly been carried out in dicot plants. Recently, we started work on ISR in our lab using the monocot rice as a model plant. We were interested to see whether the same bacterial determinants are involved in ISR in mono- and dicotyledon plants. As challenging pathogens we used the major pathogens of rice: *Pyricularia grisea*, the causal agent of rice blast; *Xanthomonas oryzae* pv. *oryzae*, the causal agent of bacterial blight and *Rhizoctonia solani*, the causal agent of sheath blight. *P. aeruginosa* 7NSK2 was able to induce resistance to rice blast, but was not effective against sheath blight or blight. We tested all available mutants of 7NSK2 for their ability to induce resistance to blast and sheath blight. Pyocyanin appeared to be the main metabolite responsible for induced resistance to blast, while there was no role for SA or pyochelin. SA-deficient mutants were in general even more effective in inducing resistance than the wild type strain (De Vleesschauwer et al., in preparation). The situation appeared to be entirely different for sheath blight. While the wild type strain 7NSK2 was not effective against *R. solani*, the pyocyanin mutants 7NSK2-*phzM* and KMPCH-*phzM* were able to induce resistance (De Vleesschauwer et al. 2006). Transient generation of H_2O_2 by redox-active pyocyanin in planta most likely accounts for the dual role of the latter compound in 7NSK2-mediated ISR in rice since exogenous application of sodium ascorbate alleviated the opposite effects of pyocyanin on *P. grisea* and *R. solani* pathogenesis (De Vleesschauwer et al. 2006). Resistance could also be induced with pure SA applied to rice roots in a gnotobiotic system (De Vleesschauwer et al., in preparation).

We conclude that while in bean, tomato and tobacco SA/pyochelin and pyocyanin act synergistically to induce resistance, in the monocot rice SA or pyocyanin alone are sufficient to induce resistance. The bacterial metabolite involved, however, depends on the challenging pathogen. It is known that pyocyanin can undergo redox-cycling, resulting in the generation of superoxide and H_2O_2 (Britigan et al. 1997). These active oxygen species (AOS) are apparently sufficient to induce resistance to blast in rice. In dicot plants such as bean and tomato, however, these AOS have to be converted to the very reactive OH-radical through the Haber-Weiss reaction in the presence of an iron-chelating compound such as Fe-pyochelin or Fe-SA.

6.2.2
Pseudomonas fluorescens CHA0 and *P. fluorescens* P3

P. fluorescens CHA0 is an effective biocontrol agent of various soil borne pathogens (see Haas and Defago 2005 for a review). This strain produces pyoverdine, salicylic acid and various antimicrobial compounds such as hydrogen cyanide, 2,4-diacetylphloroglucinol (DAPG) and pyoluteorin. *P. fluorescens* CHA0 can induce resistance against leaf necrosis caused by Tobacco Necrosis Virus (TNV) in tobacco plants. A *gacA* mutant, in which the production of hydrogen cyanide, DAPG and pyoluteorin is blocked, had the same capacity to induce resistance against TNV as did the wild type strain. The pyoverdine-negative mutant CHA400, however, was significantly less able to protect tobacco leaves against TNV (Maurhofer et al. 1994). The transposon insertion in mutant CHA400 was not localized and it is not clear whether the pyoverdine mutation is the only mutation in strain CHA400. More recently, ISR with *P. fluorescens* CHA0 was studied in *Arabidopsis thaliana* with *Peronospora parasitica* as the challenging pathogen. In this study it was shown that DAPG is required for the induction of ISR to *Peronospora parasitica*, since only mutations interfering with DAPG, including the *gacA* mutation, led to a significant decrease in ISR (Iavicoli et al. 2003). In this study, mutant CHA400 was as effective as the wild-type strain. Similarly, ISR in *A. thaliana* against *P. syringae* pv *tomato* by *P. fluorescens* Q2-87 is triggered by DAPG (Weller et al. 2004). In the latter case the possible involvement of siderophores was not studied.

Introduction of the SA biosynthesis genes *pchBA* from *P. aeruginosa* PA01 (Serino et al. 1995) into *P. fluorescens* P3, which does not produce SA, rendered this strain capable of SA production in vitro and significantly improved its ability to induce systemic resistance in tobacco against TNV (Maurhofer et al. 1998).

6.2.3
Pseudomonas fluorescens WCS417 and WCS374

P. fluorescens WCS417 was one of the first PGPR strains for which ISR was found to be an important mode of action. When strain WCS417 was applied to roots of carnation it protected the treated plants significantly from wilting caused by *F. oxysporum* f.sp. *dianthi* that was slash inoculated into the stem (Van Peer et al. 1991). Purified lipopolysaccharides of this strain also induced resistance in carnation (Van Peer and Schippers 1992), suggesting LPS as a trigger of ISR in this plant species. The pseudobactin (pyoverdin) siderophore of WCS417 seems not to be involved in ISR, since a pseudobactin negative Tn5 insertion mutant was as effective as the parental strain in protecting carnation from fusarium wilt (Duijff et al. 1993). The involvement of LPS and siderophores in ISR against fusarium wilt were investigated further for *P. fluorescens* strains WCS417 and WCS374 in a bioassay with radish specifically designed to study induced resistance (Leeman et al. 1995a). The purified LPS of both strains triggered ISR in radish; moreover, mutants of the strains lacking the 0-antigenic side chain were not able to induce resistance against fusarium wilt (Leeman et al. 1995b). However, the latter experiments were performed under conditions of high iron availability for the PGPR strains, thereby excluding a possible role for siderophores, which are produced only upon iron limitation. When iron availability for the bacteria was lowered, strains WCS374 and WCS417 reduced disease to a much lower level compared to high iron availability, and more striking, the 0-antigen mutants of both strains did trigger ISR under conditions of low iron availability (Leeman et al. 1996). These results suggest that iron-regulated metabolites of the strains are also involved in triggering ISR. Purified pseudobactin of WCS374, but not that of WCS417, induced ISR when applied to the roots of radish. Interestingly, pseudobactin mutants of both strains were as effective as the parental strains, with again additional ISR induction activity when iron availability was lowered. Both strains produce SA at low iron availability and this metabolite was suggested to be a determinant of WCS374 and WCS417 triggering ISR in radish (Leeman et al. 1996). However, in *Arabidopsis thaliana* WCS417 can induce resistance whereas WCS374 cannot, despite the fact that WCS374 produces six to seven times more SA than does WCS417. In the *A. thaliana* system ISR by WCS417 is independent of SA accumulation in the plant (Pieterse et al. 1996 1998), excluding a role for bacterially produced SA. The LPS of WCS417 does seem to be involved in ISR *in A. thaliana* (Van Wees et al. 1997). So far the involvement of pseudobactin in ISR by WCS417 in Arabidopsis has not been investigated. The inability of WCS374 to trigger ISR in *Arabidopsis* may be due to the observation that upon iron limitation this strain produces not only pseudobactin and SA, but also pseudomonine, a siderophore containing a SA moiety (Mercado-Blanco et al. 2001). Possibly in the Arabidopsis rhizosphere all SA produced by WCS374 is channelled into pseudomonine that cannot trigger ISR in this plant species. This hypothesis will be investigated using purified pseudomonine and mutants defective in pseudomonine production.

Strain WCS417 induces resistance in all plant species it has been tested in (Van Loon et al. 1998). However, recently it was reported that this strain could not trigger ISR in *Eucalyptus urophylla* (Ran et al. 2005) or rice (De Vleesschauwer, unpublished). *P. fluorescens* WCS374 does induce ISR against *Ralstonia solanacearum* in Eucalyptus and whereas a pseudobactin mutant was as effective as the parental strain in disease suppression, purified pseudobactin did trigger ISR. These results suggest that ISR by WCS374 in *E. urophylla* is triggered by its pseudobactin siderophore and as yet unknown additional determinant(s) (Ran et al. 2005). Interestingly, WCS374 could also induce systemic resistance in rice against both rice blast and sheath blight. A pseudobactin mutant lost its ability to suppress disease, while purified pseudobactin from strain WCS374 triggered ISR to rice blast and sheath blight (De Vleesschauwer, unpublished).

6.2.4
Pseudomonas putida WCS358

Strain WCS358 was originally isolated from potato tuber surface and the involvement of its fluorescent pseudobactin siderophore in disease suppression has been studied in a variety of plant-pathogen systems. Mutants defective in pseudobactin biosynthesis were isolated and analysed after Tn5 transposon mutagenesis (Marugg et al. 1985). Mutants were compared to the parental strain regarding their ability to increase potato plant growth in pot experiments and in the field (Schippers et al. 1987), and to suppress fusarium wilt in carnation (Duijff et al. 1993), and radish (Raaijmakers et al. 1995). In all cases it was observed that the parental strain was more effective than the mutant strain suggesting that pseudobactin is the key biocontrol compound in WCS358. Effective competition for ferric iron was suggested to be the main mode of action of WCS358 (Bakker et al. 1993). *P. putida* WCS358 cannot trigger ISR in carnation (Duijff et al. 1993) or radish (Leeman et al. 1995a), although it does in *A. thaliana* (Van Wees et al. 1997), *E. urophylla* (Ran et al. 2005) bean, and tomato (Meziane et al. 2005). In these plants species the involvement of the siderophore of WCS358 as a trigger of ISR was studied using both purified pseudobactin and mutants defective in pseudobactin biosynthesis. Whereas in Arabidopsis the purified siderophore induces ISR, the pseudobactin mutant was still as effective as the wild type strain (Bakker et al. 2003; Meziane et al. 2005). In addition to the siderophore, the LPS and the flagella of WCS358 also play a role in triggering ISR, indicating redundancy for ISR triggering traits of this strain in the *A. thaliana - P. syringae* pv *tomato* model system. In Eucalyptus the situation is different; the mutant no longer induces resistance and the purified siderophore does trigger ISR, suggesting that pseudobactin is the sole determinant of ISR in this plant species (Ran et al. 2005). In tomato a similar situation was observed, ISR against *B. cinerea* was triggered by the purified siderophore and the pseudobactin mutant no longer induced resistance (Meziane et al. 2005). In bean more than one determinant seems to be involved in WCS358 mediated ISR against *B. cinerea* and *C. lindemuthianum*. Both the pseudobactin mutant and the purified compound induced resistance; an additional role for LPS was suggested in this case (Meziane et al. 2005). Thus, for strain WCS358 multiple traits can be involved in ISR depending on the host plant. In previous studies in which a role for siderophore mediated competition for iron by this strain was suggested the possible involvement of ISR should be evaluated.

6.2.5
Pseudomonas putida BTP1

P. putida BTP1, obtained from barley roots, was originally selected for its specific features regarding pyoverdine-mediated iron transport (Jacques et al. 1995). *P. putida* BTP1 was shown to enhance the resistance of cucumber to root rot caused by *Pythium aphanidermatum* (Ongena et al. 1999). Results from split root experiments suggested that the protective effect was due to ISR, since systemic accumulation of antifungal compounds was observed in the host plant after treatment with BTP1 or with M3, its siderophore deficient mutant (Ongena et al. 1999). *P. putida* BTP1 is also able to protect bean against leaf infection with *Botrytis cinerea*. Biocontrol assays carried out with cell-free culture fluids of BTP1 clearly indicated that ISR was mostly induced by one or several metabolite(s) excreted by the strain under iron-limited in vitro growth conditions. In vivo assays with samples from successive fractionation steps of the BTP1 supernatant led to the isolation of fractions containing one main metabolite that retained most of the resistance-inducing activity in bean (Ongena et al. 2002). Recently, this metabolite was structurally characterized as an N-trialkylated benzylamine derivative (Ongena et al. 2005). Although the production of this metabolite is iron-regulated in strain BTP1, the compound itself has apparently no siderophore activity.

6.2.6
Serratia marcescens 90-166

Serratia marcescens 90-166 can induce resistance to fungal, viral and bacterial pathogens in cucumber such as *Colletotrichum orbiculare*, *Fusarium oxysporum*

f. sp. *cucumerinum*, Cucumber Mosaic Virus, *P. syringae* pv. *lachrymans*, and *Erwinia tracheiphila*. Strain 90-166-mediated ISR is dependent on iron concentration. The capacity of strain 90-166 to induce resistance is diminished under iron-replete conditions. In addition, suppression of cucumber anthracnose by strain 90-166 was significantly improved when the iron concentration of the planting mix was decreased by addition of the iron-chelator EDDHA (Press et al. 2001). *S. marcescens* 90-166 is known to produce SA, but mutants deficient in SA production retained ISR activity in cucumber against *C. orbiculare* (Press et al. 1997). In addition to SA, *S. marcescens* produces a catechol-type siderophore, which is probably identical to enterobactin. A siderophore-deficient mutant of *S. marcescens* was identified, in which a homologue of the *entA* gene is inactivated. *EntA* encodes 2,3-dihydro-2,3-dihydroxybenzoate dehydrogenase, an enzyme in the enterobactin biosynthesis pathway. The *entA* mutant of *S. marcescens* 90-166 was no longer able to induce resistance to *C. orbiculare* in cucumber. It was observed that there was a significant decrease in internal root population sizes of the *entA* mutant of *S. marcescens* compared with wild type strain 90-166. Press et al. (2001) hypothesized that siderophore production by strain 90-166 serves to detoxify the active oxygen species produced by the plant in response to the bacterium as was reported for *Erwinia amylovora* siderophores (Dellagi et al. 1998). The lack of enterobactin production in the *entA* mutant may render this strain more susceptible to active oxygen species and result in lowered internal populations. It was not determined, however, whether changes in internal colonization by the *entA* mutant contributed to changes in the ISR phenotype of this strain (Press et al. 2001).

6.3
Conclusions

The production of siderophores occurs under conditions of iron-limitation. Such conditions are likely to prevail in the rhizosphere (Loper and Henkels 1999), and siderophore mediated competition for iron is one of the mechanisms of bacterial antagonism against soil-borne pathogens (Loper and Buyer 1991). However, siderophore production can also trigger ISR and it can therefore play a dual role in disease suppression by depriving resident pathogens from iron locally and by inducing resistance in the plant systemically. The observation that not all siderophores induce ISR can be explained by the fact that siderophores produced by different bacteria have very different chemical structures (Höfte 1993). How siderophores are perceived by plants is presently completely unknown, but there is crop specificity as specific siderophores trigger ISR in one plant species but not another.

Several ISR-eliciting PGPR are able to produce SA in vitro, whereas others are not. Induction of systemic resistance in *NahG*-transformed plants demonstrated that ISR against TMV and *Botrytis cinerea* in tobacco and tomato by

7NSK2 (De Meyer et al. 1999a; Audenaert et al. 2002) and in Arabidopsis against *P. syringae* pv. *maculicola* by *B. pumilus* SE34 (Ryu et al. 2003) depends on SA accumulation in the plant. In other cases, PGPR still effectively induce ISR in *NahG* plants. Moreover, mutants of *S. marcescens* 90-166 that do not produce SA were as effective as the parental strain in triggering ISR in tobacco against *P. syringae* pv. *tomato* and in cucumber against *C. orbiculare* (Press et al. 1997). Thus, the importance of bacterially produced SA in PGPR-mediated ISR appears to be limited.

Whereas SA triggers a signal transduction pathway in the plant that depends on SA, ISR triggered by several PGPR strains is independent of SA but relies on jasmonic acid (JA) and ethylene signalling in the plant (Pieterse et al. 1996, 1998). Interestingly, simultaneous activation of the SA dependent and the JA/ethylene dependent pathways leads to an enhanced level of protection against pathogens (Van Wees et al. 2000). Whereas there is a wide range of pathogens against which both SA dependent and JA/ethylene dependent induced resistance are effective, some are only affected by one type of induced resistance (Ton et al. 2002). Therefore simultaneous activation of the SA dependent and independent pathways may increase the range of pathogens that are effectively suppressed after treatment with PGPR.

Redundancy in ISR eliciting determinants in PGPR on the one hand hampers studies to elucidate the involvement of these determinants; on the other hand it may give ISR robustness. If one determinant fails to elicit ISR or is not produced under certain conditions, other traits can still be effective. In these cases it would be favourable if the different traits were also differentially regulated.

Increased knowledge on the variety of bacterial determinants of ISR and their regulation in the rhizosphere will not only increase our fundamental understanding of interactions in this highly dynamic environment, but it will also increase possibilities to apply this mode of action of PGPR in crop protection strategies.

References

Abeysinghe S (1999) Investigation of metabolites of the rhizobacterium *Pseudomonas aeruginosa* strain SA44 involved in plant protection. Ph.D. thesis, Vrije Universiteit Brussel, Belgium

Audenaert K, Pattery T, Cornelis P, Höfte M (2002) Induction of systemic resistance to *Botrytis cinerea* in tomato by *Pseudomonas aeruginosa* 7NSK2: role of salicylic acid, pyochelin and pyocyanin. Mol Plant-Microbe Interact 15:1147–1156

Bakker PAHM, Raaijmakers JM, Schippers B (1993) Role of iron in the suppression of bacterial plant pathogens by fluorescent pseudomonads In: Barton LL, Hemming BC (eds) Iron chelation in plants and soil microorganisms. Academic Press, San Diego, pp 269–278

Bakker PAHM, Ran LX, Pieterse CMJ, Van Loon LC (2003) Understanding the involvement of rhizobacteria-mediated induction of systemic resistance in biocontrol of plant diseases. Can J Plant Pathol 25:5–9

Bigirimana J, Höfte M (2002) Induction of systemic resistance to *Colletotrichum lindemuthianum* in bean by a benzothiadiazole derivative and rhizobacteria. Phytoparasitica 30:159–168

Britigan BE, Roeder TL, Rasmussen GT, Shasby DM, McCormick ML, Cox CD (1992) Interaction of the *Pseudomonas aeruginosa* secretory products pyocyanin and pyochelin generates hydroxyl radicals and causes synergistic damage to endothelial cells. J Clin Invest 90:2187–2196

Britigan BE, Rasmusse GT, Cox, CD (1997) Augmentation of oxidant injury to human pulmonary epithelial cells by the *Pseudomonas aeruginosa* siderophore pyochelin. Infect Immun 65:1071–1076

Buysens S, Heungens K, Poppe J, Höfte M (1996) Involvement of pyochelin and pyoverdin in suppression of Pythium-induced damping-off of tomato by *Pseudomonas aeruginosa* 7NSK2. Appl Environ Microbiol 62: 865–871

Dellagi A, Brisset MN, Paulin JP, Expert D (1998) Dual role of desferrioxamine in *Erwinia amylovora* pathogenicity. Mol Plant-Microbe Interact 11:734–742

De Meyer G, Höfte M (1997) Salicylic acid produced by the rhizobacterium *Pseudomonas aeruginosa* 7NSK2 induces resistance to leaf infection by *Botrytis cinerea* on bean. Phytopathology 87:588–593

De Meyer G, Audenaert K, Höfte M (1999a) *Pseudomonas aeruginosa* 7NSK2-induced systemic resistance in tobacco depends on *in planta* salicylic acid accumulation but is not associated with PR1a expression. Eur J Plant Pathol 105:513–517

De Meyer G, Capieau K, Audenaert K, Buchala A, Métraux JP, Höfte M (1999b) Nanogram amounts of salicylic acid produced by the rhizobacterium *Pseudomonas aeruginosa* 7NSK2 activate the systemic acquired resistance pathway in bean. Mol Plant-Microbe Interact 12:450–458

De Vleesschauwer D, Cornelis P, Höfte M (2006) Redox-active pyocyanin secreted by *Pseudomonas aeruginosa* 7NSK2 triggers systemic resistance to *Magnaporthe grisea* but enhances *Rhizoctonia solani* susceptibility in rice. Mol Plant-Microbe Interact 19:1406–1419

Duijff BJ, Meijer JW, Bakker PAHM, Schippers B (1993) Siderophore-mediated competition for iron and induced resistance in the suppression of fusarium wilt of carnation by fluorescent *Pseudomonas* spp. Neth J Plant Pathol 99:277–289

Gaffney T, Friedrich L, Vernooij B, Negrotto D, Nye G, Uknes S, Ward E, Kessmann H, Ryals J (1993) Requirement of salicylic acid for the induction of systemic acquired resistance. Science 261:754–756

Haas D, Defago G (2005) Biological control of soil-borne pathogens by fluorescent pseudomonads. Nature Rev Microbiol 3:307–319

Höfte M (1993) Classes of microbial siderophores. In: Barton LL, Hemming BC (eds) Iron chelation in plants and soil microorganisms. Academic Press, San Diego, pp 3–26

Iavicoli A, Boutet E, Buchala A, Métraux JP (2003) Induced systemic resistance in *Arabidopsis thaliana* in response to root inoculation with *Pseudomonas fluorescens* CHA0. Mol Plant-Microbe Interact 16:851–858

Iswandi A, Bossier P, Vandenabeele J, Verstraete W (1987) Effect of seed inoculation with the rhizopseudomonas strain 7NSK2 on the root microbiota of maize (*Zea mays*) and barley (*Hordeum vulgare*). Biol Fertil Soil 3:153–158

Jacques P, Ongena M, Gwose I, Seinsche D, Schröder H, Delfosse P, Thonart P, Taraz K, Budzikiewicz H (1995) Structure and characterization of isopyoverdin from *Pseudomonas putida* BTP1 and its relation to the biogenetic pathway leading to pyoverdines. Z Naturforsch C 50:622–629

Kloepper JW, Leong J, Teintze M, Schroth MN (1980) Enhanced plant growth by siderophores produced by plant growth-promoting rhizobacteria. Nature 286:885–886

Leeman M, Van Pelt JA, Den Ouden FM, Heinsbroek M, Bakker PAHM, Schippers B (1995a) Induction of systemic resistance by *Pseudomonas fluorescens* in radish cultivars differing in susceptibility to fusarium wilt, using a novel bioassay. Eur J Plant Pathol 101:655–664

Leeman M, Van Pelt JA, Den Ouden FM, Heinsbroek M, Bakker PAHM, Schippers B (1995b) Induction of systemic resistance against fusarium wilt of radish by lipopolysaccharides of *Pseudomonas fluorescens*. Phytopathology 85:1021–1027

Leeman M, Den Ouden FM, Van Pelt JA, Dirkx FPM, Steijl H, Bakker PAHM, Schippers B (1996) Iron availability affects induction of systemic resistance against fusarium wilt of radish by *Pseudomonas fluorescens*. Phytopathology 86:149–155

Loper JE, Buyer JS (1991) Siderophores in microbial interactions on plant surfaces. Mol Plant-Microbe Interact 4:5–13

Loper JE, Henkels MD (1999) Utilization of heterologous siderophores enhances levels of iron available to *Pseudomonas putida* in the rhizosphere. Appl Environ Microbiol 65:5357–5363

Marugg JD, Van Spanje M, Hoekstra WPM, Schippers B, Weisbeek PJ (1985) Isolation and analysis of genes involved in siderophore biosynthesis in plant-growth-stimulating *Pseudomonas putida* WCS358. J Bacteriol 164:563–570

Maurhofer M, Hase C, Meuwly P, Métraux J-P, Défago G (1994) Induction of systemic resistance of tobacco to tobacco necrosis virus by the root-colonizing *Pseudomonas fluorescens* strain CHA0: influence of the *gacA* gene and of pyoverdine production. Phytopathology 84:139–146

Maurhofer M, Reimmann C, Schmidli-Sacherer P, Heeb S, Haas D, Défago G (1998) Salicylic acid biosynthetic genes expressed in *Pseudomonas fluorescens* strain P3 improve the induction of systemic resistance in tobacco against tobacco necrosis virus. Phytopathology 88:678–684

Mercado-Blanco J, Van der Drift KMGM, Olsson PE, Thomas-Oates JE, Van Loon LC, Bakker PAHM (2001) Analysis of the *pmsCEAB* gene cluster involved in biosynthesis of salicylic acid and the siderophore pseudomonine in the biocontrol strain *Pseudomonas fluorescens* WCS374. J Bacteriol 183:1909–1920

Meziane H, Van der Sluis I, Van Loon LC, Höfte M, Bakker PAHM (2005) Determinants of *Pseudomonas putida* WCS358 involved in inducing systemic resistance in plants. Mol Plant Pathol 6:177–185

Ongena M, Daayf F, Jacques P, Thonart P, Benhamou N, Paulitz T, Cornelis P, Koedam N, Belanger RR (1999) Protection of cucumber against Pythium root rot by fluorescent pseudomonads: predominant role of induced resistance over siderophores and antibiosis. Plant Pathol 48:66–76

Ongena M, Giger A, Jacques P, Dommes J, Thonart P (2002) Study of bacterial determinants involved in the induction of systemic resistance in bean by *Pseudomonas putida* BTP1. Eur J Plant Pathol 108:187–196

Ongena M, Jourdan E, Schäfer M, Kech C, Budzikiewicz H, Luxen A, Thonart P (2005) Isolation of an N-alkylated benzylamine derivative from *Pseudomonas putida* BTP1 as elicitor of induced systemic resistance in bean. Mol Plant-Microbe Interact 18:562–569

Pieterse CMJ, Van Wees SCM, Hoffland E, Van Pelt JA, Van Loon LC (1996) Systemic resistance in *Arabidopsis* induced by biocontrol bacteria is independent of salicylic acid accumulation and pathogenesis-related gene expression. Plant Cell 8:1225–1237

Pieterse CMJ, Van Wees SCM, Van Pelt JA, Knoester M, Laan R, Gerrits H, Weisbeek PJ, Van Loon LC (1998) A novel signaling pathway controlling induced systemic resistance in *Arabidopsis*. Plant Cell 10:1571–1580

Press CM, Wilson M, Tuzun S, Kloepper JW (1997) Salicylic acid produced by *Serratia marcescens* 90-166 is not the primary determinant of induced systemic resistance in cucumber or tobacco. Mol Plant-Microbe Interact 10:761–768

Press CM, Loper JE, Kloepper JW (2001) Role of iron in rhizobacteria-mediated induced systemic resistance of cucumber. Phytopathology 91:593–598

Raaijmakers JM, Leeman M, Van Oorschot MPM, Van der Sluis I, Schippers B, Bakker PAHM (1995) Dose-response relationships in biological control of fusarium wilt of radish by *Pseudomonas* spp. Phytopathology 85:1075–1081

Ran LX (2005) No role for bacterially produced salicylic acid in rhizobacterial induction of systemic resistance in *Arabidopsis*. Phytopathology 95:1349–1355

Ran LX, Li ZN, Wu GJ, Van Loon LC, Bakker PAHM (2005) Induction of systemic resistance against bacterial wilt in *Eucalyptus urophylla* by fluorescent *Pseudomonas* spp. Eur J Plant Pathol 113:59–70

Ryu CM, Hu CH, Reddy MS, Kloepper JW (2003) Different signaling pathways of induced resistance by rhizobacteria in *Arabidopsis thaliana* against two pathovars of *Pseudomonas syringae*. New Phytol 160:413–420

Ryu CM, Farag MA, Hu CH, Reddy MS, Kloepper JW, Paré PW (2004) Bacterial volatiles induce systemic resistance in *Arabidopsis*. Plant Physiol 134:1017–1026

Schippers B, Bakker AW, Bakker PAHM (1987) Interactions of deleterious and beneficial rhizosphere micro-organisms and the effect of cropping practices. Annu Rev Phytopathol 25:339–358

Serino L, Reimmann C, Baur H, Beyeler M, Visca P, Haas D (1995) Structural genes for salicylate biosynthesis from chorismate in *Pseudomonas aeruginosa*. Mol Gen Genet 249:217–228

Sticher L, Mauch-Mani B, Métraux JP (1997) Systemic acquired resistance. Annu Rev Phytopathol 35:235–270

Ton J, Van Pelt JA, Van Loon LC, Pieterse CMJ (2002) Differential effectiveness of salicylate-dependent and jasmonate/ethylene-dependent induced resistance in *Arabidopsis*. Mol Plant-Microbe Interact 15:27–34.

Van Loon LC, Bakker PAHM, Pieterse CMJ (1998) Systemic resistance induced by rhizosphere bacteria. Annu Rev Phytopathol 36:453–483

Van Peer R, Schippers B (1992) Lipopolysaccharides of plant-growth promoting *Pseudomonas* sp. strain WCS417r induce resistance in carnation to fusarium wilt. Neth J Plant Pathol 98:129–139

Van Peer R, Niemann GJ, Schippers B (1991) Induced resistance and phytoalexin accumulation in biological control of fusarium wilt of carnation by *Pseudomonas* sp. strain WCS417r. Phytopathology 81:728–734

Van Wees SCM, Pieterse CMJ, Trijssenaar A, Van 't Westende Y, Hartog F, Van Loon LC (1997) Differential induction of systemic resistance in *Arabidopsis* by biocontrol bacteria. Mol Plant-Microbe Interact 10:716–724

Van Wees SCM, De Swart EAM, Van Pelt JA, Van Loon LC, Pieterse CMJ (2000) Enhancement of induced disease resistance by simultaneous activation of salicylate- and jasmonate-dependent defense pathways in *Arabidopsis thaliana*. Proc Natl Acad Sci USA 97:8711–8716

Wei G, Kloepper JW, Tuzun S (1991) Induction of systemic resistance of cucumber to *Colletotrichum orbiculare* by select strains of plant growth-promoting rhizobacteria. Phytopathology 81:1508–1512

Weller DM, Van Pelt JA, Mavrodi DV, Pieterse CMJ, Bakker PAHM, Van Loon LC (2004) Induced systemic resistance (ISR) in *Arabidopsis* against *Pseudomonas syringae* pv. tomato by 2,4-diacetylphloroglucinol (DAPG)-producing *Pseudomonas fluorescens*. Phytopathology 94:108

7 Pyoverdine Synthesis and its Regulation in Fluorescent Pseudomonads

Paolo Visca, Franceso Imperi and Iain L. Lamont

7.1
Introduction

7.1.1
Historical Perspective

The genus *Pseudomonas* (γ-subclass of *Proteobacteria*) comprises a number of species belonging to the rRNA homology group I of *Pseudomonas sensu lato* (De Vos et al. 1989). These species are also referred to as the fluorescent pseudomonads group, which include *Pseudomonas aeruginosa, Pseudomonas fluorescens, Pseudomonas putida, Pseudomonas syringae*, and some minor species. Fluorescent pseudomonads are Gram-negative motile rods that usually grow aerobically and show remarkable nutritional and ecological versatility (Palleroni 1992). They are typical inhabitants of water and soil that can very commonly be isolated from the rhizosphere of many plants and have been well studied as biocontrol agents (Haas and Defago 2005). The type species of the genus, *P. aeruginosa*, is a widespread bacterium that has gained increasing medical significance as an opportunistic pathogen, though it normally occupies soil and water habitats (Hofte et al. 1990; Lyczak et al. 2000). The fluorescent pseudomonads are usually distinguished by the production of diffusible yellow-green fluorescent pigments, although some of the so-called fluorescent pseudomonads (e.g., *Pseudomonas stutzeri, Pseudomonas mendocina* and *Pseudomonas alcaligenes*) do not actually produce such pigments (Palleroni and Moore 2004). A number of terms have been used for the fluorescent compounds released by *Pseudomo-*

Paolo Visca, Franceso Imperi: Dipartimento di Biologia, Università di Roma Tre, Viale G. Marconi 446, 00146, Rome, Italy

Iain L. Lamont, Department of Biochemistry, University of Otago, PO Box 56. Dunedin, New Zealand, email: iain.lamont@stonebow.otago.ac.nz

Soil Biology, Volume 12
Microbial Siderophores
A. Varma, S.B. Chincholkar (Eds.)
© Springer-Verlag Berlin Heidelberg 2007

nas spp., including fluorescein, pseudobactin and pyoverdine (sometimes spelt pyoverdin). The last of these is now in general use and will be used here.

Pyoverdines were first studied in the late nineteenth century and attracted considerable interest in the first part of the last century, in particular with regard to the growth media and conditions that promoted their production (reviewed in Budzikiewicz 2004); however, their biological function was not known. An important advance was the development of a minimal medium and methodology for the purification of pyoverdines which permitted subsequent chemical characterization (Meyer and Abdallah 1978). In the same paper, the authors recognized that the biosynthesis of pyoverdine by *P. fluorescens* is suppressed by the presence of free iron in the growth medium and that the pyoverdine they were studying had a very high affinity for iron Fe^{3+} ions. It was proposed that pyoverdines act as siderophores (iron transporters) for the pseudomonads and this was demonstrated in an accompanying paper (Meyer and Hornspreger 1978).

A large number of pyoverdines have now been purified from different strains and species of *Pseudomonas*, with each strain making a single (and characteristic) form of pyoverdine although they can utilize many different pyoverdines (see chapters by Meyer and by Cornelis, this volume). *P. aeruginosa* strains, for example, secrete one of three distinct pyoverdines (Types I–III). Pyoverdines are composed of three parts: (i) a fluorescent dihydroxyquinoline chromophore that is common to all pyoverdines; (ii) an acyl side chain (either dicarboxylic acid or amide) bound to the amino group of the chromophore; and (iii) a strain-specific peptide chain linked *via* an amide bound to the C1 (rarely C3) carboxyl group

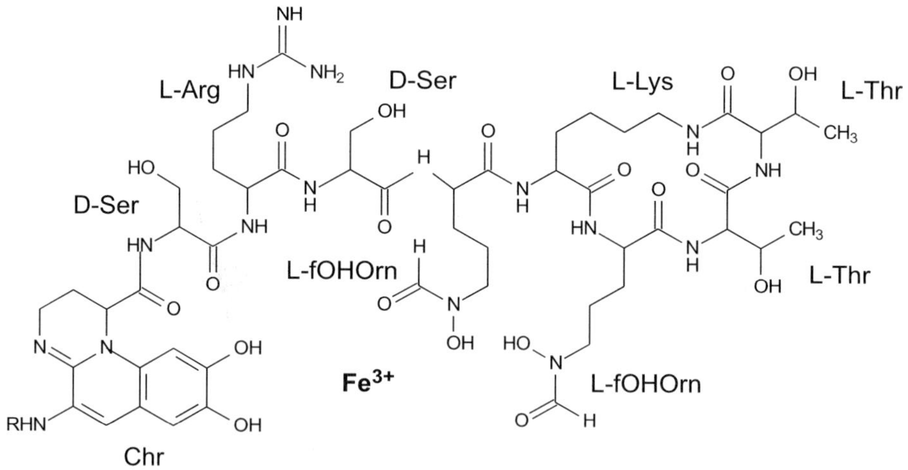

Fig. 7.1. Pyoverdine made by *P. aeruginosa* strain PAO1. The chemical structure of this pyoverdine was determined by Briskot et al. (1989). R is a variable acyl side chain (dicarboxylic acid or amide; usually α-ketoglutarate, succinate or glutamate). The catecholate group on the chromophore (Chr) and the hydroxamate groups on the L-N^5-formyl-N^5-hydroxyornithine (L-fOHOrn) residues constitute a high-affinity binding site for an Fe^{3+} ion. Amino acids are abbreviated with the conventional three-letter code. The structures of pyoverdines synthesised by other fluorescent pseudomonads are described elsewhere in this book (Meyer, this volume)

of the chromophore (Fig. 7.1). The catecholate and the hydroxamate (or sometimes β-hydroxy acid) groups provide a high-affinity binding site ($K_f \sim 10^{24}$ M^{-1}) for Fe^{3+} ions. Pyoverdines are of considerable scientific interest because of their contribution to pathogenicity of *P. aeruginosa* (Meyer et al. 1996; Takase et al. 2000) and because of their role in the ecology of soil pseudomonads and their possible involvement in biological control of plant pathogens (O'Sullivan and O'Gara 1992; Loper and Henkels 1999; Weller et al. 2002; Haas and Défago 2005). Consequently, there has been considerable research into the genetics and biochemistry of pyoverdine synthesis and into the regulation of pyoverdine production. Most research has been carried out with a strain of *P. aeruginosa* (strain PAO1, ATCC 15692) and this strain will be the main focus of this chapter. Studies with other pseudomonads are also reviewed and indicate that *P. aeruginosa* PAO1 is a suitable model for understanding pyoverdine synthesis and regulation.

7.2
Pyoverdine Genes in *Pseudomonas aeruginosa* PAO1

The structure of pyoverdine synthesized by the PAO1 strain of *P. aeruginosa* is shown in Fig. 7.1. Initial genetic studies of mutations causing pyoverdine deficiency showed that most of the genes specific to this process clustered at a single chromosomal locus, although a few mutations were mapped to a second minor locus (Ankenbauer et al. 1986; Hohnadel et al. 1986; Visca et al. 1992). A subsequent study of 24 transposon-insertion mutations, all of which mapped to the major pyoverdine locus, showed that pyoverdine synthesis genes (*pvd* genes) were present over a chromosomal region of about 90 kbp (Tsuda et al. 1995). However, some mutations in this region did not affect pyoverdine synthesis, showing that genes with other functions were also located in this part of the genome.

These gene mapping studies were followed by DNA cloning approaches that allowed the molecular characterization of genes required for synthesis of pyoverdine (Visca et al. 1994; Cunliffe et al. 1995; Merriman et al. 1995; Miyazaki et al. 1995; McMorran et al. 1996, 2001; Stintzi et al. 1996, 1999; Lehoux et al. 2000; Mossialos et al. 2002; Vandenende et al. 2004). The functions of individual genes are described below. A major breakthrough in this field was the publication of the complete genome sequence of *P. aeruginosa* strain PAO1 (Stover et al. 2000) and the availability of a web-based genome database (www.pseudomonas. com). Further genes that are required for pyoverdine synthesis were identified in the genome using micro array and candidate gene approaches (Ochsner et al. 2002; Lamont and Martin 2003). A full-list of genes required for synthesis of pyoverdine by strain PAO1 is given in Table 7.1. The organization of these genes in the genome is shown in Fig. 7.2. It is likely that most if not all of the genes required for synthesis of pyoverdine by strain PAO1 have now been identified. Different genes or groups of genes are considered below.

Table 7.1. Genes associated with pyoverdine synthesis in *P. aeruginosa* PAO1

Gene	Function/homologous genes	Phenotype of mutant strain[a]
pv-cABCD	Enzymes associated with synthesis of the pyoverdine chromophore (Stintzi et al. 1996, 1999)	Pvd[-b]
ptxR	Transcriptional regulator required for expression of *pvc* genes (Stintzi et al. 1999)	Pvd[-b]
pvdQ	38% identity with Aculeacin A acylase from *Actinoplanes utahensis* (Inokoshi et al. 1992)	Pvd[-]
pvdA	L-Ornithine hydroxylase (Visca et al. 1994)	Pvd[-]
fpvI	ECF sigma factor required for expression of *fpvA* (Beare et al. 2003)	Pvd[+]
fpvR	Antisigma factor for PvdS and FpvI (Lamont et al. 2002, Beare et al. 2003)	Pvd[+]
PA2389	Over 30% identity with periplasmic membrane fusion proteins (MFP) of RND/MFP/OMF-type efflux systems (Poole 2001; Zgurskaya and Nikaido 2000)	Pvd[±]
PA2390	Over 40% identity with resistance-nodulation-division (RND)-type transporter components of RND/MFP/OMF-type efflux systems (Poole 2001; Zgurskaya and Nikaido 2000)	Pvd[±]
PA2391	Over 30% identity with outer membrane factor (OMF) proteins of RND/MFP/OMF-type efflux systems (Poole 2001; Zgurskaya and Nikaido 2000)	Pvd[±]
pvdP	No known function	Pvd[-]
pvdM	23% identity with porcine dipeptidase (Rached et al. 1990)	Pvd[-]
pvdN	26% identity with isopenicillin N epimerase from *Streptomyces clavuligerus* (Kovacevic et al. 1990)	Pvd[-]
pvdO	No known function	Pvd[-]
pvdF	N^5-hydroxyornithine transformylase (McMorran et al. 2001)	Pvd[-]
pvdE	ABC transporter (secretion) (McMorran et al. 1996)	Pvd[-]
fpvA	Ferripyoverdine receptor protein (Poole et al. 1993b)	Pvd[+]
pvdD	Pyoverdine peptide synthetase (Merriman et al. 1995)	Pvd[-]
pvdJ	Pyoverdine peptide synthetase (Lehoux et al. 2000)	Pvd[-]
pvdI	Pyoverdine peptide synthetase (Lehoux et al. 2000)	Pvd[-]
PA2403–PA2410	Expression of these genes is co-regulated with pyoverdine synthesis genes; their roles in pyoverdine synthesis (if any) are not known. Mutations in PA2403 and PA2407 do not prevent pyoverdine synthesis (Ochsner et al. 2002)	Pvd[+]
PA2411	36% identity with thioesterase GrsT from *Bacillus brevis* (Kratzschmar et al. 1989)	Pvd[+]
PA2412	No known function	Pvd[-]
pvdH	Aminotransferase (Vandenende et al. 2004)	Pvd[-]
pvdL	33% identity with TycB peptide synthetase from *Bacillus brevis* (Mootz and Marahiel 1997)	Pvd[-]
pvdG	34% identity with GrsT thioesterase from *Bacillus brevis* (Kratzschmar et al. 1989)	Pvd[-]

[a] For *pvd* genes, as determined by Ochsner et al. (2002) and Lamont and Martin (2003) unless otherwise indicated

[b] The effects of mutations in the *pvc* genes on pyoverdine synthesis is dependent on the growth medium (see text)

Table 7.1. *(continued)* Genes associated with pyoverdine synthesis in *P. aeruginosa* PAO1

Gene	Function/homologous genes	Phenotype of mutant strain[a]
pvdS	ECF iron sigma factor (Cunliffe et al. 1995; Miyazaki et al. 1995)	Pvd⁻
pvdY	No known function (Vasil and Ochsner 1999)	Pvd±
pvdX	No known function (Vasil and Ochsner 1999)	Pvd⁺

[a] For *pvd* genes, as determined by Ochsner et al. (2002) and Lamont and Martin (2003) unless otherwise indicate

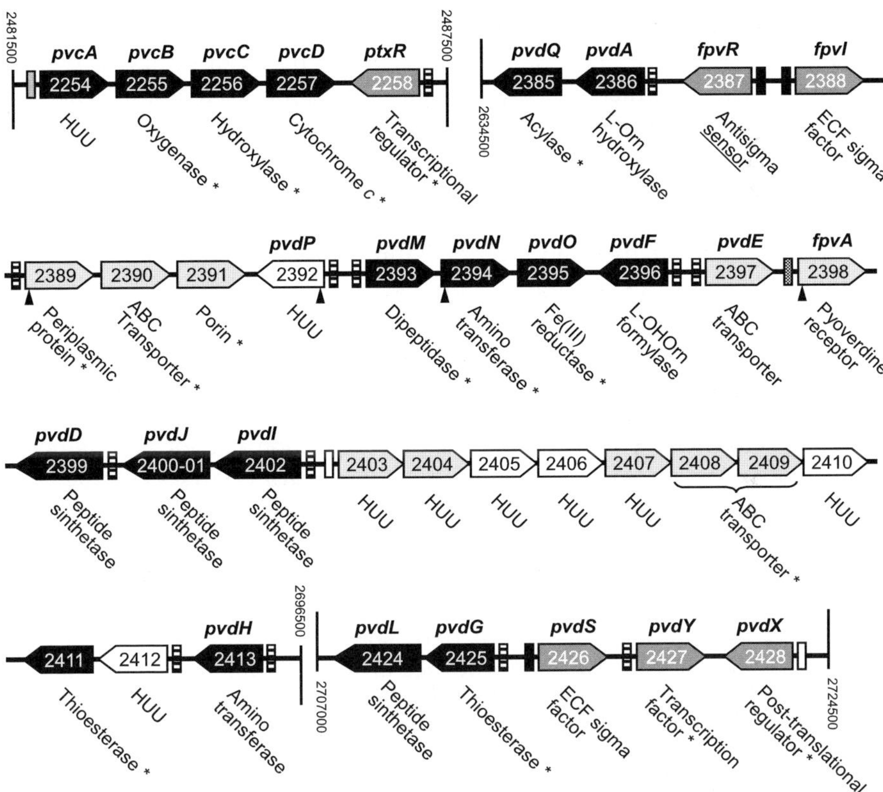

Fig. 7.2. Pyoverdine synthesis genes in *P. aeruginosa* PAO1. Genes (not drawn to scale) are oriented according to the direction of transcription. Gene numbers and map positions are according to the Pseudomonas Genome Project (www.pseudomonas.com). Gene names and function (if known) are also shown; *asterisks* denote protein functions inferred from in silico prediction. Uncharacterised gene products are indicated as HUU (hypothetical, unclassified, unknown). The *grey scale* differentiates predicted or confirmed protein functions, as follows: *black*, biosynthetic enzymes; *dark grey*, regulatory proteins; *grey*, membrane or transport proteins; *white*, HUU. Binding sites for Fur repressor protein are shown as *black boxes*, PvdS-dependent promoters as *dashed boxes*, the PtxR-dependent promoter as a *grey box*, the FpvI-dependent promoter as a *dotted box*, and uncharacterised promoters as *white boxes*. Note that PA2403–PA2410 is transcribed from an iron-regulated promoter lacking an obvious IS box. For further details and references, see Table 7.1 and the text

7.2.1
Peptide Synthetase Genes *pvdD, pvdI, pvdJ* and *pvdL*

One of the first pyoverdine synthesis genes to be characterized was *pvdD* (Merriman et al. 1995; Ackerley et al. 2003) with a mutation in this gene preventing pyoverdine synthesis. DNA sequencing showed that this gene encodes an enzyme that is part of a large family of enzymes, the non-ribosomal peptide synthetases (NRPSs). Enzymes in this class catalyze the formation of peptide bonds between substrate amino acids, and are responsible for the synthesis of a very wide range of peptides and peptide-like secondary metabolites including siderophores (reviewed in Crosa and Walsh 2002; Finking and Marahiel 2004). Although enzyme-catalyzed peptide bond formation is energetically inefficient compared to ribosomal peptide synthesis (due to the need to have a specific enzyme for each product) it enables peptide formation from amino acids that cannot be joined ribosomally, such as non-proteinogenic, D- and methylated amino acid residues.

NRPSs are very large enzymes that have a modular architecture, with each module (~1000 amino acid residues) catalyzing the incorporation of one substrate amino acid into the peptide product through a carrier thiotemplate mechanism (Kleinkauf and von Dohren 1996; Lautru and Challis 2004; Finking and Marahiel 2004). Different modules can be part of the same or different polypeptides. Each module contains a domain that is responsible for recognition and adenylation (activation) of the substrate amino acid (the A domain); a thiolation (T) domain that is the covalent attachment site for a phosphopantetheine cofactor; and, with the exception of the module responsible for incorporation of the first amino acid, a condensation (C) domain that catalyses peptide bond formation. Modules may also contain auxiliary domains that catalyze epimerization or methylation of the substrate amino acids. Each module adenylates the cognate amino acid which is then transferred to the phosphopantetheine cofactor; the C domains catalyze sequential peptide bond formation between the carboxyl group of the nascent peptide and the amino acid carried by the flanking module. The peptide is synthesized in an N-terminal to C-terminal direction. The last module is followed by a thioesterase domain that releases the assembled peptide from the phosphopantetheine cofactor. This reaction is often accompanied by cyclisation of the peptide. In most NRPSs, the organisation and the order of the modules corresponds to the amino acid sequence of the peptide product (the so-called *co-linearity* rule). The different peptides present in different pyoverdines are due to the different substrate specificities of pyoverdine-synthesising NRPSs in different strains and species (Ravel and Cornelis 2003).

Sequence analysis showed that PvdD contained two almost-identical peptide synthetase modules and the second of these was followed by a thioesterase domain (Merriman et al. 1995). This led to the hypothesis that PvdD is responsible for incorporating two L-threonine (L-Thr) residues at the carboxyl terminus of the pyoverdine peptide with the thioesterase domain catalysing release of the nascent peptide from the enzyme, accompanied by partial cyclisation (Fig. 7.1). The first part of this hypothesis was tested in subsequent biochemical studies with recombinant protein containing the first module of PvdD. The enzyme showed high substrate specificity, having high activity with L-Thr but not with D-Thr, L-allo-Thr or L-serine (L-Ser) in enzyme assays (Ackerley et al. 2003).

Genes encoding other NRPSs required for pyoverdine synthesis were identified immediately upstream of *pvdD* (Lehoux et al. 2000). Initially, three such genes were thought to be present but this was due to an error in the draft *P. aeruginosa* genome sequence, and when this error was corrected two genes (*pvdI* and *pvdJ*) were present. Mutations in either of these genes prevent pyoverdine synthesis (Lehoux et al. 2000; Ochsner et al. 2002; Lamont and Martin 2003). Sequence analysis showed that the PvdI and PvdJ proteins contain four and two peptide synthetase modules, respectively. The substrate amino acids for different peptide synthetase modules can be predicted through the *co-linearity* rule and through bioinformatic methods (Stachelhaus et al. 1999; Challis et al. 2000). PvdI is predicted to direct incorporation of (in order) D-Ser, L-arginine (L-Arg), D-Ser, and L-N^5-formyl-N^5-hydroxyornithine into the pyoverdine peptide, with PvdJ predicted to incorporate L-lysine (L-Lys) and the second L-N^5-formyl-N^5-hydroxyornithine residue (Ravel and Cornelis 2003).

Another NRPS, called PvdL, is also required for pyoverdine synthesis and is involved in synthesis of the chromophore group (Mossialos et al. 2002). The *pvdL* gene is located in the pyoverdine gene cluster though it is not adjacent to the other NRPS genes (Fig. 7.2). PvdL is composed of four modules. The first module is predicted to be an acyl CoA ligase, as it has sequence similarities with likely acyl CoA ligase domains in the first modules of NRPSs involved in synthesis of bleomycin (Du et al. 2000) and saframycin (Pospiech et al. 1996). The predicted substrates of the remaining modules are L-glutamate (L-Glu), L-tyrosine (L-Tyr) and/or L/D-tri-hydroxyphenylalanine, and L-2,4-diaminobutyric acid (L-Dab) (Mossialos et al. 2002).

Prior to the availability of the *P. aeruginosa* PAO1 genome sequence, Georges and Meyer (1995) identified high-molecular weight proteins in a number of *Pseudomonas* species that were synthesized under conditions of iron limitation and that they termed iron-regulated cytoplasmic proteins (IRCPs). Mutant strains unable to synthesize pyoverdine had altered IRCPs. This led the authors to propose that the IRCPs are NRPSs required for pyoverdine synthesis. The estimated sizes of the four largest IRCPs for *P. aeruginosa* PAO1 (approximately 550, 480, 290 and 250 kDa) correlate very well with the sizes of PvdL, PvdI, PvdD and PvdJ predicted from the genome sequence (Table 7.1), validating the proposal that the IRCPs are indeed NRPSs. These are among the largest proteins encoded by the *P. aeruginosa* PAO1 genome (www.pseudomonas.com).

As indicated above, NRPSs require a phosphopantetheine cofactor and covalent attachment of the cofactor is catalyzed by a phosphopantetheine transferase. A phosphopantetheine transferase, PcpS, has been identified in *P. aeruginosa* and shown to be required for pyoverdine synthesis (Finking et al. 2002; Barekzi et al. 2004). Other bacterial species have at least two phosphopantetheine transferases, one for primary metabolism (fatty acid synthesis) and one for secondary metabolism, but PcpS is apparently the only phosphopantetheine transferase in *P. aeruginosa*.

7.2.2
The *pvdA* and *pvdF* Genes

The *pvdA* gene was first identified via a mutation *pvd-1* that was characterized during analysis of pyoverdine-deficient mutants of *P. aeruginosa* PAO1 (Visca et al. 1992). Assays of cell-free extracts of the *pvd-1* (synonym of *pvdA*) mutant showed that it was unable to produce L-N^5-hydroxyornithine and its L-N^5-formyl derivative. Addition of L-N^5-hydroxyornithine to the growth medium restored the ability of the mutant strain to make pyoverdine and to grow under iron-limiting conditions. These findings provided strong evidence that the *pvd-1* mutation prevents synthesis of the L-N^5-formyl-N^5-hydroxyornithine residues that are present in pyoverdine. The *pvdA* gene was subsequently cloned through complementation of the *pvd-1* mutation (Visca et al. 1994). A mutation was engineered in the chromosomal *pvdA* gene and resulted in the same phenotype as that of the *pvd-1* mutant, confirming identification of the correct gene. Analysis of the deduced PvdA sequence predicted a high degree of similarity to two enzymes that catalyze similar reactions during siderophore synthesis, an L-lysine-N^6-hydroxylase from *Escherichia coli* and an L-ornithine-N^5-oxygenase from the fungus *Ustilago maydis*. A *pvdA* homologue, called *psbA*, was later identified in the rhizobacterium *Pseudomonas* B10 and found to be essential for pyoverdine synthesis (Ambrosi et al. 2000). Recently, PvdA was successfully purified and its biochemical characterization confirmed that PvdA is an L-ornithine-N^5-oxygenase that is required for conversion of L-ornithine to L-N^5-hydroxyornithine (Ge and Seah 2006).

The *pvdF* gene was identified through sequencing of cloned DNA that originated from the pyoverdine locus (McMorran et al. 2001), with a mutation that was engineered in this gene preventing pyoverdine synthesis. The sequence showed that the PvdF enzyme had some similarities to glycinamide ribonucleotide transformylases, a group of enzymes that catalyze a formylation reaction as part of the purine synthesis pathway. This led to the hypothesis that PvdF catalyses formylation of L-N^5-hydroxyornithine to form L-N^5-formyl-N^5-hydroxyornithine that is then incorporated into pyoverdine. Chemical analyses of extracts from wild-type and PvdF mutant bacteria were consistent with this hypothesis (McMorran et al. 2001).

These data indicate that PvdA and PvdF catalyse sequential reactions to synthesize L-N^5-formyl-N^5-hydroxyornithine from L-ornithine. The L-N^5-formyl-N^5-hydroxyornithine is then available for incorporation into pyoverdine by the PvdI and PvdJ NRPSs (see above).

7.2.3
The *pvdH* Gene

The *pvdH* gene was identified through bioinformatic and gene-expression approaches and mutations that were engineered into this gene showed that it is required for pyoverdine synthesis (Ochsner et al. 2002; Lamont and Martin 2003). Sequence analysis suggested that it was an aminotransferase. This was confirmed by Vandenende and co-workers (Vandenende et al. 2004) who carried out an extensive characterization of the PvdH enzyme. It was shown to catalyze an aminotransferase reaction interconverting aspartate β-semialdehyde and L-Dab. The latter amino acid is predicted to be one of the substrates for the PvdL NRPS that is almost certainly required for synthesis of the chromophore component, as described above.

7.2.4
Other *pvd* Genes

Mutations in several other genes have been shown to prevent pyoverdine synthesis (Table 7.1). Biochemical studies have not yet been carried out on the corresponding proteins but bioinformatics analyses can give insights into the possible roles of some of these proteins/enzymes in pyoverdine production.

The first gene to fall into this category is *pvdE* (McMorran et al. 1996). The predicted sequence of the PvdE protein has all the characteristics of ATP-binding-cassette (ABC) membrane transporter proteins. Such proteins are found in many species (from microbes to humans) and are responsible for transport of a very wide range of substrates including proteins, polysaccharides, peptides, drugs, sugars, amino acids, and metal ions (Higgins 2001). ABC transporters can catalyze either import or export of their substrates and PvdE has most similarity to family members that are required for export (Saurin et al. 1999). The substrate that is transported by PvdE has not been determined; it may be pyoverdine or a pyoverdine precursor, or a protein or other factor that is required extra-cytoplasmically for pyoverdine synthesis.

Mutations in *pvdG* prevent pyoverdine synthesis (Lamont and Martin 2003) although it is possible that this is because of the effect of the mutation on expression of the downstream *pvdL* gene that is predicted to be operonic with *pvdG* (Fig. 7.2). PvdG is predicted to be a thioesterase. A second predicted thioesterase in *P. aeruginosa* PAO1 is PA2411, but a mutation in this ORF does not prevent pyoverdine synthesis (Ochsner et al. 2002). Thus, three predicted thioesterases (including the thioesterase domain present in PvdD) could be involved in pyoverdine synthesis. Other NRPS systems also have more than one thioesterase. There is evidence that the additional thioesterases may act as proofreaders, removing incorrect substrates or aberrant intermediates that have been covalently linked to an NRPS enzyme (Heathcote et al. 2001; Schwarzer et al. 2002; Reimmann et al. 2004). Alternatively, PvdG and PA2411 may assist with release of the pre-pyoverdine peptide from PvdD, accompanied by partial cyclisation of the peptide.

The *pvdM*, *pvdN* and *pvdO* genes are likely to form an operon (Lamont and Martin 2003; Ochsner et al. 2002). Mutations in all three of these genes prevent pyoverdine synthesis. While it cannot be ruled out that the effects of the *pvdM* and *pvdN* mutations are polar on *pvdO*, it seems most likely that all three genes are required for pyoverdine synthesis. Amongst biochemically-characterised proteins, PvdM has highest sequence similarity with mammalian dipeptidases (Table 7.1). It is possible that PvdM catalyses processing of a pyoverdine precursor. Intriguingly, a gene encoding a putative dipeptidase with sequence homology to PvdM is also clustered with NRPS genes involved in synthesis of the fungal secondary metabolite sirodesmin (Gardiner et al. 2004), although for many other NRPS-generated peptides there is no evidence of a requirement for a (di)peptidase. Amongst characterized proteins, PvdN has highest sequence similarity to an isopenicillin N epimerase from *Streptomyces clavuligerus* (Kovacevic et al. 1990). However, the exact role of PvdN in pyoverdine synthesis has yet to be determined. PvdO is suggested to be an Fe^{3+} reductase (Ochsner et al. 2002) and its exact role in pyoverdine synthesis is also unclear.

The *pvdP* gene is expressed divergently from *pvdMNO*. Homologues of PvdP protein are restricted to the fluorescent pseudomonads; the exact role of this protein in pyoverdine synthesis is not known. The *pvdQ* gene is downstream from *pvdA*. PvdQ has sequence similarity to aculeacin A acylase from *Actinoplanes utahensis* as well as other penicillin amidases. The PvdQ enzyme has been shown to hydrolyse the amide bond in an acyl homoserine lactone that acts as a quorum-sensing signaling molecule in *P. aeruginosa* (Huang et al. 2003). This provides evidence that the enzyme is indeed an acylase although it presumably has a different substrate during pyoverdine synthesis; the physiological relevance of PvdQ-mediated hydrolysis of acyl homoserine lactone is not clear (Roche et al. 2004).

7.2.5
The *pvc* Genes

A cluster of four genes, the *pvc* genes, that is located about 150 kb away from the *pvd* genes have been implicated in the maturation of the pyoverdine chromophore as mutations in *pvc* genes prevented synthesis of pyoverdine (Stintzi et al. 1996, 1999). These genes are likely to be expressed as an operon, with regulation of expression having similarities to that of the *pvd* genes. However, under some conditions, *pvc* mutants are able to make pyoverdine (P. Cornelis, personal communication) and homologues of the *pvc* genes are not present in other *Pseudomonas* species that make pyoverdines (Baysse et al. 2002; Mossialos et al. 2002). The exact role of the *pvc* genes in pyoverdine synthesis is therefore not clear.

7.2.6
Intermediate-Based Studies of Pyoverdine Synthesis

Studies on the biosynthesis of pyoverdine have been carried out using classical biochemical approaches based on the identification and characterization of intermediate compounds in the biosynthetic pathway, in particular in the laboratory of H. Budzikiewicz (reviewed in Budzikiewicz 2004). A detailed description of these studies is beyond the scope of this book. Instead, we describe two particularly important intermediates that were purified and characterized and shed light on synthesis of the pyoverdine chromophore. These are ferribactin and 5,6-dihydropyoverdine. Ferribactin is pyoverdine in which, instead of the dihydroxyquinoline chromophoric group, there is the tripeptide γ–L-Glu – D-Tyr – L-Dab (Hohlneicher et al. 2001). Dihydropyoverdine is identical to pyoverdine except that two carbons in the chromophore are not saturated; consequently this form of pyoverdine does not fluoresce. The characterisation of these compounds, in conjunction with other studies, has led to a proposed pathway for chromophore biosynthesis that is described below.

7.2.7
The Biosynthetic Pathway of Type I Pyoverdine
in *P. aeruginosa* PAO1

The biochemical pathway of pyoverdine synthesis is understood only in part. Genetic and biochemical studies described above show that a variety of different enzymes are involved in this process. There is good evidence that both the chromophore and the peptide moiety of pyoverdine are synthesized in a multistep reaction by the NRPSs, namely, PvdL, PvdI, PvdJ and PvdD. The pyoverdine chromophore backbone is the condensation product of L-Glu, D-Tyr and L-Dab (Bockman et al. 1997), while the pyoverdine PAO1 (Type I) peptide chain results from the condensation and partial cyclisation of eight amino acids (Fig. 7.1) (Briskot et al. 1989). The composition of pyoverdine PAO1 corresponds very well with the predicted substrates of the peptide synthetases (see above) with each NRPS module catalysing incorporation of one amino acid into the pyoverdine precursor. Furthermore, because PvdL is the only NRPS lacking an initial C-domain, its role in the first step of pyoverdine biogenesis has presumptively been assigned (Mossialos et al. 2002). The architecture of this enzyme consists of two basic modules (II and IV), a third central module with an additional epimerization domain and a first, unusual short module with a T-domain preceded by an acyl-CoA ligase-like domain, whose role has not yet been elucidated in non-ribosomal peptide synthesis (Mossialos et al. 2002). On the other hand, PvdD is the only NRPS which ends with a thioesterase domain, arguing for a function in siderophore release at the end of the assembly step (Ackerley et al. 2003). The other two synthetases, PvdI and PvdJ, are composed of four and two basic modules respectively, with two supplementary epimerization domains in the first and third modules of PvdI, almost certainly involved in epimerization of two substrate L-Ser residues (Ravel and Cornelis 2003).

Except for PvdD, which was found to activate two L-Thr residues (Ackerley et al. 2003), the biochemistry of these peptide synthetases has not yet been addressed. However, their established modular organisation, together with in silico prediction of the substrate specificity of each A domain (Stachelhaus et al. 1999; Challis et al. 2000), makes possible the prediction of a nearly complete NRPS pathway for pyoverdine biogenesis (Fig. 7.3). Briefly, PvdL catalyzes the formation of the chromophore backbone by condensation of L-Glu, D-Tyr and

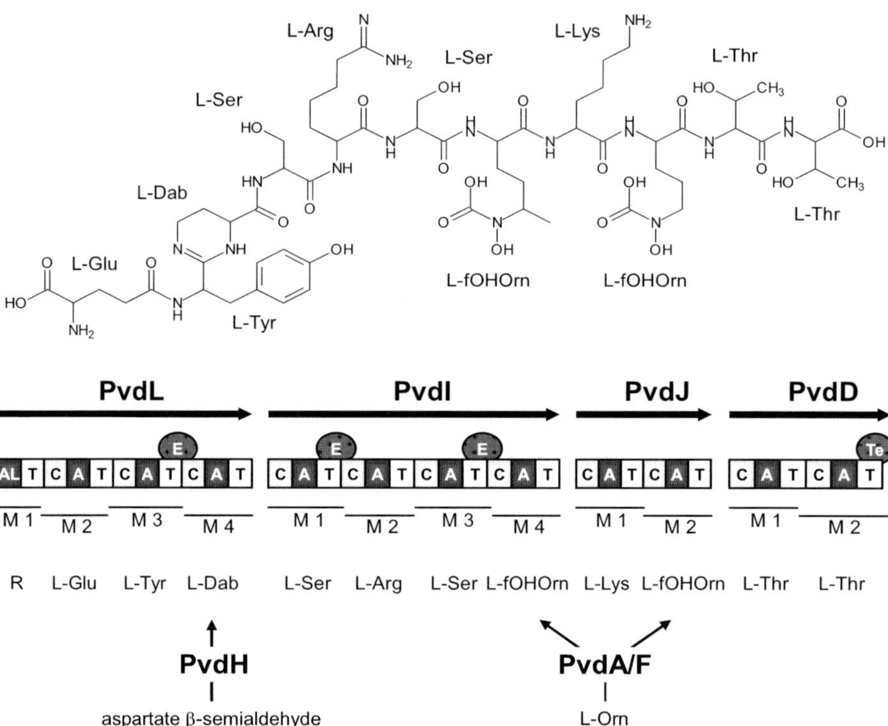

Fig.7.3. Pyoverdine biosynthesis in *P. aeruginosa* PAO1: predicted amino acid specificity of non-ribosomal peptide synthetases. NRPS enzymes are shown as *black arrows* with the respective modular domains as *adjacent squares*. Protein domains are: AL, acyl CoA ligase; A, adenylation; T, thiolation; C, condensation. Each enzyme module is indicated as M with progressive numbering. *Circles* indicate auxiliary modules: E, epimerisation; Te, thioesterase. The immature pyoverdine backbone (ferribactin) resulting from the activity of NRPSs modules is shown on the *upper part*. Amino acid substrates recognized by each module are indicated: L-Dab, L-2,4-diaminobutyrate; L-fOHOrn, L-N^5-formyl-N^5-hydroxyornithine. Other amino acids are abbreviated with conventional three-letter code. Note that the existence of E domains in PvdL (M3) and PvdI (M1 and M3) results in the incorporation of D-isomers in the pyoverdine precursor peptide. The substrate R of the putative AL domain (M1) is not known. Precursor-generating enzymes PvdH, PvdA and PvdF are shown in the *lower part*

L-Dab (Mossialos et al. 2002); the C-domain of module one of PvdI then promotes the attachment of this precursor to the first amino acid residue (D-Ser) of the growing chain. Further elongation of the peptide is provided by the linear progression of the PvdI, PvdJ and PvdD modules, while release of the peptide arises from the PvdD thioesterase domain activity. According to this model, the order of the peptide chain assembly would exactly match the physical order and orientation of NRPS gene modules responsible for peptide moiety biosynthesis, from the N-terminus of PvdI to the C-terminus of PvdD. The structure of ferribactin (see above and Fig. 7.3) exactly matches the predicted structure of this proposed precursor peptide.

In support of the proposed pathway, the first A domain of the NRPSs cluster is always specific for the first amino acid of the peptide moiety, i.e. the one joined to the chromophore, as also documented for other *P. aeruginosa* pyoverdine types (Smith et al. 2005). Moreover, PvdL is the only pyoverdine-related NRPS that is encoded by the genomes of all fluorescent *Pseudomonas* (Ravel and Cornelis 2003). Conversely, NRPSs responsible for peptide chain elongation are highly variable among different pseudomonads, consistent with the amino acid variability of the peptide moiety (Ravel and Cornelis 2003).

Although the specific modular organization of the pyoverdine NRPSs corresponds well with the amino acid sequence of the pyoverdine peptide backbone, there is no evidence for a putative cyclisation domain (Ravel and Cornelis 2003) that would be expected from the partially cyclic peptide chain structure (Fig. 7.1). Given that isolated thioesterase domains have been involved in peptide cyclisation (Kohli et al. 2001), this enzymatic activity could be provided either *in cis* by the thioesterase domain of PvdD or *in trans* by PA2411 and/or PvdG.

Genetic and biochemical characterization of PvdA and PvdF, described above, shows that they are necessary for synthesis of L-N^5-formyl-N^5-hydroxyornithine that is present in the pyoverdine peptide (Fig. 7.1). Whether these modification events occur before or during the amino acid incorporation into the peptide chain has still to be fully determined. However, both *pvdA* and *pvdF* mutants are non-fluorescent (Visca et al. 1992, 1994; McMorran et al. 2001), suggesting that N^5-formyl-N^5-hydroxyornithine formation is a pre-requisite for chromophore assembly or maturation.

The events leading to maturation of the pyoverdine chromophore are not so fully understood. PvdH catalyses the formation of L-Dab, one of the predicted substrates of PvdL. The isolation of ferribactins (described above) is consistent with the proposal that PvdL catalyses condensation of L-Dab, L-Glu and D-Tyr (although a subsequent reaction must take place so that L-Glu is joined through the γ- rather than the α-carbon). Transformation of ferribactins to pyoverdines requires a series of redox reactions followed by tautomerization leading to dihydroxyquinoline ring formation (Dorrestein et al. 2003). Furthermore, the L-Glu attached to the original D-Tyr by its γ-carboxyl group can be transformed to α-ketoglutaric acid, succinamide, malamide or hydroxylated to free acid (Schafer et al. 1991). It has been proposed that some modifications of the chromophore moiety could take place in the periplasm by the action of hemoproteins (Baysse et al. 2002), since heme has been identified as a necessary component for pyoverdine biosynthesis (Baysse et al. 2001). It may be that the *pvc* genes contribute to this step. Additionally, *pvd* genes that have not yet been assigned biochemical functions (see above; Table 7.1) may catalyse reactions involved in chromophore maturation.

Following (or concomitant with) synthesis, pyoverdine must be secreted into the extra cellular milieu. The mechanism of secretion is not known although one obvious candidate protein that may be involved is PvdE (see above). PA2389–PA2391 are co-regulated with pyoverdine genes (Ochsner et al. 2002), encode proteins with the characteristics of a secretion system and are located at the pyoverdine locus (Fig. 7.2) so that it is tempting to speculate that they are involved in pyoverdine synthesis. However, mutations in these genes do not prevent secretion of pyoverdine (Ochsner et al. 2002; Lamont and Martin 2003). A different secretary system has also been implicated in pyoverdine secretion (Poole et al. 1993a) but the possible contribution of this system has not been further studied.

7.3
Iron Regulation of Pyoverdine Synthesis: The Master Roles of Fur and PvdS

As for all siderophores, pyoverdines are produced in response to nutritional iron deficiency. This is inferred by the evidence that pyoverdine synthesis is repressed in suspension cultures grown in chemically-defined laboratory media containing more than 5 µM $FeCl_3$ (Meyer and Abdallah 1978). Above this concentration, Fe(III) is likely to diffuse passively into the cell where it is reduced to Fe(II) in the cytoplasm, thereby acting as a co-repressor for the global regulatory protein Fur (ferric uptake regulator) (Escolar et al. 1999). Iron-dependent transcriptional repression is primarily dependent on the binding of the Fur-Fe(II) holorepressor complex to the promoters of iron-regulated genes. In the case of pyoverdine genes Fur is not the only regulatory protein. Positive control of siderophore synthesis also occurs through a membrane-spanning signalling mechanism (Sect. 7.4).

Fur is an essential protein in *P. aeruginosa* PAO1 and in other fluorescent pseudomonads which tolerate only partial loss of Fur function through amino acid substitutions resulting from point mutations in the *fur* gene (reviewed in Vasil and Ochsner 1999; Venturi et al. 1995b). *P. aeruginosa fur* mutants show pleiotropic phenotypes and deregulation of siderophore synthesis (Vasil and Ochsner 1999). The iron regulon has been extensively investigated in *P. aeruginosa* PAO1, confirming the role of Fur as the master regulatory protein in iron metabolism. The expression of more than 100 genes shows strict dependence on iron deficiency (Ochsner et al. 2002; Palma et al. 2003), but only a minority of these genes contain a Fur-binding sequence (the Fur box, GATAATGATAAT-

CATTATC) (Lavrrar and McIntosh 2003) in their promoter. This apparent discrepancy is due to the occurrence of Fur-controlled transcriptional regulators that regulate subsets of genes, most of which are involved in the uptake of endogenous or exogenous iron chelates (Visca et al. 2002). The crystal structure of *P. aeruginosa* Fur has been solved (Pohl et al. 2003). The protein consists of two domains: the dimerisation domain containing the regulatory Fe(II)-binding site and the DNA-binding domain containing a structural Zn(II) atom. Fur-Fe(II) associates as a dimer, and two dimers have been proposed to recognise 5-bp spaced operator sites located on opposite sides along the DNA sequence, thereby hindering access of vegetative (RpoD-dependent) RNA polymerase (RNAP) to the promoter. Within the pyoverdine region of the *P. aeruginosa* PAO1 chromosome, Fur-binding sequences were identified in the promoter regions of the *pvdS*, *fpvI* and *fpvR* genes (Beare et al. 2003; Cunliffe et al. 1995; Lamont et al. 2002; Miyazaki et al. 1995; Ochsner and Vasil 1996).

Fur can also act as a positive regulator of the expression of some genes involved in oxidative stress response (superoxide dismutase, *sodB*), Krebs cycle (succinate dehydrogenase, *sdh*) and iron storage (bacterioferritin, *bfrB*). Positive iron regulation of gene expression is indirect and involves two small RNAs, PrrF1 and PrrF2, tandemly arranged on the *P. aeruginosa* PAO1 chromosome (Wilderman et al. 2004). PrrF RNAs act at the post-transcriptional level to reduce the expression of target genes. They reduce mRNA stability by increasing the rate of its decay through complementary base pairing at the 5′-end of target transcript (Wilderman et al. 2004). Transcription of PrrF1 and Prrf2 only occurs in low-iron conditions as expression of these RNAs is repressed by Fur-Fe(II) binding to their promoters. This system enhances expression of target genes (including *sodB*, *sdh* and *bfrB*) in response to iron proficiency.

Despite their iron-regulated expression profile, all pyoverdine biosynthesis and uptake genes or operons lack Fur-binding sequence in their promoter region but contain the iron starvation (IS) box (originally described as [G/C]CTAAATCCC, but currently reconsidered as TAAAT), a typical sequence signature involved in the recognition and binding of the alternative sigma factor PvdS (Rombel et al. 1995; Wilson and Lamont 2000). PvdS is the prototypic member of IS sigma factors, an iron-responsive subgroup of the extracytoplasmic function (ECF) family of RpoD-like sigmas (Leoni et al. 2000). The *pvdS* promoter region contains three overlapping Fur boxes and binds Fur in vitro (Ochsner et al. 1995), thus explaining the tight iron-regulated expression of pyoverdine genes. PvdS is a 187-amino acid protein that spontaneously associates with the core fraction of RNAP to form a transcriptionally active RNAP holoenzyme (Leoni et al. 2000; Wilson and Lamont 2000). As a typical ECF sigma, PvdS is very much smaller than the primary sigma factor RpoD and has only a low amount of sequence similarity. This is consistent with its promoter recognition sequence, which is markedly different from the RpoD consensus. PvdS-dependent promoters are all characterised by an IS box at about position −33 relative to the transcription start point, followed by the CGT triplet 16/17 nt downstream (Ochsner et al. 2002; Wilson and Lamont 2000). The TAAAT-$N_{16/17}$-CGT element has been recognised with minor variation in the promoter region of 19 out of the 26 PvdS-regulated genes of *P. aeruginosa* PAO1 (Ochsner et al. 2002).

7.4
Receptor-dependent Autoregulation of Pyoverdine Synthesis

Positive control of siderophore transport has been extensively investigated in *P. aeruginosa*. In this bacterium, siderophores account for at least three levels of regulation and signalling. First of all, iron supply via siderophores enables *P. aeruginosa* to switch-on the activity of Fur repressor thereby silencing the Fur regulon. Second, receptor binding by both endogenous and exogenous siderophores communicates the real efficacy of a given siderophore in iron supply, acting as a stimulus for co-ordinate expression of the cognate uptake and, eventually, biosynthesis genes. Third, and apparently unique to the pyoverdine system, binding of the siderophore to its receptor acts as an extracytoplasmic stimulus for a signal transduction cascade leading to the overexpression of several exoproducts, including a protease, an exotoxin, and pyoverdine itself. Four protein partners located in distinct cellular compartments enable the pyoverdine signal to be transmitted from the cell surface to the cytoplasm. These are the pyoverdine receptor FpvA, the antisigma factor FpvR and two alternative sigma factors, PvdS and FpvI. An overall view of transport-coupled pyoverdine signalling is shown in Fig. 7.4.

▶ **Fig. 7.4 a,b.** Receptor-dependent pyoverdine signaling in *P. aeruginosa* PAO1: **a** crystal structure of the pyoverdine-loaded FpvA receptor (Cobessi et al. 2005). *Side* (*left*) and *top* (*right*) view of FpvA showing 22 antiparallel β strands filled by the plug domain. The pyoverdine molecule, depicted in space filling representation, is shown in the ligand binding pocket of the receptor. The structure of the periplasmic N-terminal extension implicated in signalling is not known and this is represented by a dotted line; **b** model for the mechanism of pyoverdine signalling. The *left part* of the figure shows the resting state of the signalling complex. Under low-iron conditions, the FpvA receptor is permanently engaged with iron-free pyoverdine (*white triangle*) and interaction between the membrane-spanning antisigma factor FpvR and the PvdS sigma reduces transcription of *pvd* genes. Within the FpvA schematic structure, the cork-bound signalling domain and the TonB box are depicted as a *grey plug* and a *black sphere*, respectively. The *right part* of the figure shows the model for the activated state of the signalling complex. Displacement of apo pyoverdine by ferric pyoverdine (*black circle embedded in white triangle*) causes a conformational transition of the FpvA receptor that results in TonB-dependent ferric pyoverdine transport into the periplasm followed by transport of iron into the cytoplasm (*black arrow*) via an unknown mechanism. Concomitantly, release of PvdS by the FpvR antisigma causes PvdS to bind the core enzyme of RNA polymerase (RNAPc) and direct transcription of pyoverdine biosynthetic genes (*pvd*). These are characterised by the typical TAAAT-N_{16-17}-CGT promoter motif. A similar pattern of interaction can be predicted between FpvR and the *fpvA*-specific IS sigma factor FpvI (not shown). OM, outer membrane; PP, periplasm; CM, cytoplasmic membrane. For further details see the text

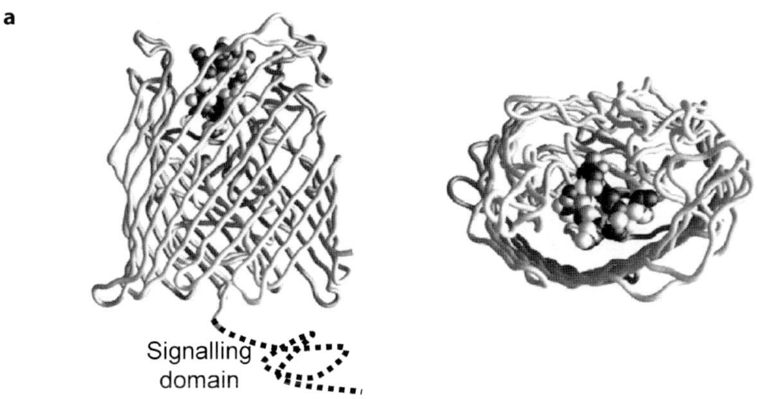

a
Signalling
domain

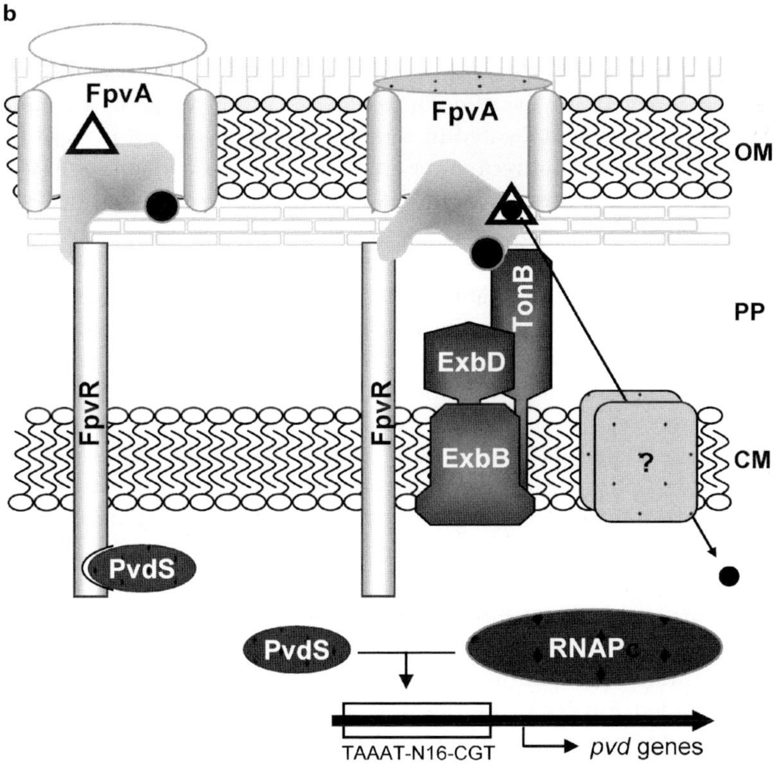

b
FpvA
FpvA
OM
TonB
ExbD
PP
FpvR
FpvR
ExbB
?
CM
PvdS
PvdS
RNAP
TAAAT-N16-CGT
pvd genes

7.4.1
The FpvA Receptor

The ferric pyoverdine receptor is an iron-regulated outer membrane protein of ca. 86 kDa, encoded by the *fpvA* gene (Poole et al. 1993b). Its crystal structure has recently been solved (Cobessi et al. 2005) (see Fig. 7.4a). FpvA has a periplasmic N-terminal domain that is responsible for signal transduction and transcriptional regulation of *pvd* genes, a typical feature of TonB-dependent receptors involved in surface signalling (Schalk et al. 2004). Consistent with its periplasmic location, this domain is sensitive to proteolysis (James et al. 2005). At the structural level, the membrane-associated portion of FpvA is folded into two domains: a transmembrane 22-stranded *beta*-barrel domain occluded by an N-terminal plug domain containing a mixed four-stranded *beta*-sheet. The *beta*-strands of the barrel are connected by long extracellular loops and short periplasmic turns. The iron-free pyoverdine is bound at the surface of the receptor in a pocket lined with aromatic residues while the extracellular loops do not completely cover the pyoverdine binding site. The N-terminal domain of ca. 70 amino acids that is implicated in signal transduction is not present in this structure. The TonB box that is involved in intermolecular contacts with the TonB protein(s) is poorly defined but could be modelled in an extended flexible conformation (Cobessi et al. 2005).

FpvA-bound ferric pyoverdine is located on the extracellular side of the outer membrane. Translocation of the siderophore into the periplasm requires the proton motive force and energy coupling in the inner membrane through the TonB complex. Three TonB homologues (*tonB1*, *tonB2* and PA0695) have been identified in the *P. aeruginosa* genome (Poole et al. 1996; Zhao and Poole 2000). The *tonB1* gene is iron-regulated, Fur-repressible, and essential for siderophore-mediated iron uptake (Poole et al. 1996). Homologues of *exbB* and *exbD* that are required for ferrisiderophore transport in *E. coli* were identified downstream of the *tonB2* gene, but the whole *tonB2* gene system seems to be dispensable for iron acquisition (Zhao and Poole 2000). Therefore, TonB1 could act in concert with the *exb* homologues encoded by PA0693 and PA0694, or with other proteins, or it could be capable of working autonomously.

Functional studies highlighted that FpvA is capable of binding both unsaturated and iron-saturated pyoverdine (Schalk et al. 1999, 2001). This indicates that pyoverdine-loaded FpvA can be the normal state of the receptor. FpvA-bound apo pyoverdine is not transported by *P. aeruginosa*, while the iron-loaded siderophore efficiently enters the cell (Schalk et al. 1999, 2001). The TonB system is essential for the efficient displacement of FpvA-bound apo pyoverdine by ferric pyoverdine, but dispensable for siderophore binding (Schalk et al. 2001). Iron-saturated pyoverdine does not exchange the metal with the receptor-bound apo siderophore in vivo but replaces it in the receptor pocket, showing fast displacement kinetics that parallel the rate of iron uptake. Thus, FpvA appears to adopt different conformations depending on its loading status, the transport (and likely signalling) competent state being achieved upon binding of the fer-risiderophore (Cobessi et al. 2005; Schalk et al. 2004).

The involvement of the flexible N-terminal extension of FpvA in pyoverdine signalling was documented by genetic studies (James et al. 2005; Shen et al. 2002). Deletion of *fpvA* curtails both pyoverdine synthesis and transport, while complementation with a truncated version of *fpvA* lacking the 5′ (N-terminal) coding region restores transport of exogenously supplied pyoverdine, but not pyoverdine synthesis (Shen et al. 2002). Mutations within the N-terminal do-main do not affect pyoverdine transport but reduce pyoverdine production, consistent with a regulatory function for this region (James et al. 2005). Mu-tational analysis also provided evidence that FpvA-dependent transport and signalling are not mutually required processes, and that either function can be independently inactivated by insertion mutagenesis (James et al. 2005). Thus, the periplasmic extension of FpvA is definitively involved in the regulation of expression of pyoverdine biosynthesis genes.

P. aeruginosa PAO1 contains an additional receptor FpvB that enables uptake of Type I pyoverdine (Ghysels et al. 2004). FpvB can substitute for pyoverdine transport in an *fpvA*-deficient background. The *fpvB* gene is iron-regulated, but is located outside of the *pvd* locus and is differentially expressed with respect to *fpvA*, depending on the carbon source. Its role in pyoverdine signalling has not yet been addressed.

7.4.2
The FpvR Antisigma

How does binding of (ferri) pyoverdine to the FpvA outer membrane recep-tor trigger the activity of sigma factor PvdS in the cytoplasm of *P. aeruginosa*? Signal transduction requires the *fpvR* gene product. FpvR is a 331 amino acid protein with a predicted single transmembrane helix (residues 93–115) and a cytoplasmic N-terminus. The overall topology of FpvR is very similar to that ob-served for two other transmembrane sensors, namely, FecR of *E.coli* and PupR of *P. putide*, and the similarity extends to the functional level (Lamont et al. 2002). The site of interaction between FpvA and FpvR is predicted in the periplasmic N-terminal part of the receptor, in a region (residues 50–59) called the FecR-box (sequence I-LL-GTGA in FpvA) (Visca 2004). There is a high conservation of Trp residues in the N-terminal region of sensor proteins implicated in sidero-phore signalling, including FpvR (Visca 2004), and the corresponding residues of FecR are known to be crucial for signal transduction to the sigma factor FecI (Stiefel et al. 2001). FpvR has typical antisigma functions (Fig. 7.4b); when over-expressed it causes transcriptional repression of *pvd* genes, while its deletion has only minor effects (Lamont et al. 2002). Genetic evidence indicates that the N-terminal part of FpvR is located in the cytoplasm and interacts with the alter-native sigma factors PvdS and FpvI (Lamont et al. 2002; Beare et al. 2003; Rédly and Poole 2005). *fpvR* is transcribed from a Fur-regulated promoter (Ochsner and Vasil 1996) and is located in opposite orientation to the *fpvI* gene that encodes a sigma factor responsible for transcription of *fpvA* (Fig. 7.2) (Beare et al. 2003; Rédly and Poole 2003).

7.4.3
A Divergent Signalling Pathway from FpvR to PvdS and FpvI Sigma Factors

Transcription of *pvd* genes is primarily dependent upon the expression of the Fur-repressible *pvdS* gene (Cunliffe et al. 1995; Miyazaki et al. 1995). As an alternative sigma factor, PvdS forms a transcriptionally active RNAP holoenzyme complex and shows fair conservation of regions implicated in promoter binding and recognition of core RNAP (reviewed in Visca et al. 2002). As anticipated, the main determinant of promoter recognition by PvdS is the IS box (see Sect. 7.3), a sequence signature also present in the promoters of genes regulated by the PvdS homologues PfrI, PbrA and PsbS in other fluorescent pseudomonads (Rombel et al. 1995).

An additional IS ECF sigma factor has been identified that is encoded by the *fpvI* gene, adjacent to *fpvR* (Fig. 7.2) (Beare et al. 2003; Rédly and Poole 2003). Transcription of both *fpvI* and *fpvR* is co-regulated by iron due to the presence of Fur boxes in their divergently oriented promoters (Ochsner and Vasil 1996; Ochsner et al. 2002). FpvI is implicated in the regulation of FpvA expression, as inferred by the reduced *fpvA* transcription in the *fpvI* mutant and by the increased *fpvA* transcription resulting from *fpvI* overdosage (Beare et al. 2003; Rédly and Poole 2003). Since FpvA is also implicated in the control of *pvd* genes through the FpvR-PvdS signalling cascade, both pyoverdine production and transcription of *pvd* genes are reduced in either *fpvI* or *fpvA* mutants (Beare et al. 2003; Shen et al. 2002).

The available evidence is consistent with a model whereby the N-terminal extension of FpvA interacts with the C-terminal domain of FpvR in the periplasm and the N-terminal portion of FpvR interacts with PvdS and FpvI in the cytoplasm. On this model, binding of (ferri) pyoverdine to FpvA on the outer membrane enables FpvR to transmit a signal through the cytoplasmic membrane to the two sigma factors PvdS and FpvI, leading to transcription of pyoverdine biosynthesis (*pvd*) and uptake (*fpvA*) genes, respectively (Fig. 7.4b). The proposed dual activity of FpvR on both PvdS and FpvI, and the response of *P. aeruginosa* PAO1 to the endogenous pyoverdine are features that differentiate the FpvA/FpvR/FpvI-PvdS system from other surface signalling devices such as the Fec and the Pup systems (Beare et al. 2003; Lamont et al. 2002).

Autogenous control of siderophore synthesis has broad biological implications. In species like *P. aeruginosa* that produce and acquire multiple siderophores, coupling between siderophore uptake and biosynthesis ensures that each siderophore is produced only when it is effective in delivering iron to the cell, namely after binding to the cognate receptor.

7.4.4
Additional Regulatory Proteins Involved in Modulation of Pyoverdine Gene Expression

Other regulators of the pyoverdine system have been identified in addition to PvdS, FpvI and FpvR. The effect of AlgQ (or AlgR2) on pyoverdine production by *P. aeruginosa* PAO1 has been elucidated (Ambrosi et al. 2005). AlgQ is a global regulatory protein that activates alginate, ppGpp, and polyP synthesis through a cascade involving nucleoside diphosphate kinase (Ndk). At the sequence level, AlgQ is similar to the *E. coli* Rsd protein that is an antisigma factor for RpoD and, like Rsd, AlgQ is capable of interacting with region 4 of RpoD (Dove and Hochschild 2001). In *P. aeruginosa* PAO1, deletion of *algQ* results in moderate but reproducible reduction in pyoverdine production as the result of diminished transcription of *pvd* genes. Complementation with wild-type *algQ* fully restores pyoverdine production and expression of *pvd* genes, while *ndk* does not. Thus, AlgQ is a functional homologue of Rsd, operating as an antisigma factor for RpoD and so assisting recruitment of core RNAP by PvdS and hence transcription of *pvd* genes. Expression of pyoverdine synthesis genes in *P. putida* requires the presence of the PfrA protein that is similar to AlgQ (Venturi et al. 1993) and could enable the *P. putida* PfrI sigma factor to compete with RpoD for RNAP core enzyme. However, AlgQ has a subtle effect on transcription of *pvd* genes in *P. aeruginosa* (Ambrosi et al. 2005), while PfrA is absolutely essential for the expression of homologous *P. putida* genes (Venturi et al. 1993).

Expression of the *pvc* genes that are associated with synthesis of the pyoverdine chromophore (see Sect. 7.2.5) is controlled by the LysR-type regulator PtxR, which in turn is controlled by PvdS (Stintzi et al. 1999). The *ptxR* gene is absent in fluorescent pseudomonads other than *P. aeruginosa* (Ravel and Cornelis 2003; Stintzi et al. 1999), consistent with the absence of the *pvc* genes from those species. Additional regulatory proteins that could affect PvdS expression or activity are the PA2427 and PA2428 gene products (PvdY and PvdX, respectively; see Table 7.1 and Fig. 7.2), though their functions are merely speculative at the moment (Vasil and Ochsner 1999). Thus, while the role of sigma factors in directing the expression of genes required for the synthesis of pyoverdines is conserved among fluorescent pseudomonads, other regulatory factors involved in modulation of siderophore gene expression can differ between *P. aeruginosa* and other fluorescent *Pseudomonas* species.

7.4.5
Strain- and Species-dependent Variability of Pyoverdine Synthesis and Regulation

The diversity of pyoverdine structures and receptors is mirrored by a huge variability of the pyoverdine genomic region. A whole-genome diversity study comparing one environmental and two clinical *P. aeruginosa* strains with the PAO1 type strain recognised the pyoverdine region as the most variable alignable locus (Spencer et al. 2003), coherent with the pyoverdine type diversity (Meyer, this volume). Comparing the pyoverdine locus from nine *P. aeruginosa* strains producing different types of pyoverdine, the highest divergence was observed in the genomic region encompassing the *pvdE*, *fpvA*, *pvdD*, *pvdI* and *pvdJ* genes (Smith et al. 2005). Three divergent sequence types were identified, corresponding to the three pyoverdine structures. In each case, NRPSs for the three pyoverdine types were co-linear and consistent with the variable peptide sequence. Moreover, strong evidence of co-evolution between the *fpvA* and *pvd* genes was observed, indicating that the receptor may have driven diversity at the *pvd* locus. Emergence of receptor variants is likely to occur in natural populations exposed to bacteriocins and phages (Baysse et al. 1999), and these must be compensated by reorganisation of biosynthesis genes through re-arrangement of NRPS modules. Most divergence arose from recombination between types leading to generation of novel pyoverdine structures (Smith et al. 2005). Such an evolutionary dynamic would be advantageous to fluorescent pseudomonads in natural communities, as it would avoid siderophore parasitism by competing microorganisms (Tummler and Cornelis 2005). Conversely, comparison of *pvd* clusters from different *P. aeruginosa* strains revealed a significant conservation of pyoverdine signalling genes *pvdS*, *fpvI* and *fpvR* (Smith et al. 2005), suggesting a positive selection for this regulatory network in *P. aeruginosa*.

Pyoverdine biosynthesis genes have also been identified in the genomes of other fluorescent pseudomonads, both experimentally (Adams et al. 1994; Devescovi et al. 2001; Mossialos et al. 2002; Lamont and Martin 2003; Putignani et al. 2004) and through genomic analysis (Lamont and Martin 2003; Ravel and Cornelis 2003). The available information indicates that the pathway of pyoverdine synthesis will be generally the same as in *P. aeruginosa*, with substrate amino acids being assembled into a precursor peptide by NRPSs and subsequent modifications resulting in formation of pyoverdine. Differences between species are likely to reflect differences in the pyoverdines that are made.

Siderophore-dependent induction of gene expression has been documented in *P. fluorescens* M114, *Pseudomonas* strain B10 and in *P. putida* WCS358 (Callanan et al. 1996; Ambrosi et al. 2002; Venturi et al. 1995a, b), raising the possibility that similar signalling pathways exist in other fluorescent pseudomonads. However, analysis of the *P. putida* KT2440, *P. syringae* DC3000 and *P. fluorescens* Pf0-1 *pvd* locus for the presence of putative PvdS, FpvR and FpvI homologues revealed significant variability. Although the *pvdS* homologue is present in a conserved genetic context for all four genomes analysed, an *fpvR*-like gene is missing in species other than *P. aeruginosa*. In fact, the gene showing highest homology with *P. aeruginosa fpvR* maps far away from the *pvd* locus of *P. putida* (PT3555) and *P. syringae* (PSPTO2358), in a genomic region apparently unrelated to iron uptake (www.tigr.org). It has been proposed that homologues of the *P. syringe* phosphorelay sensor SyrP could compensate for the absence of FpvR to drive positive control of pyoverdine synthesis in such species, and a *syrP* homologue has indeed been identified in the pyoverdine region of *P. pu-tida*, *P. syringae* and *P. fluorescens* (Ravel and Cornelis 2003). An *fpvI* homologue is present in *P. putida* KT2440 and *P. fluorescens* Pf0-1 but apparently absent in *P. syringae* DC3000, suggestive of a different mechanism of regulation of pyover-dine uptake in this species. Intriguingly, the presence of an IS box upstream of the tandemly-arranged *fpvA*-like receptor genes of *P. syringae* (Ravel and Cor-nelis 2003) suggests a role for the *pvdS* homologue (PSPTO2133) in biosynthe-sis-coupled control of pyoverdine receptor expression in this species. Given the heterogeneity of the pyoverdine genomic region in fluorescent pseudomonads, functional studies are needed to clarify the molecular partners involved in au-togenous control of pyoverdine synthesis in species other than *P. aeruginosa*.

Putative AlgQ homologues were also identified in several Pseudomonada-ceae (Ambrosi et al. 2005). The whole *algQ* genomic locus appeared remarkably conserved in *P. aeruginosa* PAO1, *P. putida* KT2440, *P. syringae* DC3000 and *P. fluorescens* PfO-1, suggesting that AlgQ homologues may play similar regula-tory functions in Pseudomonadaceae. However, the different impact of the *algQ* (*pfrA*) mutation on pyoverdine biosynthesis in different fluorescent species de-serves further investigation.

7.5
Conclusions

The synthesis of pyoverdines, and its regulation, has been a subject of research for many years. A combination of genetic and biochemical approaches has led to a good, though still incomplete, understanding of these processes in *P. aeru-ginosa*. For example, further research will be needed to determine the complete biosynthetic pathway and the roles of genes with as-yet-unassigned function. Similarly, the molecular interactions that control expression of pyoverdine syn-thesis genes are not yet fully understood. Other species have been less inten-sively studied. The data so far available show that in other fluorescent pseudo-monads there are clear parallels but also significant differences in pyoverdine synthesis and its regulation when compared to *P. aeruginosa*. A major challenge will be to relate differences in biosynthesis and regulation seen in the labora-tory to the processes that occur during growth and survival of these bacteria in complex soil ecosystems.

References

Ackerley DF, Caradoc-Davies TT, Lamont IL (2003) Substrate specificity of the nonribosomal peptide synthetase PvdD from *Pseudomonas aeruginosa*. J Bacteriol 185:2848–2855

Adams C., Dowling DN, O'Sullivan DJ, O'Gara F (1994) Isolation of a gene (*psbC*) required for siderophore biosynthesis in fluorescent *Pseudomonas* sp. strain M114. Mol Gen Genet 243:515–524

Ambrosi C, Leoni L, Putignani L, Orsi N, Visca P (2000) Pseudobactin biogenesis in the plant growth-promoting rhizobacterium *Pseudomonas* B10: identification and functional analysis of the L-ornithine N^5-oxygenase (*psbA*) gene. J Bacteriol 182:6233–6238

Ambrosi C, Leoni L, Visca P (2002) Different responses of pyoverdine genes to autoinduction in *Pseudomonasaeruginosa* and the group *Pseudomonas fluorescens-Pseudomonas putida*. Appl Environ Microbiol 68:4122–4126

Ambrosi C, Tiburzi F, Imperi F, Putignani L, Visca P (2005) Involvement of AlgQ in transcriptional regulation of pyoverdine genes in *Pseudomonas aeruginosa* PAO1. J Bacteriol 187: 5097–5107

Ankenbauer R, Hanne LF, Cox CD. (1986) Mapping of mutations in *Pseudomonas aeruginosa* defective in pyoverdin production. J Bacteriol 167:7–11

Barekzi N, Joshi S, Irwin S, Ontl T, Schweizer HP (2004) Genetic characterization of *pcpS*, encoding the multifunctional phosphopantetheinyl transferase of *Pseudomonas aeruginosa*. Microbiology 150:795–803

Baysse C, Budzikiewicz H, Uria Fernandez D, Cornelis P (2002) Impaired maturation of the siderophore pyoverdine chromophore in Pseudomonas fluorescens ATCC 17400 deficient for the cytochrome c biogenesis protein CcmC. FEBS Lett 523:23–28

Baysse C, Matthijs S, Pattery T, Cornelis P (2001) Impact of mutations in *hemA* and *hemH* genes on pyoverdine production by *Pseudomonas fluorescens* ATCC 17400. FEMS Microbiol Lett 205:57–63

Baysse C, Meyer J-M, Plesiat P, Geoffroy V, Michel-Briand Y, Cornelis P (1999) Uptake of pyocin S3 occurs through the outer membrane ferripyoverdine type II receptor of *Pseudomonas aeruginosa*. J Bacteriol 181:3849–3851.

Beare PA, For RJ, Martin LW, Lamont IL (2003) Siderophore-mediated cell signalling in *Pseudomonas aeruginosa*: divergent pathways regulate virulence factor production and siderophore receptor synthesis. Mol Microbiol 47:195–207

Bockmann M, Taraz K, Budzikiewicz H (1997) Biogenesis of the pyoverdin chromophore. Z Naturforsch C 52:319–324

Briskot G, Taraz K, Budzikiewicz H (1989) Pyoverdin-type siderophores from *Pseudomonas aeruginosa*. Liebigs Ann Chem 37:375–384

Budzikiewicz H (2004) Siderophores of the Pseudomonadaceae *sensu stricto* (fluorescent and non-fluorescent *Pseudomonas* spp.). Fortschr Chem Org Naturst 87:81–237

Callanan M, Sexton, R, Dowling DN, O'Gara F (1996) Regulation of the iron uptake genes in *Pseudomonas fluorescens* M114 by pseudobactin M114: the *pbrA* sigma factor gene does not mediate the siderophore regulatory response. FEMS Microbiol Lett 144: 61–66

Challis GL, Ravel J, Townsend CA (2000) Predictive, structure-based model of amino acid recognition by nonribosomal peptide synthetase adenylation domains. Chem Biol 7:211–224

Cobessi D, Celia H, Folschweiller N, Schalk IJ, Abdallah MA, Pattus F (2005) The crystal structure of the pyoverdine outer membrane receptor FpvA from *Pseudomonas aeruginosa* at 3.6 angstroms resolution. J Mol Biol 347:121–134

Crosa JH, Walsh CT (2002) Genetics and assembly line enzymology of siderophore biosynthesis in bacteria. Microbiol Mol Biol Rev 66:223–249

Cunliffe HE, Merriman TR, Lamont IL (1995) Cloning and characterization of *pvdS*, a gene required for pyoverdine synthesis in *Pseudomonas aeruginosa*: PvdS is probably an alternative sigma factor. J Bacteriol 177:2744–2750

De Vos P, van Landschoot A, Segers P, Tytgat R, Gillis M, Bauwens NS, Rossu R, Goor M, Pot B, Kersters K, Lizzaraga P, De Ley J (1989) Genotypic relationships and taxonomic localization of unclassified *Pseudomonas* and *Pseudomonas*-like strains by deoxyribonucleic acid: ribosomal ribonucleic acid hybridizations. Int J Syst Bacteriol 39: 35–49

Devescovi G, Aguilar C, Majolini MB, Marugg, J, Weisbeek P, Venturi V (2001) A siderophore peptide synthetase gene from plant-growth-promoting *Pseudomonas putida* WCS358. System Appl Microbiol 24:321–330

Dorrestein PC, Poole K, Begley TP (2003) Formation of the chromophore of the pyoverdine siderophores by an oxidative cascade. Org Lett 5:2215–2217

Dove SL, Hochschild A (2001) Bacterial two-hybrid analysis of interactions between region 4 of the σ^{70} subunit of RNA polymerase and the transcriptional regulators Rsd from *Escherichia coli* and AlgQ from *Pseudomonas aeruginosa*. J Bacteriol 183:6413–6421

Du L, Sanchez C, Chen M, Edwards DJ, Shen B (2000) The biosynthetic gene cluster for the antitumor drug bleomycin from *Streptomyces verticillus* ATCC15003 supporting functional interactions between nonribosomal peptide synthetases and a polyketide synthase. Chem Biol 7:623–642

Escolar L, Perez-Martin J, de Lorenzo V (1999) Opening the iron box: transcriptional metalloregulation by the Fur protein. J Bacteriol 181:6223–6229

Finking R, Marahiel MA (2004) Biosynthesis of nonribosomal peptides. Annu Rev Microbiol 58: 453–488

Finking R, Solsbacher J, Konz D, Schobert M, Schafer A, Jahn D, Marahiel MA (2002) Characterization of a new type of phosphopantetheinyl transferase for fatty acid and siderophore synthesis in *Pseudomonas aeruginosa*. J Biol Chem 277:50293–50302

Gardiner DM, Cozijnsen AJ, Wilson LM, Soledade M, Pedras C, Howlett BJ (2004) The sirodesmin biosynthetic gene cluster of the plant pathogenic fungus *Leptosphaeria maculans*. Mol Microbiol 53:1307–1318

Ge L, Seah SY (2006) Heterologous expression, purification, and characterization of an L-ornithine N5-hydroxylase involved in pyoverdine siderophore biosynthesis in *Pseudomonas aeruginosa*. J Bacteriol 188:7205–7210.

Georges C, Meyer J-M (1995) High-molecular-mass, iron-repressed cytoplasmic proteins in fluorescent *Pseudomonas*: potential peptide-synthetases for pyoverdine biosynthesis. FEMS Microbiol Lett 132:9–15

Ghysels B, Min Dieu BT, Beatson SA, Pirnay J-P, Ochsner UA, Vasil ML, Cornelis P (2004) FpvB, an alternative type I ferripyoverdine receptor of *Pseudomonas aeruginosa*. Microbiology 150:1671–1680

Haas D, Défago G (2005) Biological control of soil-borne pathogens by fluorescent pseudomonads. Nat Rev Microbiol 3: 307–319

Heathcote ML, Staunton J, Leadlay PF (2001) Role of type II thioesterases: evidence for removal of short acyl chains produced by aberrant decarboxylation of chain extender units. Chem Biol 8:207–220

Higgins CF (2001) ABC transporters: physiology, structure and mechanism – an overview. Res Microbiol 152:205–210

Hofte M, Mergeay M, Verstraete W (1990) Marking the rhizopseudomonas strain 7NSK2 with a Mu d(lac) element for ecological studies. Appl Environ Microbiol 56:1046–1052

Hohlneicher U, Schafer M, Fuchs R, Budzikiewicz H (2001) Ferribactins as the biosynthetic precursors of the *Pseudomonas* siderophores pyoverdins. Z Naturforsch C 56:308–310

Hohnadel D, Haas D, Meyer J-M (1986) Mapping of mutations affecting pyoverdine production in *Pseudomonas aeruginosa*. FEMS Microbiol Lett 36:195–199

Huang JJ, Han JI, Zhang LH, Leadbetter JR (2003) Utilization of acyl-homoserine lactone quorum signals for growth by a soil pseudomonad and *Pseudomonas aeruginosa* PAO1. Appl Environ Microbiol 69:5941–5949

Inokoshi J, Takeshima H, Ikeda H, Omura S (1992) Cloning and sequencing of the aculeacin A acylase-encoding gene from *Actinoplanes utahensis* and expression in *Streptomyces lividans*.Gene 119:29–35

James HE, Beare PA, Martin LW, Lamont IL (2005) Mutational analysis of a bifunctional ferrisiderophore receptor and signal-transducing protein from *Pseudomonas aeruginosa*. J Bacteriol 187:4514–4520

Kleinkauf H, von Dohren H (1996) A non-ribosomal system of peptide biosynthesis. Eur J Biochem 236:335–351

Kohli RM, Trauger JW, Schwarzer D, Marahiel M, Walsh MA (2001) Generality of peptide cyclization catalyzed by isolated thioesterase domains of nonribosomal peptide synthetase. Biochemistry 40:7099–7108

Kovacevic S, Tobin MB, Miller JR (1990) The beta-lactam biosynthesis genes for isopenicillin N epimerase and deacetoxycephalosporin C synthetase are expressed from a single transcript in *Streptomyces clavuligerus*. J Bacteriol 172:3952–3958

Kratzschmar J, Krause M, Marahiel MA (1989) Gramicidin S biosynthesis operon containing the structural genes *grsA* and *grsB* has an open reading frame encoding a protein homologous to fatty acid thioesterases. J Bacteriol 171:5422–5429

Lamont IL, Beare PA, Ochsner U, Vasil AI, Vasil ML (2002) Siderophore-mediated signaling regulates virulence factor production in *Pseudomonas aeruginosa*. Proc Natl Acad Sci USA 99:7072–7077

Lamont IL, Martin LW (2003) Identification and characterization of novel pyoverdine synthesis genes in *Pseudomonas aeruginosa*. Microbiology 149:833–842

Lautru S, Challis GL (2004) Substrate recognition by nonribosomal peptide synthetase multienzymes. Microbiology 150:1629–1636

Lavrrar JL, McIntosh MA (2003) Architecture of a Fur binding site: a comparative analysis. J Bacteriol 185:2194–2202

Lehoux DE, Sanschagrin F, Levesque RC (2000) Genomics of the 35-kb *pvd* locus and analysis of novel *pvdIJK* genes implicated in pyoverdine biosynthesis in *Pseudomonas aeruginosa*. FEMS Microbiol Lett 190:141–146

Leoni L, Orsi N, de Lorenzo V, Visca P (2000) Functional analysis of PvdS, an iron starvation sigma factor of *Pseudomonas aeruginosa*. J Bacteriol 182:1481–1491

Lyczak JB, Cannon CL, Pier GB (2000) Establishment of *Pseudomonas aeruginosa* infection: lessons from a versatile opportunist. Microbes Infect 2:1051–1060

Loper JE, Henkels MD (1999) Utilization of heterologous siderophores enhances levels of iron available to *Pseudomonas putida* in the rhizosphere. Appl Environ Microbiol 65:5357–5363

McMorran BJ, Kumara HMCS, Sullivan K, Lamont IL (2001) Involvement of a transformylase enzyme in siderophore synthesis in *Pseudomonas aeruginosa*. Microbiology 147:1517–1524

McMorran BJ, Merriman ME, Rombel IT, Lamont IL (1996) Characterisation of the *pvdE* gene which is required for pyoverdine synthesis in *Pseudomonas aeruginosa*. Gene 176: 55–59

Merriman TR, Merriman ME, Lamont IL (1995) Nucleotide sequence of *pvdD*, a pyoverdine biosynthetic gene from *Pseudomonas aeruginosa*: PvdD has similarity to peptide synthetases. J Bacteriol 177:252–258

Meyer J-M, Abdallah MA (1978) The fluorescent pigment of *Pseudomonas fluorescens*: biosynthesis, purification and physicochemical properties. J Gen Microbiol 107:319–328

Meyer J-M, Hornspreger J (1978) Role of pyoverdine$_{Pf}$, the iron binding fluorescent pigment of *Pseudomonas fluorescens* in iron transport. J Gen Microbiol 107:319–328

Meyer J-M, Neely A, Stintzi A, Georges C, Holder IA (1996) Pyoverdine is essential for virulence of *Pseudomonas aeruginosa*. Infect Immun 64:518–523

Miyazaki H, Kato H, Nakazawa T, Tsuda M (1995) A positive regulatory gene, *pvdS*, for expression of pyoverdine biosynthetic genes in *Pseudomonas aeruginosa* PAO. Mol Gen Genet 248:17–24

Mossialos D, Ochsner U, Baysse C, Chablain P, Pirnay JP, Koedam N, Budzikiewicz H, Fernandez DU, Schafer M, Ravel J, Cornelis P (2002) Identification of new, conserved, non-ribosomal peptide synthetases from fluorescent pseudomonads involved in the biosynthesis of the siderophore pyoverdine. Mol Microbiol 45:1673–1685

Mootz HD, Marahiel MA (1997) The tyrocidine biosynthesis operon of *Bacillus brevis*: complete nucleotide sequence and biochemical characterization of functional internal adenylation domains. J Bacteriol 179:6843–6850

Ochsner UA, Vasil ML (1996) Gene repression by the ferric uptake regulator in *Pseudomonas aeruginosa*: cycle selection of iron-regulated genes. Proc Natl Acad Sci USA 93:4409–4414

Ochsner UA, Vasil AI, Vasil ML (1995) Role of the ferric uptake regulator of *Pseudomonas aeruginosa* in the regulation of siderophore and exotoxin A expression: purification and activity on iron-regulated promoters. J Bacteriol 177:7194–7201

Ochsner UA, Wilderman PJ, Vasil AI,Vasil ML (2002) GeneChip(R) expression analysis of the iron starvation response in *Pseudomonas aeruginosa*: identification of novel pyoverdine biosynthesis genes. Mol Microbiol 45:1277–1287

O'Sullivan DJ, O'Gara F (1992) Traits of fluorescent *Pseudomonas* spp. involved in suppression of plant root pathogens. Microbiol Rev 56:662–676

Palma M, Worgall S, Quadri LEM (2003) Transcriptome analysis of the *Pseudomonas aeruginosa* response to iron. Arch Microbiol 180:374–379

Palleroni, NJ (1992) Introduction to the family Pseudomonadaceae. In: A. Balows, H. G. Trüper, M. Dworkin, W. Harder and K-H Schleifer (ed). The Prokaryotes (Second edition), volume III. Springer-Verlag, New York, pp. 3071–3085

Palleroni NJ, Moore ERB (2004) Taxonomy of pseudomonads: experimental approaches. In: J.L. Ramos (ed). *Pseudomonas* vol. I. Kluwer Academic/Plenum Publisher, New York, pp 3–44

Pohl E, Haller JC, Mijovilovich A, Meyer-Klaucke W, Garman E, Vasil ML (2003) Architecture of a protein central to iron homeostasis: crystal structure and spectroscopic analysis of the ferric uptake regulator. Mol Microbiol 47:903–915

Poole K (2001) Multidrug efflux pumps and antimicrobial resistance in *Pseudomonas aeruginosa* and related organisms. J Mol Microbiol Biotechnol 3:255–264

Poole K, Heinrichs D, Neshat S (1993a) Cloning and sequence analysis of an EnvCD homologue in *Pseudomonas aeruginosa*: regulation by iron and possible involvement in the secretion of the siderophore pyoverdine. Mol Microbiol 10:529–544

Poole K, Neshat S, Krebes K, Heinrichs DE (1993b) Cloning and nucleotide sequence analysis of the ferripyoverdine receptor gene *fpvA* of *Pseudomonas aeruginosa*. J Bacteriol 175:4597–4604

Poole K, Zhao Q, Neshat S, Heinrichs DE, Dean CR (1996) The *Pseudomonas aeruginosa tonB* gene encodes a novel TonB protein. Microbiology 142:1449–1458

Pospiech A, Bietenhader J, Schupp T (1996) Two multifunctional peptide synthetases and an O-methyltransferase are involved in the biosynthesis of the DNA-binding antibiotic and antitumour agent saframycin Mx1 from *Myxococcus xanthus*. Microbiology 142:741–746

Putignani L, Ambrosi C, Ascenzi P, Visca P. (2004) Expression of L-ornithine N$^\delta$-oxygenase (PvdA) in fluorescent *Pseudomonas* species: an immunochemical and in silico study. Biochem Biophys Res Commun 313: 245–257

Rached E, Hooper NM, James P, Semenza G, Turner AJ, Mantei N (1990) cDNA cloning and expression in *Xenopus laevis* oocytes of pig renal dipeptidase, a glycosyl-phosphatidylinositol-anchored ectoenzyme. Biochem J 271:755–760

Ravel J, Cornelis P (2003) Genomics of pyoverdine-mediated iron uptake in pseudomonads. Trends Microbiol 11:195–200

Rédly GE, Poole K (2003) Pyoverdine-mediated regulation of FpvA synthesis in *Pseudomonas aeruginosa*: involvement of a probable extracytoplasmic-function sigma factor, FpvI. J Bacteriol 185:1261–1265

Rédly GE, Poole K (2005) FpvIR control of *fpvA* ferric pyoverdine receptor gene expression in *Pseudomonas aeruginosa*: demonstration of an interaction between FpvI and FpvR and identification of mutations in each compromising this interaction. J Bacteriol 187:5648–5657

Reimmann C, Patel HM, Walsh CT, Haas D (2004) PchC thioesterase optimizes nonribosomal biosynthesis of the peptide siderophore pyochelin in *Pseudomonas aeruginosa*. J Bacteriol 186:6367–6373

Roche DM, Byers JT, Smith DS, Glansdorp FG, Spring DR, Welch M (2004) Communications blackout? Do N-acylhomoserine-lactone-degrading enzymes have any role in quorum sensing? Microbiology 150:2023–2028

Rombel IT, McMorran BJ, Lamont IL (1995) Identification of a DNA sequence motif required for expression of iron-regulated genes in pseudomonads. Mol Gen Genet 246:519–528

Saurin W, Hofnung M, Dassa E (1999) Getting in or out: early segregation between importers and exporters in the evolution of ATP-binding cassette (ABC) transporters. J Mol Evol 48:22–41

Schafer H, Taraz K, Budzikiewicz H (1991) Zur genese der amidisch an den chromophor von pyoverdinen gebundenen dicarbonsauren Z Naturforsch 46c:398–406

Schalk IJ, Hennard C, Dugave C, Poole K, Abdallah MA, Pattus F (2001) Iron-free pyoverdin binds to its outer membrane receptor FpvA in *Pseudomonas aeruginosa*: a new mechanism for membrane iron transport. Mol Microbiol 39:351–360

Schalk IJ, Kyslik P, Prome D, van Dorsselaer A, Poole K, Abdallah MA, Pattus F (1999) Co-purification of the FpvA ferric pyoverdin receptor of *Pseudomonas aeruginosa* with its iron-free ligand: implications for siderophore-mediated iron transport. Biochemistry 38:9357–9365

Schalk IJ, Yue WW, Buchanan SK (2004) Recognition of iron-free siderophores by TonB-dependent iron transporters. Mol Microbiol 54:14–22

Schwarzer D, Mootz HD, Linne U, Marahiel MA (2002) Regeneration of misprimed nonribosomal peptide synthetases by type II thioesterases. Proc Natl Acad Sci USA 99:14083–14088

Shen J, Meldrum A, Poole K (2002) FpvA receptor involvement in pyoverdine biosynthesis in *Pseudomonas aeruginosa*. J Bacteriol 184:3268–3275

Smith EE, Sims EH, Spencer DH, Kaul R, Olson MV (2005) Evidence for diversifying selection at the pyoverdine locus of *Pseudomonas aeruginosa*. J Bacteriol 187:2138–2147

Spencer DH, Kas A, Smith EE, Raymond CK, Sims EH, Hastings M, Burns JL, Kaul R, Olson MV (2003) Whole-genome sequence variation among multiple isolates of *Pseudomonas aeruginosa*. J Bacteriol 185:1316–1325

Stachelhaus T, Mootz HD, Marahiel MA (1999) The specificity-conferring code of adenylation domains in nonribosomal peptide synthetases. Chem Biol 6:493–505

Stiefel A, Mahren S, Ochs M, Schindler P, Enz S, Braun V (2001) Control of the ferric citrate transport system of *Escherichia coli*: mutations in region 2.1 of the FecI extracytoplasmic-function sigma factor suppress mutations in the FecR transmembrane regulatory protein. J Bacteriol 183:162–170

Stintzi A, Cornelis P, Hohnadel D, Meyer JM, Dean C, Poole K, Kourambas S, Krishnapillai V (1996) Novel pyoverdine biosynthesis gene(s) of *Pseudomonas aeruginosa* PAO. Microbiology 142:1181–1190

Stintzi A, Johnson Z, Stonehouse M, Ochsner U, Meyer J-M, Vasil ML, Poole K (1999) The *pvc* gene cluster of *Pseudomonas aeruginosa*: role in synthesis of the pyoverdine chromophore and regulation by PtxR and PvdS. J Bacteriol 181:4118–4124

Stover CK, Pham XQ, Erwin AL, Mizoguchi SD, Warrener P, Hickey MJ, Brinkman FSL, Hufnagle WO, Kowalik DJ, Lagrou M, Garber RL, Goltry L, Tolentino E, Westbrock-Wadman S, Yuan Y, Brody LL, Coulter SN, Folger KR, Kas A, Larbig K, Lim R, Smith K,

Spencer D, Wong GK-S, Wu Z, Paulsen IT, Reizer J, Saier MH, Hancock REW, Lory S, Olson MV (2000) Complete genome sequence of *Pseudomonas aeruginosa* PAO1, an opportunistic pathogen. Nature 406:959–964

Takase H, Nitanai H, Hoshino K, Otani T (2000) Impact of siderophore production on *Pseudomonas aeruginosa* infections in immunocompromised mice. Infect Immun 68:1834–1839

Tsuda M, Miyazaki H, Nakazawa T (1995) Genetic and physical mapping of genes involved in pyoverdin production in *Pseudomonas aeruginosa* PAO. J Bacteriol 177:423–431

Tummler B, Cornelis P. (2005) Pyoverdine receptor: a case of positive Darwinian selection in *Pseudomonas aeruginosa*. J Bacteriol 187:3289–3292

Vandenende CS, Vlasschaert M, Seah SY (2004) Functional characterization of an aminotransferase required for pyoverdine siderophore biosynthesis in *Pseudomonas aeruginosa* PAO1. J Bacteriol 186:5596–5602

Vasil ML, Ochsner UA (1999) The response of *Pseudomonas aeruginosa* to iron: genetics, biochemistry and virulence. Mol Microbiol 34:399–413

Venturi V, Ottevanger C, Bracke M, Weisbeek PJ (1995a) Iron regulation of siderophore biosynthesis and transport in *Pseudomonas putida* WCS358: involvement of a transcriptional activator and of the Fur protein. Mol Microbiol 15:1081–1093

Venturi V, Ottevanger C, Leong J, Weisbeek PJ (1993) Identification and characterization of a siderophore regulatory gene (*pfrA*) of *Pseudomonas putida* WCS358: homology to the alginate regulatory gene *algQ* of *Pseudomonas aeruginosa*. Mol Microbiol 10:63–73

Venturi V, Weisbeek P, Koster M (1995b) Gene regulation of siderophore-mediated iron acquisition in *Pseudomonas*: not only the Fur repressor. Mol Microbiol 17:603–610

Visca P (2004) Iron regulation and siderophore signalling in virulence by *Pseudomonas aeruginosa*. In: J.L. Ramos (ed). *Pseudomonas* vol. II p. 69–123. Kluwer Academic/Plenum Publisher, New York

Visca P, Serino L, Orsi N (1992) Isolation and characterization of *Pseudomonas aeruginosa* mutants blocked in the synthesis of pyoverdine. J Bacteriol 174:5727–5731

Visca P, Ciervo A, Orsi N (1994) Cloning and nucleotide sequence of the *pvdA* gene encoding the pyoverdine biosynthetic enzyme L-Ornithine N^5-Oxygenase in *Pseudomonas aeruginosa*. J Bacteriol 176:1128–1140

Visca P, Leoni L, Wilson MJ, Lamont IL (2002) Iron transport and regulation, cell signalling and genomics: lessons from *Escherichia coli* and *Pseudomonas*. Mol Microbiol 45:1177–1190

Weller DM, Raaijmakers JM, Gardener BB, Thomashow LS (2002) Microbial populations responsible for specific soil suppressiveness to plant pathogens. Annu Rev Phytopathol 40:309–348

Wilderman PJ, Sowa NA, FitzGerald DJ, FitzGerald PC, Gottesman S, Ochsner UA, Vasil ML (2004) Identification of tandem duplicate regulatory small RNAs in *Pseudomonas aeruginosa* involved in iron homeostasis. Proc Natl Acad Sci USA 101:9792–9797

Wilson MJ, Lamont IL (2000) Characterization of an ECF sigma factor protein from *Pseudomonas aeruginosa*. Biochem Biophys Res Commun 273:578–583

Zhao Q, Poole K (2000) A second *tonB* gene in *Pseudomonas aeruginosa* is linked to the *exbB* and *exbD* genes. FEMS Microbiol Lett 184:127–132

Zgurskaya HI, Nikaido H (2000) Multidrug resistance mechanisms: drug efflux across two membranes. Mol Microbiol 37:219–225

8 Implication of Pyoverdines in the Interactions of Fluorescent Pseudomonads with Soil Microflora and Plant in the Rhizosphere

Philippe Lemanceau, Agnès Robin,
Sylvie Mazurier, Gérard Vansuyt

8.1
Introduction

Soils are known to be oligotrophic environments whereas soil microflora is mostly heterotrophic in such way that microbial growth in soil is mainly limited by the scarce sources of readily available organic compounds (Wardle 1992). Therefore, in soils, microflora is mostly in stasis (fungistasis/bacteriostasis) (Lockwood 1977). In counterpart, plants are autotrophic organisms responsible for the primary production resulting from the photosynthesis. A significant part of photosynthetates are released from plant roots to the soil through a process called rhizodeposition. These products, i.e. the rhizodeposits, are made of exudates, lysates, mucilage, secretions and dead cell material, as well as gases including respiratory CO_2 and ethylene. Depending on plant species, age and environmental conditions, rhizodeposits can account for up to 40% of net fixed carbon (Lynch and Whipps 1990). On average, 17% of net fixed carbon appears to be released by the roots (Nguyen 2003).

This significant release of organic compounds by plant roots in soil oligotrophic environments is then expected to affect strongly the heterotrophic microflora located closely to the roots. Indeed, one century ago, Hiltner (1904) observed an increased proliferation of heterotrophic bacteria in contact with the roots. This author proposed to call rhizosphere the volume of soil surrounding roots in which the microflora is influenced by these roots. Since then, further studies have shown that living roots modify the biological and physicochemical properties of rhizospheric soil determining the rhizosphere effect (Curl and Truelove 1986; Lemanceau and Heulin 1998; Lynch 1990; Rovira 1965).

Rhizodeposition affects the soil microflora and especially leads to (i) an increase of its density (Clark 1949; Rovira 1965), biomass (Barber and Lynch 1977)

INRA-CMSE, UMR 1229 INRA/Université de Bourgogne 'Microbiologie du Sol et de l'Environnement', 17 rue Sully, 21034 Dijon cedex, France, email: lemanceau@dijon.inra.fr

Soil Biology, Volume 12
Microbial Siderophores
A. Varma, S.B. Chincholkar (Eds.)
© Springer-Verlag Berlin Heidelberg 2007

and activity (Soderberg and Bååth 1998) and to (ii) variations of the structure and diversity of their populations (Edel et al. 1997; Lemanceau et al. 1995; Mavingui et al. 1992). In the rhizosphere, trophic interactions and communications lead indeed to (i) selection of the most adapted microbial groups and populations, to (ii) variations of their physiology, and then to (iii) shift in the structure, diversity and activity of the microbial community compared to that of the bulk soil. Some microbial groups and populations appear to be preferentially adapted and favored in the rhizosphere. Among them, fluorescent pseudomonads were shown to have a higher density and activity in the rhizosphere than in bulk soil and are considered as rhizobacteria.

Besides impacting soil borne microflora, plant roots also modify physicochemical properties of soil in the rhizosphere by changing ionic concentrations, pH, redox potential and by complexing metals including iron (Hinsinger 1998; Hinsinger et al. 2003).

Energetic metabolism of heterotrophic microorganisms is based on electron donors (organic compounds) and electron acceptors (ferric iron, oxygen and nitrogen oxides). Bacteria are usually able to use a wide range of organic compounds as for example shown for fluorescent pseudomonads (Lemanceau et al. 1995), in such a way that, according to the organic compounds present, different catabolic activities can be induced to take advantages of the nutrients available in the rhizosphere (Cheng et al. 1996). In contrast, the range of possible electron acceptors is limited (Latour and Lemanceau 1997), making the competition for them especially intense. This is the case with the competition for ferric iron. Although being the fourth element of the earth crust, at pH values compatible with plant growth, Fe(III) is mostly precipitated as hydroxides, iron is also associated with phosphorus and colloids and therefore the concentration of Fe(III) available for living organisms in cultivated soils is usually low (Lindsay 1979). In the rhizosphere, this concentration is even lower due to the uptake of this ion both by the roots and the microflora. The combined low concentration of Fe(III) in solution (offer) together with the requirements of aerobic organisms (plant and microorganisms) (demand) lead to a specially high level of competition of Fe(III) in the rhizosphere (Guerinot and Yi 1994; Loper and Buyer 1991). Since Fe(III) is an essential element for most aerobic microorganisms, this ion is often a limiting factor for microbial growth and activity in soil and rhizosphere (Loper and Buyer 1991).

Most aerobic organisms have developed an active strategy for iron uptake. For the microorganisms, this strategy is based on siderophores and on ferrisiderophore membrane receptors (Neilands 1981). As an example, in iron stress conditions, fluorescent pseudomonads synthesize siderophores, called pyoverdines, showing a high affinity for Fe(III) (Meyer and Abdallah 1978) and membrane receptors that are usually specific (Hohnadel and Meyer 1988). Pyoverdines are chromopetides composed of a quinoleinic chromophore bound together with a peptide and an acyl chain (Meyer et al. 1987). Synthesis of pyoverdines and related protein membrane receptors correspond to a significant metabolic effort for bacterial cells which is then only expressed when required through regulation processes. Their syntheses occur in response to cellular iron deficiency resulting from a low iron content of the environment (low offer) but are repressed in non-iron stress conditions (Meyer and Abdallah 1978); similarly pyoverdine synthesis was shown to be regulated by the phenomenon of Quorum Sensing through the production of acyl homoserines lactones (AHLs) when the bacterial density is high and corresponds to a significant demand in iron (Stintzi et al. 1998).

For the plants, two types of strategies have been described for the active iron uptake (Marschner et al. 1986, Marschner and Römheld 1994). The first strategy, employed by dicotyledons and non-graminaceaous monocotyledons plants, involves (i) the excretion of protons (Bienfait 1985; Guerinot and Yi 1994), (ii) the reduction of Fe(III) by reductases to the more soluble Fe(II) form (Robinson et al. 1999; Ying and Guerinot 1996) and (iii) plasmalemma transport of Fe(II) by iron transporters (Curie et al. 2000; Eide et al. 1996; Vert et al. 2001). The second strategy, employed by grasses, involves (i) the synthesis of phytosiderophores forming a complex with Fe(III) (Von Wirén et al. 2000) and (ii) the incorporation in the roots by a transporter specific for the Fe(III)-phytosiderophore complex (Curie et al. 2001). For more details concerning iron plant assimilation and trafficking, excellent reviews were recently published (Curie and Briat 2003; Hell and Stephan 2003; Schmidt 2003).

In the rhizosphere, there is both (i) a strong competition for iron resulting from the increased microbial biomass, activity and then uptake and from the plant uptake and (ii) a high density of fluorescent pseudomonads leading to a significant frequency of populations producing AHLs (Elasri et al. 2001). Rhizosphere conditions appear to be favorable to pyoverdine synthesis. This synthesis was indeed shown to occur in the rhizosphere by the use of monoclonal antibodies against ferri-pyoverdine (Buyer et al. 1990) and by the use of an ice-nucleation reporter gene *inaZ* under the control of an iron-regulated promoter involved in pyoverdine synthesis (Duijff et al. 1999; Loper and Lindow 1994). Use of *P. fluorescens* Pf-5 containing *pvd-inaZ* further allowed Loper and Henkels (1999) to show the low biological availability of iron in soil and rhizosphere.

Taking into account the strong iron competition in the rhizosphere and the high affinity of pyoverdines for iron, these molecules are expected to interact with the iron nutrition of other organisms in soil: microbes, including plant pathogens, and plants. Indeed, since the 1980s many studies have been dedicated to these interactions. These studies were stimulated by the potential of some fluorescent pseudomonads to suppress soil borne diseases and to promote plant growth (Cook et al. 1995; Haas and Défago 2005; Lemanceau 1992; O'Sullivan and O'Gara 1992; Weller 1988). The aims of the present chapter are (i) to present the state of the art on these interactions in the rhizosphere in relation with the possible application of fluorescent pseudomonads for promoting plant health and growth, and (ii) to propose some prospects for further research.

8.2
Contribution of Pyoverdines to Microbial Interactions in the Rhizosphere

The combined low biological availability and high demand of ferric iron in the rhizosphere determine a strong competition for this ion between microorganisms (Loper and Buyer 1991). In this competition context, microorganisms with the most efficient siderophore-mediated iron uptake are expected to be favoured (competitive advantage), whereas the further iron depletion determined by the efficient iron-competitors is expected to reduce saprophytic growth of microorganisms with a less efficient iron-uptake system through iron starvation (microbial antagonism). This assumption has been supported and illustrated by several studies, some of them being summarized here below.

8.2.1
Competitive Advantage Given by Pyoverdine-Mediated Iron Uptake

Various studies have been conducted at the population level and at the strain level to assess the possible competitive advantage given by pyoverdine-mediated iron uptake to fluorescent pseudomonads.

Population studies carried in our group consisted in comparing the susceptibility to iron stress of fluorescent pseudomonads isolated from flax rhizosphere in two different soils (Carquefou and Châteaurenard, France) (Lemanceau et al. 1988b). The strategy followed, consisted in determining, for each strain, the Minimal Inhibitory Concentration (MIC) of 8-hydroxyquinoline (8HQ), a strong iron chelator (Geels et al. 1985); the higher the MIC, the stronger the ability of the strain to grow in iron stress conditions and the lower its susceptibility to iron starvation. The distributions of the soil and rhizosphere strains in the different MIC classes were significantly different, the rhizosphere strains being more represented in the classes with high MIC values than the soil strains. Recently, we have further shown that there is a gradient between the bulk soil, rhizosphere soil, rhizoplane and root tissues, the MIC values of pseudomonad populations increasing when isolated closer to the roots (Robin et al., in press). Altogether, these data indicate that populations of fluorescent pseudomonads from rhizosphere are less susceptible to iron starvation than those from bulk soil suggesting that (i) they benefit from a more efficient iron-uptake than those from bulk soil, and that (ii) this efficient iron-uptake might be implicated in their rhizosphere competence (Lemanceau et al. 1988b).

Since pyoverdine is the major class of siderophores produced by fluorescent pseudomonads, the above hypothesis was tested by evaluating the possible implication of pyoverdine in the rhizosphere competence of a model strain (*P. fluorescens* C7R12) using a pyoverdine minus mutant (pvd–). The strain C7R12 was chosen since it was previously shown to be competitive in the rhizosphere (Eparvier et al. 1991) and to be a biocontrol agent (Lemanceau and Alabouvette 1991). The survival kinetics of the wild-type strain and of the pvd– mutant in the rhizosphere were compared in competition, in the absence (gnotobiotic conditions) and in the presence of the indigenous microflora (Mirleau et al. 2000, 2001). In the absence of the indigenous microflora, bacterial competition was favourable to the pyoverdine producer wild-type, whereas in non-gnotobiotic conditions the survival of both strains was similar. In the latter conditions, the fitness of the pvd– mutant in competition with the wild-type strain was assumed to be related to its ability to take-up, as the wild-type strain, pyoverdines of foreign origin (Mirleau et al. 2000). Altogether, these results suggest that pyoverdine-mediated iron uptake is involved in the rhizosphere competence of *P. fluorescens* C7R12.

The possible ecological advantage conferred to bacteria by their ability to utilize a large variety of siderophores was previously proposed by Jurkevitch et al. (1992). More specifically, the assumed importance of the ability to incorporate foreign siderophores in the rhizosphere competence of a model strain of fluorescent pseudomonads (*P. aeruginosa* 7NSK2) was suggested by Höfte et al. (1992) by showing a reduced rhizosphere competence of a mutant impaired in the synthesis of membrane protein receptors. The role of siderophore receptor in rhizosphere competence was further supported by experiments performed with a mutant of *P. fluorescens* WCS374 that harboured siderophore receptor PupA from *P. putida* WCS358 and was then able to utilize ferri-pyoverdine from WCS358. This ability was shown to give the WCS374 mutant a competitive advantage over the corresponding wild-type strain when co-inoculated with WCS358 (Raaijmakers et al. 1995). Following a similar approach with a different strain, Moënne-Loccoz et al. (1996) came to a different conclusion, but a heterologous membrane receptor harboured in the mutant was shown to incorporate a pyoverdine which was widely produced by indigenous fluorescent pseudomonads and then expected to be incorporated by a high proportion of these populations, reducing then the competitive advantage given to the mutant over the wild-type. However, this report points out possible differences in traits accounting for rhizosphere competence according to bacterial strains and environmental conditions as previously stressed (Latour et al. 2003).

This statement has stimulated us to go back to a population approach to identify traits that would be shared by rhizosphere competent populations. The adaptation to the rhizosphere of 21 strains, representative of the diversity of a larger collection (340) previously characterized (Latour et al. 1996; Lemanceau et al. 1995) plus two reference strains, was evaluated by measuring their survival rate in the rhizosphere of tomato grown in soil in non-gnotobiotic conditions. Siderophore typing of the strains was performed according to Meyer et al. (2002) by combining (i) iso-eletrophoretic patterns of pyoverdine iso-forms determined by iso-electrofocusing and (ii) pyoverdine-mediated iron uptake of each strain with ^{59}Fe-pyoverdine complexes. The survival rate of the strains varied from 61.4% for the most competitive down to 0.11% for the less competitive, indicating clearly the high diversity of the strains tested. Nine different siderotypes were identified but only one of them was shared by eight of the nine most competitive in the rhizosphere, suggesting again the importance of the pyoverdine-mediated iron uptake in the rhizosphere competence (Delorme et al., unpublished data).

8.2.2
Implication of Pyoverdines in Microbial Antagonism

Studies on possible involvement of pyoverdines in suppression of soil borne diseases were stimulated by the early report of Kloepper et al. (1980a) indicating that addition of *Pseudomonas* sp. B10 or its pyoverdine to soils conducive to fusarium wilts and to take-all rendered them suppressive. They further showed that supplementation of the soils with iron overcame the positive effects of the bacterial inoculation and pyoverdine addition.

Possible involvement of pyoverdine in microbial antagonism was then supported by a series of observations: (i) in vitro antagonism by some fluorescent pseudomonads against specific pathogens only occurring when pyoverdines were produced (Kloepper et al. 1980a; Misaghi et al. 1982), (ii) positive correlation between the intensity of siderophore synthesis in vitro by different pseudomonads isolates and their ability to reduce chlamydospore germination of pathogenic *F. oxysporum* in soil (Elad and Baker 1985b; Sneh et al. 1984), (iii) in vitro antagonistic activity of purified pyoverdine against *Pythium* (Meyer et al. 1987) and *F. oxysporum* (Lemanceau et al. 1992), and (iv) reduced chlamydospore germination of pathogenic *F. oxysporum* in soil upon pyoverdine introduction (Elad and Baker 1985a).

The demonstration of the involvement of pyoverdines in the antagonism achieved by some fluorescent pseudomonads against plant pathogens (*F. oxysporum*, *Pythium* spp.) and so-called deleterious microorganisms (fluorescent pseudomonads, *Pythium* spp.) was fulfilled by the use of mutants impaired in their ability to synthesize such siderophore (Bakker et al. 1986, 1987; Becker and Cook 1988; Buysens et al. 1996; De Boer et al. 2003; Duijff et al. 1993; Leeman et al. 1996; Loper 1988; Raaijmakers et al. 1995). Antagonistic activity against plant pathogens lead to an improvement of plant health (Loper and Buyer 1991), and that against deleterious microorganisms to an enhanced plant growth (Becker and Cook 1988; Kloepper et al. 1980b; Schippers et al. 1987). Examples of these pyoverdine-mediated antagonistic activities and consequences to plant health and growth are listed in Table 8.1. As an example of the strategy based on the use of pvd− mutant, Lemanceau et al. (1992) showed that *P. putida* WCS358, but not its pvd− mutant, was able to improve the control of fusarium wilt determined by non-pathogenic *F. oxysporum* Fo47. Furthermore, iron availability to *P. putida* WCS358 was shown to be low enough in the rhizosphere to allow pyoverdine production as assessed by ice-nucleation conferred from a transcriptional fusion (*pvd-inaZ*) of an ice-nucleation reporter gene to an iron-regulated promoter. The pyoverdine-mediated improvement of the control by *P. putida* WCS358 was related to a reduced saprophytic density and activity of the pathogenic *F. oxysporum* as assessed by β-glucoronidase reporter gene in *gusA*-marked derivative of the pathogen (Duijf et al. 1999).

Implication of iron competition in the pyoverdine-mediated antagonism was first suggested by using EDDHA to mimic the effect of bacterial siderophore on iron availability (Scher and Baker 1982; Elad and Baker 1985b). FeEDDHA shows a stability constant ($K = 10^{33.9}$) (Lindsay 1979) close to that of ferri-pyoverdines ($K = 10^{32}$) (Meyer and Abdallah 1978) and significantly higher than that of siderophores of *Fusarium*, called fusarinines (Emery 1965; Lemanceau et al. 1986), chelated with iron ($K = 10^{29}$) (Scher and Baker 1982). As expected by the comparison of the K values, the synthetic ligand decreased the iron availability to *F. oxysporum* as suggested by the reduced germination of chlamydospores and germ tube length in vitro and in soil, whereas FeEDDHA did not. Lemanceau et al. (1993) further showed that the antagonistic activity of increasing concentrations of purified pyoverdine from *P. putida* WCS358 was related to iron competition, since the same concentrations of Fe-pyoverdine did not have any deleterious effect; the highest concentrations had even a promoting effect probably due to the dissociation of the chelate leading to an iron release. These data are in agreement with the proposal made by Loper and Buyer (1991) that pyoverdine may make Fe(III) unavailable to the target pathogens.

The importance of iron competition in microbial antagonism was further supported by studies on natural soil suppressiveness to fusarium wilts which was ascribed at least partly to fluorescent pseudomonads (Lemanceau et al. 1988b; Scher and Baker 1980). In these soils, introduction of FeEDTA, increasing the availability of Fe(III) to pathogenic *F. oxysporum*, resulted in a lower iron competition and consequently higher fusarium wilt severity, whereas introduction of FeEDDHA, lowering the availability of Fe(III) to pathogenic *F. oxysporum* resulted in a stronger iron competition and consequently lower fusarium wilt severity (Lemanceau et al. 1988a; Scher and Baker 1982).

Naturally suppressive soils to fusarium wilts are known to have physico-chemical properties (high pH and $CaCO_3$ content) contributing to a very low solubility of ferric iron, accounting for the strong iron competition in these soils (Alabouvette et al. 1996). Simeoni et al. (1987) reported that the critical level of Fe(III) concentration in nutrient solution, below which chlamydospore germination was suppressed, was between 10^{-19} and 10^{-22} M, and that optimal suppression took place between Fe(III) concentrations of 10^{-22} and 10^{-27} M. Duijff et al. (1994a) further showed that the pyoverdine producing strain *P. putida* WCS358 only decreased fusarium wilt incidence when the iron availability of the nutrient solution was reduced by EDDHA supplementation. On top of their low Fe(III) solubility, suppressive soils are characterized by their high biomass contributing to a high demand for this ion and then enhancing the intensity of iron competition. This may explain why the suppression of fusarium wilt based on pyoverdine-mediated iron competition was reported to be mostly effective when the microbial biomass is high (Duijff et al. 1991). Another hypothesis for this observation could be related to the interaction between carbon and iron competition as described in the suppressive soil of Châteaurenard (Lemanceau 1989) and in the synergistic interaction between *P. putida* WCS358 and non-pathogenic *F. oxysporum* Fo47 (Lemanceau et al. 1992, 1993). Experiments with combined supplementation of the Châteaurenard suppressive soil with glucose and EDDHA clearly pointed out the interaction between competition for carbon and for iron (Lemanceau 1989). This type of interaction was further elucidated when studying the modes of action of the protection by combined inoculation of WCS358+Fo47. The efficacy of this protection was ascribed to a synergistic effect between (i) carbon competition by Fo47, reducing the amount of carbohydrates available for the pathogen and (ii) pyoverdine-mediated iron competition by WCS358, reducing the efficacy of the energetic metabolism of the pathogen and making it more susceptible to carbon competition (Lemanceau et al. 1993).

Pyoverdines do not have a universal role in biological control activity of *Pseudomonas* spp. (Gutterson 1990; Loper and Buyer 1991). The role of pyoverdine in microbial antagonism and disease suppression varies according to the target pathogens and to the bacterial strains (Table 8.1). Variations according to the target pathogen may be related to the endogenous iron content of the pathogenic fungal spores. It has been suggested that the lower susceptibility of *F. solani* to iron-mediated antagonism compared to that of *F. oxysporum* could be ascribed to the higher iron content of *F. solani* chlamydospores making them less susceptible to iron deprivation (Baker et al. 1986). Similarly, *Thielavopsis basicola* was shown to be not susceptible to pyoverdine-mediated iron competition. A pvd− mutant of *P. fluorescens* CHA0 was reported to suppress black root-rot as efficiently as the wild-type (Keel et al. 1989). Even more, high concentrations of ferri-siderophore appeared to have an antagonistic effect (Ahl et al. 1986) and iron sufficiency was described as being a pre-requisite for disease suppression by CHA0 (Keel et al. 1989). For other target pathogens, contribution of pyoverdine-mediated iron competition seems to play a minor role compared to other bacterial traits. This is the case of *Gaeummanomyces graminis* var. *tritici* responsible for take-all. Although early works had suggested or indicated the involvement of this mode of action for some fluorescent pseudomonads (Kloepper et al. 1980a; Weller et al. 1988; Wong and Baker 1984), it was then nicely demonstrated that its relative contribution to the antagonism and disease suppression by *P. fluorescens* 2-79 and M4-80R was minor or even nil compared to those by phenazine antibiotic (Hamdan et al. 1991). Thomashow and Weller (1990) suggested that the effects attributed in the early studies to pyoverdine in the antagonism against *G. graminis* var. *tritici* could be related to an artefact linked to the production of iron-regulated antibiotics, this type of compound being already described by Gill and Warren (1988).

Table 8.1. Examples showing evidence or lack of evidence for pyoverdine-mediated antagonistic activity and consequences on plant health and growth

Consequences on plant health and growth	Pathogen	Bacterial strain	Strategy	Reference
Non-tested				
	Geotrichum candidum *Pythium aphanidermatum*	Collection of *Pseudomonas* spp. A strain of *P. fluorescens*	Crude fluorescent pigment Partially purified fluorescent pigment	Misaghi et al. 1982
	Alternaria solani, Colletrotichum coccodes, Fusarium tabacinum, Phoma exigua, Verticillium albo-atrum, Rhizoctonia solani, Erwinia carotovora, Streptomyces scabies	Collection of *Pseudomonas* spp.	Iron-regulated antagonism	Geels and Schippers 1983a
	Pseudomonas spp. A214, A225, B117, 7SR1 (Deleterious RhizoBacteria, DRB)	Collection of *Pseudomonas* spp. including B10	Purified pyoverdine	Buyer and Leong 1986
	Fusarium oxysporum f. sp. *cucumerinum*	*Pseudomonas* sp. 346	Compared effect of purified pyoverdine with supplementation of iron or not	Elad and Baker 1985a
	F. oxysporum f. sp. *lycopersici*	*P. putida* PPU3	WT/pvd– mutant Purified pyoverdine	Vandenbergh et al. 1983
	Phytophthora nicotianae var. *parasitica*	*P. putida* 06909, *P. fluorescens* 09906	Bacterial inoculation/ iron supplementation WT/pvd– mutant	Yang et al. 1994
	Pythium ultimum	*P. tolaasi* 2068	Purified pyoverdine	Meyer et al. 1987
Disease suppression				
Charcoal rot in peanut	*Macrophomina phaseolina*	*Pseudomonas* sp. GRC$_2$	Correlation between in vitro production of siderophore and disease suppression by bacteria	Gupta et al. 2002

Table 8.1. *(continued)* Examples showing evidence or lack of evidence for pyoverdine-mediated antagonistic activity and consequences on plant health and growth

Consequences on plant health and growth	Pathogen	Bacterial strain	Strategy	Reference
Disease suppression *(continued)*				
Verticillium wilt in olive	*Verticillium dahliae*	Collection of *P. fluorescens* and *P. putida*	Correlation between in vitro production of siderophore and disease suppression by bacteria	Mercado-Blanco et al. 2004
Fusarium wilt in cucumber, flax and radish	*F. oxysporum* f. sp. *cucumerinum*, f. sp. *lini*, f. sp. *conglutinans*	*P. putida* A12	Compared effect of bacteria and EDDHA on disease severity	Scher and Baker 1982
Fusarium wilt in cucumber	*F. oxysporum* f. sp. *cucumerinum*	Collection of *Pseudomonas* spp.	Correlation between in vitro production of siderophore and disease suppression by bacteria	Sneh et al. 1984
Fusarium wilt in cucumber, radish and pea	*F. oxysporum* f. sp. *cucumerinum*, f. sp. *conglutinans*, f. sp. *pisi*	*Pseudomonas* sp. 346	Correlation between in vitro production of siderophore and disease suppression by bacteria	Elad and Baker 1985b
Fusarium wilt in carnation	*F. oxysporum* f. sp. *dianthi*	*P. fluorescens* WCS417	Purified pyoverdine	Van Peer et al. 1990
Fusarium wilt in carnation	*F. oxysporum* f. sp. *dianthi*	*P. putida* WCS358	WT/pvd− mutant	Lemanceau et al. 1992
Fusarium wilt in carnation	*F. oxysporum* f. sp. *dianthi*	*P. putida* WCS358	Purified pyoverdine	Lemanceau et al. 1993
Fusarium wilt in carnation	*F. oxysporum* f. sp. *dianthi*	*P. putida* WCS358, WCS417	WT/pvd− mutant	Duijff et al. 1993
Fusarium wilt in carnation	*Fusarium oxysporum* f. sp. *dianthi*	*P. putida* WCS358	WT/pvd− mutant	Duijff et al. 1994a
Fusarium wilt in flax	*F. oxysporum* f. sp. *lini*	*P. putida* WCS358	WT/pvd− mutant	Duiff et al. 1999
Fusarium wilt in radish	*F. oxysporum* f. sp. *raphani*	*P. putida* WCS358	WT/pvd− mutant, positive effect differing according to the experiment	De Boer et al. 2003

Table 8.1. (*continued*) Examples showing evidence or lack of evidence for pyoverdine-mediated antagonistic activity and consequences on plant health and growth

Consequences on plant health and growth	Pathogen	Bacterial strain	Strategy	Reference
Disease suppression (*continued*)				
Fusarium wilt in flax Take-all in barley	*F. oxysporum* f. sp. *lini Gaeumannomyces graminis* var. *tritici*	*Pseudomonas* sp. B10	Purified pyoverdine Bacterial inoculation	Kloepper et al. 1980a
Take-all in wheat	*G. graminis* var. *tritici*	*P. fluorescens* 2-79RN10, R1a-80R, R7z-80R	WT/pvd– mutant Supplementation of soil with FeEDTA	Weller et al. 1988
Damping-off in cotton	*P. ultimum*	*P. fluorescens* 3551	WT/pvd– mutant	Loper 1988
Damping-off in bean	*F. oxysporum, F. solani, P. ultimum, R. solani, Sclerotinia sclerotiorum*	Collection of *Pseudomonas* spp.	Correlation between microbial antagonism in iron stress conditions and disease suppression by bacteria	Lemanceau and Samson 1983
Damping-off in tomato	*Pythium*	*P. aeruginosa* 7NSK2	WT/pvd– mutant	Buysens et al. 1996
Grey mould in eucalyptus	*Botrytis cinerea*	*P. fluorescens* WCS374, WCS417 *P. putida* WCS358	WT/pvd– mutant	Ran et al. 2005a
Bacterial wilt in eucalyptus	*Ralstonia solanacaearum*	*P. putida* WCS358	WT/pvd– mutant	Ran et al. 2005b

Table 8.1. *(continued)* Examples showing evidence or lack of evidence for pyoverdine-mediated antagonistic activity and consequences on plant health and growth

Consequences on plant health and growth	Pathogen	Bacterial strain	Strategy	Reference
Plant growth promotion				
Bean	*F. oxysporum, F. solani, P. ultimum, R. solani, S. sclerotiorum*	Collection of *Pseudomonas* spp.	Correlation between microbial antagonism in iron stress conditions and plant growth promotion	Lemanceau and Samson 1983
Potato	Reduced densities of fungi in the rhizosphere	*Pseudomonas* sp. B10	Purified pyoverdine Bacterial inoculation WT/pvd– mutant	Kloepper et al. 1980b
Potato	DRB	Collection of *Pseudomonas* spp.	Correlation between production of pyoverdine in vitro and plant growth promotion	Geels and Schippers 1983b
Potato	DRB	*P. putida* WCS358 *P. putida* WCS358, *P. fluorescens* WCS374	WT/pvd– mutant	Bakker et al. 1986 Bakker et al. 1987
Wheat	*P. ultimum* var. *sporangiiferum*	*Pseudomonas* sp. B324R	WT/pvd– mutant	Becker and Cook 1988
Cucumber, maize, spinach	DRB	*P. aeruginosa* 7NSK2	WT/pvd– mutant	Höfte et al. 1991

Table 8.1. *(continued)* Examples showing evidence or lack of evidence for pyoverdine-mediated antagonistic activity and consequences on plant health and growth

Consequences on plant health and growth	Pathogen	Bacterial strain	Strategy	Reference
Lack of evidence for pyoverdine effect				
Fusarium wilt in bean	*F. oxysporum* f. sp. *phaseoli*	*Pseudomonas* sp. 346	Correlation between production of pyoverdine in vitro and disease suppression by bacteria	Elad and Baker 1985b
Fusarium wilt in radish	*F. oxysporum* f. sp. *raphani*	*P. putida* WCS358 *P. putida* RE8	WT/pvd– mutant	De Boer et al. 2003
Inhibition of chlamydo-spores and mycelial growth	*Thielaviopsis basicola*	*P. fluorescens* CHA0	Purified pyoverdine	Ahl et al. 1986
Damping-off in cucumber	*P. ultimum*	*P. putida* N1R	WT/pvd– mutant	Paulitz and Loper 1991
Damping-off in tomato	*Pythium sporangia*	*P aeruginosa* 7NSK2	WT/pvd– mutant	Heungens et al. 1992
Damping-off in cucumber	*P. ultimum*	*P. fluorescens* Pf-5	WT/pvd– mutant	Kraus and Loper 1992
Root rot in cucumber	*Pythium aphanidermatum*	*P putida* BTP1	WT/pvd– mutant	Ongena et al. 1999
Take-all in wheat	*G. graminis* var. *tritici*	*P. fluorescens* 2-79 *P. fluorescens* M4-80R	WT/pvd– mutant	Hamdan et al. 1991
Bacterial wilt in eucalyptus	*Ralstonia solanacaearum*	*P. fluorescens* WCS417	WT/pvd– mutant	Ran et al. 2005a

8.3
Contribution of Pyoverdines to Plant-Microbe Interactions in the Rhizosphere

8.3.1
Induced Systemic Resistance

Induced resistance has been defined by Van Loon et al. (1998) as being a state of enhanced defensive capacity by a plant when appropriately stimulated. Various fluorescent pseudomonads strains have been shown to enhance in a systemic way this capacity (Bakker et al. 2003; Van Loon et al. 1998).

The corresponding demonstration is commonly based on the physical separation of the pathogen and biocontrol agent, disease suppression being then not related to microbial antagonism but to induced systemic resistance. This has been achieved in the so-called split-root system by separating the root system in two independent parts, one side being infested with the pathogen and the other side being inoculated with the biocontrol agent (Leeman et al. 1995a; Van Peer et al. 1991; Van Wees et al. 1997); another experimental set-up consisted in spraying the pathogen on the leaves while inoculating the roots with the biocontrol agent (Maurhofer et al. 1994; Pieterse et al. 1996).

Using the latter, *P. fluorescens* CHA0 was shown to induce systemically resistance of tobacco against TNV, whereas its pvd– mutant was les efficient than the wild-type, suggesting that pyoverdine was playing a role in the ISR by CHA0 (Maurhofer et al. 1994). However, with another pathosystem (*Arabidopsis thaliana/Peronospora parasitica*), this pvd– mutant protected the host-plant as efficiently as the wild-type strain (Iavicoli et al. 2003). Applying the split-root set-up, Duijff et al. (1993) and Leeman et al. (1996) have also reported that, in *P. fluorescens* WCS417, pyoverdine does not seem to be involved in ISR of radish against fusarium wilt since (i) the control by the wild-type strain and a pvd– mutant were as efficient and (ii) no protection was obtained with the purified pyoverdine. The lack of contribution of pyoverdine in ISR by WCS417 is in agreement with the major role given to the O-antigenic side chain of the lipopolysaccharides in that strain (Leeman et al. 1995b).

Further examples showing evidence or lack of evidence for pyoverdine involvement in induced systemic resistance are presented in Table 8.2.

Meziane et al. (2005) have suggested a redundancy of bacterial traits involved in ISR including, amongst others, pyoverdine. Indeed, although the purified pyoverdine of *P. fluorescens* WCS374 induced resistance in radish against fusarium wilt, a pvd– mutant of this strain was able to induce resistance (Leeman et al. 1996). Similarly, purified pyoverdine from *P. putida* WCS358 induced resistance in *Arabidopsis* but a pvd– mutant remained able to induce resistance. However,

Table 8.2. Examples showing evidence or lack of evidence for pyoverdine involvement in induced systemic resistance

Pyoverdine involvement in ISR	Pathogen	Bacterial strain	Strategy	References
Evidence for pyoverdine involvement				
Tobacco necrosis	Tobacco Necrosis Virus	*P. fluorescens* CHA0	WT/pvd– mutant	Maurhofer et al. 1994
Fusarium wilt in radish	*F. oxysporum* f. sp. *raphani*	*P. fluorescens* WCS374	Purified pyoverdine	Leeman et al. 1996
Gummy stem rot in watermelon	*Didymella bryoniae*	*P. fluorescens* WR8-3m *P. putida* WR9-16m	WT/pvd– mutant	YongHoon et al. 2001
Grey mould in tomato Bacterial speck in *Arabidospsis thaliana*	*Botrytis cinerea* *Pseudomonas. syringae* pv. *tomato*	*P. putida* WCS358	WT/pvd– mutant Purified pyoverdine	Meziane et al. 2005
Lack of evidence for pyoverdine involvement				
Fusarium wilt in radish	*F. oxysporum* f. sp. *raphani*	*P. fluorescens* WCS358, WCS417 *P. fluorescens* WCS374, WCS417	Purified pyoverdine WT/pvd– mutant	Leeman et al. 1996
Downy mildew in *Arabidospsis thaliana*	*Peronospora parasitica*	*P. fluorescens* CHA0	WT/pvd– mutant	Iavicoli et al. 2003
Grey mould in bean	*Botrytis cinerea*	*P. putida* BTP1	Supernatant from bacteria grown in iron stress conditions	Ongena et al. 2002
Gey mould in bean Athracnose in bean Bacterial speck *Arabidospsis thaliana*	*B. cinerea* *Colletotrichum lindemutthianum* *P. syringae* pv. *tomato*	*P. putida* WCS358	WT/pvd– mutant	Meziane et al. 2005
Fusarium wilt in carnation	*F. oxysporum* f. sp. *dianthi*	*P. fluorescens* WCS417	WT/pvd– mutant	Duijff et al. 1993
Grey mould in bean	*B. cinerea*	*P. aeruginosa* 7NSK2	WT/pvd– mutant	De Meyer and Höfte 1997

pyoverdine seems to plays a major role in the ISR determined by *P. putida* WCS358 in tomato, since this purified siderophore induced resistance and the pvd– mutant did not (Meziane et al. 2005). Altogether, these data indicate that traits accounting for ISR not only differ according to the bacterial strains but also to the plant species.

Demonstration of the possible involvement of pyoverdine in ISR should prompt us to modulate the initial conclusions made, from the early works on the use of pvd– mutants, on the sole contribution of iron competition in the pyoverdine-mediated disease suppression (see Sect. 8.2.2). Experiments must be performed to assess the relative importance of antagonism and ISR in the disease suppression. Such an approach was followed by Duijff et al. (1998) who have shown that the control of fusarium wilt of tomato by *P. fluorescens* WCS417 was ascribed both to ISR and microbial antagonism by comparing the level of suppressiveness in split (resulting only from ISR) and non split root (resulting both from antagonism and ISR) systems.

More generally, application of experimental set-ups, physically separating pathogens and biocontrol agents, has allowed demonstrating that various iron-regulated metabolites are implicated in the induced systemic resistance (Bakker et al. 2003, Van Loon et al. 1998). This topic is further detailed in the present book (Chap. 13).

8.3.2
Implication of Pyoverdine in Plant Nutrition

Contribution of soil microflora to plant iron nutrition was first shown following global approaches. When cultivated under axenic conditions, strategy I and strategy II plants had a significant reduced Fe and chlorophyll contents compared to plants cultivated in the presence of indigenous microflora (Masalha et al. 2000; Rroço et al. 2003). Chen et al. (1998) further showed that addition of compost, including a complex microflora, to nutrient solution contributed to improve iron nutrition of strategy I (*Glycine max*) and II (*Avena sativa*) plants.

More specifically, possible effects of pyoverdine and ferri-pyoverdine on plant iron nutrition have been evaluated over the last decade (Table 8.3). Supernatant of *Pseudomonas* sp. GRP3, including pyoverdine, enhanced iron content in shoots of *Vignata radiata* (strategy I) when cultivated in the presence of Fe-citrate as iron source (Sharma et al. 2003). Iron chlorosis of *Arachis hypogea* (strategy I) was corrected by application of purified Fe-pyoverdine (Jurkevitch et al. 1988).

Different reports indicated the presence of low quantities of ^{59}Fe or ^{55}Fe in several strategy I (*Arabidopsis thaliana, Arachis hypogeae, Cucumis sativus, Dianthus caryophyllus, Gossypium hirsutum*) and strategy II (*Avena sativa, Hordeum vulgare, Sorghum bicolor, Zea mays*) plants when cultivated in nutrient solution or soil supplemented with ^{59}Fe-pyoverdine or ^{55}Fe-pyoverdine (Bar-Ness et al.

Table 8.3. Examples of effects of pyoverdine and ferri-pyoverdine on plant iron nutrition

Plant	Growing conditions	Strain	Type of supplementation	Effect on plant	Reference
Strategy I					
Arachis hypogeae L.	Calcareous soil	*P. putida*	Fe-pyoverdine	Enhanced chlorophyll content	Jurkevitch et al. 1988
Arachis hypogeae L. *Gossypium hirsutum* L.	Nutrient solution	*P. putida* P-3	^{59}Fe-pyoverdine	Enhanced chlorophyll content Presence of ^{59}Fe in roots	Bar-Ness et al. 1991
Cucumis sativus L.	Soil	*P. putida*	^{59}Fe-pyoverdine	Presence of ^{59}Fe in the roots	Walter et al. 1994
Dianthus caryophyllus L.	Nutrient solution	*P. putida*	Fe-pyoverdine	Enhanced chlorophyll content and ferric reductase activity	Duijff et al. 1994b
Pisum sativum L.	Nutrient solution	*Pseudomonas* sp. B10	Pyoverdine + ^{59}FeCl$_3$	Reduced plant iron uptake and chlorophyll content	Becker et al. 1985
Vigna radiata L.	Quartz matrix	*Pseudomonas* sp.GRP3	Bacterial supernatant supplemented with Fe-citrate, Fe-EDTA or Fe(OH)$_3$	Enhanced iron content in shoots only when supplemented with Fe-citrate	Sharma et al. 2003
Strategy II					
Hordeum vulgare L.	Nutrient solution	*P. putida* WCS358	^{59}Fe-pyoverdine Pyoverdine + Fe(OH)$_3$	Enhanced chlorophyll content Presence of ^{59}Fe in the roots Enhanced chlorophyll content	Duijff et al. 1994c
Sorghum bicolor L.	Nutrient solution	*P. putida* P-3	^{59}Fe-pyoverdine	Enhanced chlorophyll content Uptake of ^{59}Fe by the host plant	Bar-Ness et al. 1991
Zea mays L.	Nutrient solution	*Pseudomonas* sp. B10	Pyoverdine + ^{59}FeCl$_3$	Reduced plant iron uptake and chlorophyll content	Becker et al. 1985
Zea mays L. *Avena sativa* L.	Nutrient solution	*P.putida*	^{55}Fe-pyoverdine	Presence of ^{55}Fe in the host plant	Bar-Ness et al. 1992
Zea mays L.	Soil	*P. putida*	^{59}Fe-pyoverdine	Presence of ^{59}Fe in the roots	Walter et al. 1994

1991, 1992; Duijff et al. 1994b, c; Vansuyt et al. (in press); Walter et al. 1994). This enhanced iron accumulation in the presence of ferri-pyoverdine appears to be more significant in root that in shoot (Bar-Ness et al. 1991; Walter et al. 1994, Vansuyt et al., in press). In strategy II plants, Von Wirén et al. (1993, 1995) showed that in plants producing low amounts of phytosiderophores (*Sorghum bicolor, Zea mays*), the degradation of these plant siderophores by the rhizospheric microflora could affect negatively the plant iron content, whereas such negative effect was not recorded in *Hordeum vulgare* which produces higher amount of phytosiderophores. Non-chelated purified pyoverdine was shown to either increase the chlorophyll content of barley (strategy II) or decrease that of pea (strategy I) when cultivated in nutrient solution supplemented with iron (Becker et al. 1985; Duijff et al. 1994c).

Contribution of pyoverdine to plant iron nutrition is in agreement with studies performed with other microbial siderophores such as ferrioxamine and rhodothorulic acid (Cline et al. 1984; Crowley and Gries 1994; Crowley et al. 1988, 1992; Fett et al. 1998; Hördt et al. 2000; Jonhson et al. 2002; Reid et al. 1984; Siebner-Freibach et al. 2003; Wang et al. 1993; Yehuda et al. 1996, 2000, 2003).

On the other way round, thanks to the reporter gene *pvd-inaZ*, phytosiderophores were shown to be a possible source of iron for *P. fluorescens* (Marschner and Crowley 1998). Similar observation was made by Jurkevitch et al. (1993) using ^{55}Fe uptake experiment from ^{55}Fe-mugeneic acid with a *P. putida* strain.

Mechanism by which plants acquire iron from ferri-pyoverdines remain basically to be elucidated even if some hypotheses have been proposed and recent observations have been made. The ability of strategy I plants to reduce ferri-pyoverdines is expected to depend on soil pH, iron availability and uptake; however, plant reductase activity is unlikely to reduce ferri-pyoverdines due to their low standard redox potential (Bar-Ness et al. 1991; Duijff et al. 1994b). Recent data from Vansuyt et al. (in press) indicate that incorporation of iron by *Arabidopsis thaliana* Col. from ferri-pyoverdine did not either rely on reductase activity or on the major IRT1 transporter, suggesting that the strategy I does not mediate this incorporation. This is supported by the presence of pyoverdine in planta as shown by ELISA and by tracing ^{15}N of ^{15}N-pyoverdine'. For strategy II plants, only indirect mechanisms may account for improved plant nutrition by pyoverdines due to their significant higher affinity for iron compared to phytosiderophores (Meyer and Abdallah 1978; Sugiura et al. 1981). Duijff et al. (1994c) have proposed that possible degradation of ferri-pyoverdines by microflora could lead to the release of iron that would then be available for phytosiderophores. Diurnal production cycle of phytosiderophores is submitted to pulses (Crowley and Gries 1994). During these pulses, phytosiderophore concentration in rhizosphere might be higher than that of microbial siderophores, therefore affecting the ligand exchange in favour ofphytosiderophores (Jurkevitch et al. 1993; Yehuda et al. 1996).

8.4
Conclusions

The above overview clearly underlines the significant contribution of pyoverdines to the interactions of fluorescent pseudomonads among themselves, with other microorganisms, and with the host-plant. These interactions are diverse. They are mostly related to the ability of these pyoverdines to chelate strongly iron and to contribute to (i) iron nutrition of the producing bacteria and/or having the corresponding protein membrane receptors and to (ii) iron deprivation of microorganisms producing siderophores with a lower affinity for iron. Whether the ability of pyoverdines to elicitate induced systemic resistance of specific host-plants is related to their ability to chelate iron or to their peptidic chain remains to be evaluated (Van Loon et al., personal communication).

So far, strategies followed to assess the role of iron-mediated competition in regards to microbial interactions were based on modification of the iron availability by the use of (i) synthetic ligands and purified pyoverdines compared to their corresponding iron chelates and of (ii) pvd– mutants compared to their corresponding wild-type strains. Another strategy recently proposed by Robin et al. (2006b) relies on modification of iron availability in the rhizosphere by cultivation of plant transgene that over express ferritin and then over accumulate iron (Goto et al. 1998; Van Wuytswinkel et al. 1999). This deregulation was indeed shown to induce an iron deprivation in the rhizosphere when the transgene is cultivated in soil with a low iron content (Robin et al. 2006a). Some populations of fluorescent pseudomonads selected in the rhizosphere of the ferritin over expressing mutant show siderotypes differing from that selected by the wild-type, and express a more intense in vitro antagonism in low iron medium than populations selected by the wild-type (Robin et al., in press).

As often with studies on rhizosphere ecology, we tend to draw general conclusions from data yielded on the whole root system, whereas it is now well known that there is spatial and temporal heterogeneity in the rhizosphere. Indeed, the distribution of fluorescent pseudomonads was shown to be not even (Gamalero et al. 2004). Similarly, physico-chemical properties at the root surface vary along the root system (Hinsinger et al. 2003). The intensity of pyoverdine synthesis and its biological meaning are then expected to vary along the root. Microscopical observations aiming at localizing pyoverdine in the rhizosphere in relation with the distribution of fluorescent pseudomonads and of plant pathogens should allow us to take into account better the heterogeneity of the rhizosphere and consequences on pyoverdine-mediated antagonism.

The relative importance of microbial siderophores in plant iron nutrition and possible mechanisms by which plants make use of Fe-siderophore have been assessed by a number of investigators over the last decades. However, the ecological relevance of microbial siderophores in plant nutrition remains controversial and mechanisms by which plants acquire iron from these compounds remain to be further explored. Using plant mutants affected in iron transporters (Curie and Briat, 2003), Vansuyt et al. (in press) have recently shown that iron from ferri-pyoverdine was incorporated by *A. thaliana* Col. and that this incorporation did not rely on the strategy I but on a pathway which remains to be described.

The complexity of pyoverdine-mediated interactions in the rhizosphere resulting from both abiotic and biotic interactions, the relevance of these interactions for plant growth and health, the sum of information yielded by the international scientific community and finally the variety of tools and methods available make these pyoverdine-mediated interactions a choice model for studying rhizosphere ecology and its modeling.

References

Ahl P, Voisard C, Défago G (1986) Iron bound-siderophores, cyanic acid, and antibiotics involved in the suppression of *Thielaviopsis basicola* of a *Pseudomonas fluorescens* strain. J Phytopathol 116:121–134

Alabouvette C, Höper H, Lemanceau P, Steinberg C (1996) Soil suppressiveness to diseases induced by soil-borne plant pathogens. In: Stotzky G, Bollag J-M (eds) Soil biochemistry. Marcel Dekker, New York, pp 371–413

Baker R, Elad Y, Sneh B (1986) Physical, biological and host factors in iron competition in soils. In: Swinburne TR (ed) Iron, siderophores and plant diseases. Plenum Press, New York London, pp 77–84

Bakker PAHM, Lamers JG, Bakker AW, Marugg JD, Weisbeek PJ, Schippers B (1986) The role of siderophores in potato tuber yield increase by *Pseudomonas putida* in a short rotation of potato. Neth J Plant Pathol 92:249–256

Bakker PAHM, Bakker AW, Marugg JD, Weisbeek PJ, Schippers B (1987) Bioassay for studying the role of siderophores in potato growth stimulation by *Pseudomonas* spp in short potato rotations. Soil Biol Biochem 4:443–449

Bakker PAHM, Ran LX, Pieterse CMJ, Van Loon LC (2003) Understanding the involvement of rhizobacteria-mediated induction of systemic resistance in biocontrol of plant diseases. Can J Plant Pathol 25:5–9

Barber DA, Lynch JM (1977) Microbial growth in the rhizosphere. Soil Biol Biochem 9:305–308

Bar-Ness E, Chen Y, Hadar Y, Marschner H, Römheld V (1991) Siderophores of *Pseudomonas putida* as an iron source for dicot and monocot plants. Plant Soil 130:231–241

Bar-Ness E, Hadar Y, Chen Y, Romheld V, Marschner H (1992) Short-term effects of rhizosphere microorganisms on Fe uptake from microbial siderophores by maize and oat. Plant Physiol 100:451–456

Becker JO, Cook RJ (1988) Role of siderophores in suppression of *Pythium* species and production of increased-growth response of wheat by fluorescent pseudomonads. Phytopathology 78:778–782

Becker JO, Hedges RW, Messens E (1985) Inhibitory effect of pseudobactin on the uptake of iron by higher plants. Appl Environ Microbiol 49:1090–1093

Bienfait HF (1985) Regulated redox process at the plasmalemma of plant root cells and their function in iron uptake. J Bioenerg Biomembr 17:73–83

Buyer JS, Leong J (1986) Iron transport-mediated antagonism between plant-growth promoting and plant-deleterious *Pseudomonas* strains. J Biol Chem 261:791–794

Buyer JS, Sikora LJ, Kratzke MG (1990) Monoclonal antibodies to ferric pseudobactin, the siderophore of plant growth-promoting *Pseudomonas putida* B10. Appl Environ Microbiol 56:419–424

Buysens S, Heungens K, Poppe J, Höfte M (1996) Involvement of pyochelin and pyoverdin in suppression of *Pythium*-induced damping-off of tomato by *Pseudomonas aeruginosa* 7NSK2. Appl Environ Microbiol 62:865–871

Chen L, Dick WA, Streeter JG, Hoitink HAJ (1998) Fe chelates from compost microorganisms improve Fe nutrition of soybean and oat. Plant Soil 139:139–147

Cheng W, Zhang Q, Coleman, DC, Carroll CR, Hoffman CA (1996) Is available carbon limiting microbial respiration in the rhizosphere? Soil Biol Biochem 28:1283-1288

Clark FE (1949) Soil microorganisms and plant roots. Adv Agron 1:241–248

Cline GR, Reid CPP, Powell PE, Szaniszlo PJ (1984) Effects of a hydroxamate siderophore on iron absorption by sunflower and sorghum. Plant Physiol 76:36–39

Cook JR, Thomashow LS, Weller DM, Fujimoto D, Mazzola M, Bangera G, Kim DS (1995) Molecular mechanisms of defense by rhizobacteria against root disease. Proc Natl Acad Sci USA 92:4197–4201

Crowley DE, Gries D (1994) Modelling of iron availability in the plant rhizosphere. In: Manthey JA, Crowley DE, Luster DG (ed) Biochemistry of metal micronutrients in the rhizosphere. CRC Press, Boca Raton, Florida, pp 199–220

Crowley DE, Reid CPP, Szaniszlo PJ (1988) Utilization of microbial siderophores in iron acquisition by oat. Plant Physiol 87:680–685

Crowley DE, Romheld V, Marschner H, Szaniszlo PJ (1992) Root-microbial effects on plant iron uptake from siderophores and phytosiderophores. Plant Soil 142:1–7

Curie C, Briat JB (2003) Iron transport and signalling in plants. Annu Rev Plant Biol 54:183–206

Curie C, Alonso JM, Le Jean M, Ecker JR, Briat JF (2000) Involvement of NRAMP1 from *Arabidopsis thaliana* in iron transport. Biochem J 347:749–755

Curie C, Panaviene Z, Loulergue C, Dellaporta SL, Briat JF, Walker EL (2001) Maize yellow stripe1 encodes a membrane protein directly involved in Fe(III) uptake. Nature 409:346–349

Curl EA, Truelove B (1986) The rhizosphere. Springer, Berlin Heidelberg New York

De Boer M, Bom P, Kindt F, Keurentjes JJB, Von der Sluis I, Van Loon LC, Bakker PAHM (2003) Control of Fusarium wilt of radish by combining *Pseudomonas putida* strains that have different disease-suppressive mechanisms. Phytopathology 93:626–632

De Meyer G, Höfte M (1997) Salicylic acid produced by the rhizobacterium *Pseudomonas aeruginosa* 7NSK2 induces resistance to leaf infection by *Botrytis cinerea* on bean. Phytopathology 87:588–593

Duijff BJ, Bakker, PAHM, Schippers B (1991) Suppression of fusarium wilt of carnation by *Pseudomonas* in soil: mode of action. In: Keel C, Knoller, B, Défago G (ed) Plant-Growth Promoting Rhizobacteria, IOBC/WPRS, Interlaken, Switzerland, pp 152-157

Duijff BJ, Meijer JW, Bakker PAHM, Schippers B (1993) Siderophore-mediated competition for iron and induced resistance in the suppression of Fusarium wilt of carnation by fluorescent *Pseudomonas* spp. Neth Plant Pathol 99:277–289

Duijff BJ, Bakker PAHM, Schippers B (1994a) Suppression of fusarium wilt of carnation by *Pseudomonas putida* WCS358 at different levels of disease incidence and iron availability. Biocontrol Sci Technol 4:279–288

Duijff BJ, Bakker PAHM, Schippers B (1994b) Ferric pseudobactin 358 as an iron source for carnation. J Plant Nutr 17:2069–2078

Duijff BJ, De Kogel WJ, Bakker PAHM, Schippers B (1994c) Influence of pseudobactin 358 on the iron nutrition of barley. Soil Biol Biochem 26:1681–1994

Duijff BJ, Pouhair D, Olivain C, Alabouvette C, Lemanceau P (1998) Implication of systemic induced resistance in the suppression of fusarium wilt of tomato by *Pseudomonas fluorescens* WCS417r and nonpathogenic *Fusarium oxysporum* Fo47. Eur J Plant Pathol 104:903–910

Duijff BJ, Recorbet G, Bakker PAHM, Loper J, Lemanceau P (1999) Microbial antagonism at the root level is involved in the suppression of fusarium wilt by the combination of nonpathogenic *Fusarium oxysporum* Fo47 and *Pseudomonas putida* WCS358. Phytopathology 89:1073–1079

Edel V, Steinberg C, Gautheron N, Alabouvette C (1997) Populations of nonpathogenic *Fusarium oxysporum* associated with roots of four plant species compared to soilborne populations. Phytopathology 87:693–697

Eide D, Broderius M, Fett J, Guerinot ML (1996) A novel iron-regulated metal transporter from plants identified by functional expression in yeast. Proc Natl Acad Sci USA 93:5624–5628

Elad Y, Baker R (1985a) Influence of trace amounts of cations and siderophore-producing pseudomonads on chlamydospore germination of *Fusarium oxysporum*. Phytopathology 75:1047–1052

Elad Y, Baker R (1985b) The role of competition for iron and carbon in suppression of chlamydospore germination of *Fusarium* spp. by *Pseudomonas* spp. Phytopathology 75:1053–1059

Elasri M, Delorme S, Lemanceau P, Stewart G, Laue B, Glickmann E, Oger PM, Dessaux Y (2001) Acyl-homoserine lactone production is more common amongst plant-associated than soil-borne *Pseudomonas* spp. Appl Environ Microbiol 67:1198–1209

Emery T (1965) Isolation, characterization, and properties of fusarinine, a hydroxamic derivative of ornithine. Biochem 4:1410–1417

Eparvier A, Lemanceau P, Alabouvette C (1991) Population dynamics of non-pathogenic *Fusarium* and fluorescent *Pseudomonas* strains in rockwool, a substratum for soilless culture. FEMS Microbiol Ecol 86:177–184

Fett J, LeVier K, Guerinot ML (1998) Soil microorganisms and iron uptake by higher plants. In: Sigel A, Sigel H (eds) Iron transport and storage in microorganisms, plants and animals, vol 35. Marcel Dekker, New York, pp 187–214

Gamalero E, Lingua G, Capri FG, Fusconi A, Berta G, Lemanceau P (2004) Colonization pattern of tomato primary roots by *Pseudomonas fluorescens* A6RI characterized by dilution plating, flow cytometry, fluorescence, confocal and scanning electron microscopy. FEMS Microbiol Ecol 48:79–87

Geels FP, Schippers B (1983a) Reduction of yield depressions in high frequency potato cropping soil after seed tuber treatments with antagonistic fluorescent *Pseudomonas* spp. Phytopath Z 108:207–214

Geels FP, Schippers B (1983b) Selection of antagonistic fluorescent *Pseudomonas* spp. and their root colonization and persistence following treatment of seed potatoes. Phytopath Z 108:193–206

Geels FP, Schmidt EDL, Schippers B (1985) The use of 8-hydroxyquinoline for the isolation and prequalification of plant growth-stimulating rhizosphere pseudomonads. Biol Fertil Soil 1:167–173

Gill PR, Warren GJ (1988) An iron-antagonized fungistatic agent that is not required for iron assimilation from a fluorescent rhizosphere pseudomonad. J Bacteriol 170:163–170

Goto F, Yoshihara T, Saiki H (1998) Iron accumulation in tobacco plants expressing soybean ferritin gene. Transgenic Res 7:173–180

Guerinot ML, Yi Y (1994) Iron: nutritious, noxious, and not readily available. Plant Physiol 104:815–820

Gupta CP, Dubey RC, Maheshwari DK (2002) Plant growth enhancement and suppression of *Macrophomina phaseolina* causing charcoal rot of peanut by fluorescent *Pseudomonas*. Biol Fertil Soils 35:399–405

Gutterson N (1990) Microbial fungicides: recent approaches to elucidating mechanisms. Crit Rev Biotechnol 10:69–91

Haas D, Défago G (2005) Biological control of soil-borne pathogens by fluorescent pseudomonads. Nat Rev Microbiol 3:307–319

Hamdan H, Weller DM, Thomashow LS (1991) Relative importance of fluorescent siderophores and other factors in biological control of *Gaeumannomyces graminis* var. *tritici* by *Pseudomonas fluorescens* 2-79 and M4-80R. Appl Environ Microbiol 57:3270–3277

Hell R, Stephan UW (2003) Iron uptake, trafficking and homeostasis in plants. Planta 216:541–551

Heungens K, Höfte M, Buysens S, Poppe J (1992) Role of siderophores in biological control of *Pythium* spp. by *Pseudomonas* strain 7NSK2. Medede Fac Landbouwwet Univ Gent 57:365–372

Hiltner L (1904) Über neuere erfahrungen und problem auf dem gebeit der bodenbakteriologie und unter besonderer berucksichtigung der grundungung und brache. Arbeit und Deutsche Landwirschaft Gesellschaft 98:59–78

Hinsinger P (1998) How do plant roots acquire mineral nutrients? Chemical processes involved in the rhizosphere. Adv Agron 64:225–264

Hinsinger P, Plassard C, Tang C, Jaillard B (2003) Origins of root-mediated pH changes in the rhizosphere and their responses to environmental constraints: a review. Plant Soil 248:43–59

Höfte M, Seong KY, Jurkevitch E, Verstraete W (1991) Pyoverdin production by the plant growth beneficial *Pseudomonas* strain 7NSK2: Ecological significance in soil. Plant Soil 130: 249–257

Höfte M, Boelens J, Verstraete W (1992) Survival and root colonization of mutants of plant growth-promoting pseudomonads affected in siderophore biosynthesis or regulation of siderophore production. J Plant Nutr 15:2253–2262

Hohnadel D, Meyer JM (1988) Specificity of pyoverdine-mediated iron uptake among fluorescent *Pseudomonas* strains. J Bacteriol 170:4865–4873

Hördt W, Römheld V, Winkelmann G (2000) Fusarinines and dimerum acid, mono- and dihydroxamate siderophores from *Penicillium chrysogenum*, improve iron utilization by strategy I and strategy II plants. BioMetals 13:37–46

Iavicoli A, Boutet E, Buchala A, Metraux JP (2003) Induced systemic resistance in *Arabidopsis thaliana* in response to root inoculation with *Pseudomonas fluorescens* CHA0. Mol Plant-Microbe Interact 16:851–858

Jonhson GV, Lopez A, La Valle Foster N (2002) Reduction and transport of Fe from siderophores. Plant Soil 241:27–33

Jurkevitch E, Hadar Y, Chen Y (1988) Involvement of bacterial siderophores in the remedy of lime-induced chlorosis in peanut. Soil Sci Soc Am J 52:1032–1037

Jurkevitch E, Hadar Y, Chen Y (1992) Utilization of the siderophores FOB and pseudobactin by rhizosphere microorganisms of cotton plants. J Plant Nutr 15:2183–2192

Jurkevitch E, Hadar Y, Chen Y, Chino M, Mori S (1993) Indirect utilization of the phytosiderophore mugineic acid as an iron source to rhizosphere fluorescent *Pseudomonas*. BioMetals 6:119–123

Keel C, Voisard C, Berling CH, Kahr G, Défago G (1989) Iron sufficiency, a prerequisite for the suppression of tobacco black root rot by *Pseudomonas fluorescens* strain CHA0 under gnotobiotic conditions. Phytopathology 79:584–589

Kloepper JW, Leong J, Teintze M, Schroth MN (1980a) *Pseudomonas* siderophores: a mechanism explaining disease-suppressive soils. Current Microbiol 4:317–320

Kloepper JW, Leong J, Teintze M, Schroth MN (1980b) Enhanced plant growth by siderophores produced by plant growth-promoting rhizobacteria. Nature 286:885–886

Kraus J, Loper JE (1992) Lack of evidence for the role of antifungal metabolite production by *Pseudomonas fluorescens* Pf-5 in biological control of Pythium damping-off of cucumber. Phytopathology 82:264–271

Latour X, Lemanceau P (1997) Métabolisme carboné et énergétique des *Pseudomonas* spp. fluorescents saprophytes à oxydase positive. Agronomie 17:427–433

Latour X, Corberand T, Laguerre G, Allard F, Lemanceau P (1996) The composition of fluorescent pseudomonad populations associated with roots is influenced by plant and soil type. Appl Environ Microbiol 62:2449–2456

Latour X, Delorme S, Mirleau P, Lemanceau P (2003) Identification of traits implicated in the rhizosphere competence of fluorescent pseudomonads: description of a strategy based on population and model strain studies. Agronomie 23:397–405

Leeman M, Van Pelt JA, Den Ouden FM, Heinsbroek M, Bakker PAHM, Schippers B (1995a) Induction of systemic resistance by *Pseudomonas fluorescens* in radish cultivars differing in susceptibility to Fusarium wilt, using a novel bioassay. Eur J Plant Pathol 101:655–664

Leeman M, Van Pelt JA, Den Ouden FM, Heinsbroek M, Bakker PAHM, Schippers B (1995b) Induction of systemic resistance against Fusarium wilt of radish by lipopolysaccharides of *Pseudomonas fluorescens*. Phytopathology 85:1021–1027

Leeman M, Den Ouden FM, Van Pelt JA, Dirkx FPM, Steijl H, Bakker PAHM, Schippers B (1996) Iron availability affects induction of systemic resistance to fusarium wilt of radish by *Pseudomonas fluorescens*. Phytopathology 86:149–155

Lemanceau P (1989) Role of competition for carbon and iron in mechanisms of soil suppressiveness to fusarium wilts. In: Tjamos EC, Beckman CH (ed) Vascular wilt diseases of plants, basic studies and control. Springer, Berlin Heidelberg New York, pp 386–396

Lemanceau P (1992) Effets bénéfiques de rhizobactéries sur les plantes: exemple des *Pseudomonas* spp fluorescent. Agronomie 12:413–437

Lemanceau P, Alabouvette C (1991) Biological control of fusarium diseases by fluorescent *Pseudomonas* and non-pathogenic *Fusarium*. Crop Protect 10:279–286

Lemanceau P, Heulin T (1998) La rhizosphère. In: Stengel P, Gelin S (eds) Sol: interface fragile. INRA Edition, Paris, pp 93–106

Lemanceau P, Samson R (1983) Relations entre quelques caractéristiques *in vitro* de 10 *Pseudomonas* fluorescents et leur effet sur la croissance du haricot (*Phaseolus vulgaris*). In: Dubos B, Olivier JM (eds) Les antagonismes microbiens, INRA, Paris pp 327-328

Lemanceau P, Alabouvette C, Meyer JM (1986) Production of fusarinine and iron assimilation by pathogenic and non-pathogenic *Fusarium*. In: Swinburne TR (ed) Iron, siderophores and plant diseases. Plenum Press, New York London, pp 251–259

Lemanceau P, Alabouvette C, Couteaudier Y (1988a) Recherches sur la résistance des sols aux maladies. XIV. Modification du niveau de réceptivité d'un sol résistant et d'un sol sensible aux fusarioses vasculaires en réponse à des apports de fer ou de glucose. Agronomie 8:155–162

Lemanceau P, Samson R, Alabouvette C (1988b) Recherches sur la résistance des sols aux maladies. XV. Comparaison des populations de *Pseudomonas* fluorescents dans un sol résistant et un sol sensible aux fusarioses vasculaires. Agronomie 8:243–249

Lemanceau P, Bakker PAHM, De Kogel WJ, Alabouvette C, Schippers B (1992) Effect of pseudobactin 358 production by *Pseudomonas putida* WCS358 on suppression of fusarium wilt of carnations by nonpathogenic *Fusarium oxysporum* Fo47. Appl Environ Microbiol 58:2978–2982

Lemanceau P, Bakker PAHM, De Kogel WJ, Alabouvette C, Schippers B (1993) Antagonistic effect of nonpathogenic *Fusarium oxysporum* Fo47 and pseudobactin 358 upon pathogenic *Fusarium oxysporum* f. sp. *dianthi*. Appl Environ Microbiol 59:74–82

Lemanceau P, Corberand T, Gardan L, Latour X, Laguerre G, Boeufgras JM, Alabouvette C (1995) Effect of two plant species, flax (*Linum usitatissinum* L.) and tomato (*Lycopersicon esculentum* Mill.), on the diversity of soilborne populations of fluorescent pseudomonads. Appl Environ Microbiol 61:1004–1012

Lindsay WL (1979) Chemical equilibria in soils. Wiley, New York

Lockwood JL (1977) Fungistasis in soils. Biol Rev 52:1–43

Loper JE (1988) Role of fluorescent siderophore production in biological control of *Pythium ultimum* by a *Pseudomonas fluorescens* strain. Phytopathology 78:166–172

Loper JE, Buyer JS (1991) Siderophores in microbial interactions on plant surfaces. Mol Plant-Microbe Interact 4:5–13

Loper JE, Henkels MD (1999) Utilization of heterologous siderophores enhances levels of iron available to *Pseudomonas putida* in the rhizosphere. Appl Environ Microbiol 65:5357–5363

Loper JE, Lindow SE (1994) A biological sensor for iron available to bacteria in their habitats on plant surfaces. Appl Environ Microbiol 60:1934–1941

Lynch JM (1990) The rhizosphere. In: Lynch JM (ed) The rhizosphere. Wiley, Chichester, pp 59–97

Lynch JM, Whipps JM (1990) Substrate flow in the rhizosphere. In: Keister DL, Cregan JP (eds) The rhizosphere and plant growth. Kluwer Academic Publishers, Dordrecht, pp 15–24

Marschner H, Römheld V (1994) Strategies of plants for acquisition of iron. Plant Soil 165:261–274

Marschner P, Crowley DE (1998) Phytosiderophores decrease iron stress and pyoverdine production of *Pseudomonas fluorescens* Pf-5 (*pvd-inaZ*). Soil Biol Biochem 30:1275–1280

Marschner H, Römheld V, Kissel M (1986) Different strategies in higher plants in mobilization and uptake of iron. J Plant Nutr 9:695–713

Masalha J, Kosegarten H, Elmaci O, Mengel K (2000) The central role of microbial activity for iron acquisition in maize sunflower. Biol Fertil Soil 30:433–439

Maurhofer M, Hase C, Meuwly P, Metraux JP, Défago G (1994) Induction of systemic resistance of tobacco to tobacco necrosis virus by the root-colonizing *Pseudomonas fluorescens* strain CHA0: influence of the *gacA* gene and of pyoverdine production. Phytopathology 84:139–146

Mavingui P, Laguerre G, Berge O, Heulin T (1992) Genetic and phenotypic diversity of *Bacillus polymyxa* in soil and in the wheat rhizosphere. Appl Environ Microbiol 58:1894–1903

Mercado-Blanco J, Rodriguez-Jurado D, Hervas A, Jimenez-Diaz RM (2004) Suppression of *Verticillium wilt* in olive planting stocks by root-associated fluorescent *Pseudomonas* spp. Biol Cont 30:474–486

Meyer JM, Abdallah MA (1978) The fluorescent pigment of *Pseudomonas fluorescens*: biosynthesis, purification and physico-chemical properties. J Gen Microbiol 107:319–328

Meyer JM, Hallé F, Hohnadel D, Lemanceau P, Ratefiarivelo H (1987) Siderophores of *Pseudomonas* - biological properties. In: Winkelmann G, Van der Helm D, Neilands JB (eds) Iron transport in microbes, plants and animals. VCH, Weinheim, pp 189–205

Meyer JM, Geoffroy VA, Baida N, Gardan L, Izard D, Lemanceau P, Achouak W, Palleroni NJ (2002) Siderophore typing, a powerful tool for the taxonomy of fluorescent and non-fluorescent *Pseudomonas*. Appl Environ Microbiol 68:2745–2453

Meziane H, Van der Sluis I, Van Loon LC, Höfte M, Bakker PAHM (2005) Determinants of *Pseudomonas putida* WCS358 involved in inducing systemic resistance in plants. Mol Plant Pathol 6:177–185

Mirleau P, Delorme S, Philippot L, Meyer JM, Mazurier S, Lemanceau P (2000) Fitness in soil and rhizosphere of *Pseudomonas fluorescens* C7R12 compared with a C7R12 mutant affected in pyoverdine synthesis and uptake. FEMS Microbiol Ecol 34:35–44

Mirleau P, Philippot L, Corberand T, Lemanceau P (2001) Involvement of nitrate reductase and pyoverdine in competitiveness of *Pseudomonas fluorescens* strain C7R12 in soil. Appl Environ Microbiol 67:2627–2635

Misaghi IJ, Stowell LJ, Grogan RG, Spearman LC (1982) Fungistatic activity of water-soluble fluorescent pigments of fluorescent pseudomonads. Phytopathology 72:33–36

Moënne-Loccoz Y, McHugh B, Stephens PM, McConnell FI, Glennon JD, Dowling DN, O'Gara F (1996) Rhizosphere competence of fluorescent *Pseudomonas* sp. B24 genetically modified to use additional ferric siderophores. FEMS Microbiol Ecol 19:215–225

Neilands JB (1981) Iron absorption and transport in microorganisms. Ann Rev Nut 1:27–46

Nguyen C (2003) Rhizodeposition of organic C by plants: mechanisms and controls. Agronomie 23:375–396

Ongena M, Daayf F, Jacques P, Thonart P, Benhamou N, Paulitz TC, Cornelis P, Koedam N, Belanger RR (1999) Protection of cucumber against Pythium root rot by fluorescent pseudomonads: predominant role of induced resistance over siderophores and antibiosis. Plant Pathol 48:66–76

Ongena M, Giger A, Jacques P, Dommes J, Thonart P (2002) Study of bacterial determinants involved in the induction of systemic resistance in bean by *Pseudomonas putida* BTP1. Eur J Plant Pathol 108:187-196

O'Sullivan DJ, O'Gara F (1992) Traits of fluorescent *Pseudomonas* spp. involved in suppression of plant root pathogens. Microbiol Rev 56:662–676

Paulitz TC, Loper JE (1991) Lack of a role for fluorescent siderophore production in the biological control of Pythium damping-off of cucumber by a strain of *Pseudomonas putida*. Phytopathology 81:930–935

Pieterse CMJ, Van Wees SCM, Hoffland E, Van Pelt JA, Van Loon LC (1996) Systemic resistance in *Arabidopsis* induced by biocontrol bacteria is independent of salicylic acid accumulation and pathogenesis-related gene expression. Plant Cell 8:1225–1237

Raaijmakers JM, Van der Sluis L, Koster M, Bakker PAHM, Weisbeek PJ, Schippers B (1995) Utilisation of heterologous siderophores and rhizosphere competence of fluorescent *Pseudomonas* spp. Can J Microbiol 41:126–135

Ran L, Xiang M, Zhou B, Bakker PAHM (2005a) Siderophores are the main determinants of fluorescent *Pseudomonas* strains in suppression of grey mould in *Eucalyptus urophylla*. Acta Phytopathol Sinica 35:6–12

Ran LX, Liu CY, Wu GJ, Van Loon C, Bakker PAHM (2005b) Suppression of bacterial wilt in *Eucalyptus urophylla* by fluorescent *Pseudomonas* spp. in China. Biol Cont 32:111–120

Reid CPP, Crowley DE, Kim HJ, Powell PE, Szaniszlo PJ (1984) Utilization of iron by oat when supplied as ferrated synthetic chelate or as ferrated hydroxamate siderophore. J Plant Nutr 7:437–447

Robin A, Vansuyt G, Corberand T, Briat JF, Lemanceau P (2006a) The soil type affects both the differential accumulation of iron between wild type and ferritin over-expressor tobacco plants and the sensitivity of their rhizosphere bacterioflora to iron stress. Plant Soil 283:73–83

Robin A, Mougel C, Siblot S, Vansuyt G, Mazurier S, Lemanceau P (2006b) Effect of ferritin over-expression in tobacco on the structure of bacterial and pseudomonad communities associated with the roots. FEMS Microbiol Ecol 58:492–502

Robin A, Mazurier S, Meyer JM, Vansuyt G, Mougel C, Lemanceau P. (In press) Diversity of root-associated fluorescent pseudomonads as affected by ferritin overexpression in tobacco. Environ Microbiol

Robinson NJ, Procter CM, Connolly EL, Guerinot ML (1999) A ferric-chelate reductase for iron uptake from soils. Nature 397:694–697

Rovira AD (1965) Interactions between plant roots and soil microorganisms. Ann Rev Microbiol 19:241–266

Rroço E, Kosegarten H, Harizaj F, Imani J, Mengel K (2003) The importance of soil microbial activity for the supply of iron to sorghum and rape. Eur J Agronomy 19:487–493

Scher FM, Baker R (1980) Mechanism of biological control in a Fusarium-suppressive soil. Phytopathology 70:412–417

Scher FM, Baker R (1982) Effect of *Pseudomonas putida* and a synthetic iron chelator on induction of soil suppressiveness to Fusarium wilt pathogens. Phytopathology 72:1567–1573

Schimdt W (2003) Iron solutions: acquisition strategies and signalling pathways in plants. Trend Plant Sci 8:188–193

Schippers B, Bakker AW, Bakker PAHM (1987) Interactions of deleterious and beneficial rhizosphere microorganisms and the effect of cropping practices. Annu Rev Phytopathol 25:339–358

Sharma A, Johri BN, Sharma AK, Glick BR (2003) Plant growth-promoting bacterium *Pseudomonas* sp. strain GRP3 influences iron acquisition in mung bean (*Vigna radiata* L. Wilzeck). Soil Biol Biochem 35:887–894

Siebner-Freibach H, Hadar Y, Chen Y (2003) Siderophores sorbed on Ca-montmorillonite as an iron source for plants. Plant Soil 251:115–124

Simeoni LA, Lindsay WL, Baker R (1987) Critical iron level associated with biological control of Fusarium wilt. Phytopathology 77:1057–1061

Sneh B, Dupler M, Elad Y, Baker R (1984) Chlamydospore germination of *Fusarium oxysporum* f.sp. *cucumerinum* as affected by fluorescent and lytic bacteria from a Fusarium-suppressive soil. Phytopathology 74:1115–1124

Söderberg KH, Bååth E (1998) Bacterial activity along a young barley root measured by the thymidine and leucine incorporation techniques. Soil Biol Biochem 30:1259–1268

Stintzi A, Evans K, Meyer JM, Poole K (1998) Quorum sensing and siderophore biosynthesis in *Pseudomonas aeruginosa*: *lasR/lasI* mutants exhibit reduced pyoverdine synthesis. FEMS Microbiol Lett 166:341–345

Sugiura Y, Tanaka H, Mino Y, Ishida T, Ota N, Nomoto K, Yosioka H, Takemoto T (1981) Structure, properties, and transport mechanism of iron(III) complex of mugineic acid, a possible phytosiderophore. J Am Chem Soc 103:6979–6982

Thomashow LS, Weller DM (1990) Role of antibiotics and siderophores in biocontrol of take-all disease of wheat. Plant Soil 129:93–99

Vandenbergh PA, Gonzales CF, Wright AM, Kunka BS (1983) Iron-chelating compounds produced by soil pseudomonads: correlation with fungal growth inhibition. Appl Environ Microbiol 46:128–132

Van Loon LC, Bakker PAHM, Pieterse CMJ (1998) Systemic resistance induced by rhizosphere bacteria. Annu Rev Phytopathol 36:453–483

Van Peer R, Van Kuik AJ, Rattink H, Schippers B (1990) Control of Fusarium wilt in carnation grown on rockwool by *Pseudomonas* sp. strain WCS417r and by Fe-EDDHA. Netherlands J Plant Pathol 96:119–132

Van Peer R, Niemann GJ, Schippers B (1991) Induced resistance and phytoalexin accumulation in biological control of Fusarium wilt of carnation by *Pseudomonas* sp. strain WCS417r. Phytopathology 81:728–734

Vansuyt G, Robin A, Briat JF, Curie C, Lemanceau P (In press) Iron acquisition from Fe-pyoverdine by *Arabidopsis thaliana*. Mol Plant-Microbe Interact

Van Wees SCM, Pieterse CMJ, Trijssenaar A, Van't Westende YAM, Hartog F, Van Loon LC (1997) Differential induction of systemic resistance in *Arabidopsis* by biocontrol bacteria. Mol Plant-Microbe Interact 10:716–724

Van Wuytswinkel O, Vansuyt G, Grignon N, Fourcroy P, Briat JF (1999) Iron homeostasis alteration in transgenic tobacco overexpressing ferritin. Plant J 17:93–98

Vert G, Briat JF, Curie C (2001) *Arabidopsis IRT2* gene encodes a root-periphery iron transporter. Plant J 26:181–189

Von Wirén N, Romheld V, Morel JL, Guckert A, Marschner H (1993) Influence of microorganisms on iron acquisition in maize. Soil Biol Biochem 25:371–376

Von Wirén N, Romheld V, Shioiri T, Marschner H (1995) Competition between micro-organisms and roots of barley and sorghum for iron accumulated in the root apoplasm. New Phytol 130:511–521

Von Wirén N, Khodr H, Hider RC (2000) Hydroxylated phytosiderophore species possess an enhanced chelate stability and affinity for iron(III). Plant Physiol 124:1149–1157

Walter A, Romheld V, Marschner H, Crowley DE (1994) Iron nutrition of cucumber and maize: effect of *Pseudomonas putida* YC 3 and its siderophore. Soil Biol Biochem 26:1023–1031

Wang Y, Brown HN, Crowley DE, Szaniszlo PJ (1993) Evidence for direct utilization of a siderophore, ferrioxamine B, in axenically grown cucumber. Plant Cell Environ 16:579–585

Wardle DA (1992) A comparative assessment of factors which influence microbial biomass carbon and nitrogen levels in soil. Biol Rev 67:321–342

Weller DM (1988) Biological control of soilborne plant pathogen in the rhizosphere with bacteria. Annu Rev Phytopathol 26:379–407

Weller DM, Howie WJ, Cook RJ (1988) Relationship between in vitro inhibition of *Gaeumannomyces graminis* var. *tritici* and suppression of take-all of wheat by fluorescent pseudomonads. Phytopathology 78:1094–1100

Wong PTW, Baker R (1984) Suppression of wheat take-all and *Ophiobolus* patch by fluorescent pseudomonads from fusarium-suppressive soil. Soil Biol Biochem 16:397–403

Yang C, Menge JA, Cooksey DA (1994) Mutations affecting hyphal colonization and pyoverdine production in pseudomonads antagonistic toward *Phytophthora parasitica*. Appl Environ Microbiol 60:473–481

Yehuda Z, Shenker M, Romheld V, Marschner H, Hadar Y, Chen Y (1996) The role of ligand exchange in the uptake of iron from microbial siderophores by gramineous plants. Plant Physiol 112:1273–1280

Yehuda Z, Shenker M, Hadar Y, Chen Y (2000) Remedy of chlorosis induced by iron deficiency in plants with the fungal siderophore rhizoferrin. J Plant Nutr 23:1991–2006

Yehuda Z, Hadar Y, Chen Y (2003) Immobilized EDDHA and DFOB as iron carriers to cucumber plants. J Plant Nutr 26:2043–2056

Ying Y, Guerinot ML (1996) Genetic evidence that induction of root Fe(III) chelate reductase activity is necessary for iron uptake under iron deficiency. Plant J 10:835–844

YongHoon L, WangHyu L, Duku L, HyeongKwon S (2001) Factors relating to induced systemic resistance in watermelon by plant growth-promoting *Pseudomonas* spp. Plant Pathol J 17:174–179

9 *Pseudomonas* Siderophores and their Biological Significance

Pierre Cornelis and Sandra Matthijs

9.1 Introduction

Fluorescent pseudomonads are γ-proteobacteria known for their capacity to colonize various ecological niches (Goldberg 2000). This adaptability is reflected by the high diversity of their iron uptake systems (Cornelis and Matthijs 2002). The majority of fluorescent pseudomonads produce complex fluorescent peptidic siderophores called pyoverdines or pseudobactins, which are very efficient iron scavengers (Cornelis and Matthijs 2002). Each species or sub-species produce a different pyoverdine as described in Chaps 9 and 10. In some cases pyoverdines have been identified as being responsible for the control of plant diseases by beneficial fluorescent pseudomonads (Cornelis and Matthijs 2002; see below). Other siderophores of lower affinity for iron are also produced by pseudomonads, but only a few have been described so far. These "secondary" siderophores are sometimes endowed with interesting properties in addition to iron scavenging, such as formation of complexes with other metals or antimicrobial activity. In this chapter we will review our current knowledge about pyoverdines and secondary siderophores with emphasis on their biological effects on other organisms.

9.2 Pyoverdines

As already mentioned in other chapters, pyoverdines (PVDs) are composed of a conserved chromophore and a variable peptide chain (Meyer 2000). The

Department of Molecular & Cellular Interactions, Laboratory of Microbial Interactions, Flanders Interuniversity Institute of Biotechnology, Vrije Universiteit Brussel, Pleinlaan 2, 1050 Brussels, Belgium, email: Pierre.Cornelis@imol.vub.ac.be

Soil Biology, Volume 12
Microbial Siderophores
A. Varma, S.B. Chincholkar (Eds.)

biosynthesis of pyoverdines has been reviewed by Lamont and Visca in Chap. 7. Briefly, both for the chromophore and the peptide chain, non-ribosomal peptide synthetases (NRPS) are involved (Mossialos et al. 2002; Ravel and Cornelis 2003; Lamont and Martin 2003). As already mentioned in the introduction, the chromophore is conserved among different pyoverdines while the peptide chain is highly variable, reflecting a rapid evolution of the NRPS and the receptor since its specificity resides in the recognition of the peptide chain of pyoverdine (Smith et al. 2005; Tümmler and Cornelis 2005). Some *Pseudomonas putida* or *P. fluorescens* strains have a remarkable capacity to utilise a broad spectrum of heterologous pyoverdines while others are restricted in their use of heterologous pyoverdines (Raaijmakers et al. 1995; Ongena et al. 2001; Mirleau et al. 2000; Mossialos et al. 2000; Ghysels et al. 2004). Sometimes, pyoverdine receptors are redundant, meaning that a particular strain has more than one receptor which is able to recognize its own pyoverdine as described recently for FpvB, the second type I receptor of *P. aeruginosa* (Ghysels et al. 2004). Interestingly, some bacteriocins produced by *Pseudomonas* species seem to enter cells of closely related bacteria via receptors for PVD (Baysse et al. 1999; Michel-Briand and Baysse 2002). It has recently been proposed that this state of warfare between strains producing pyocins and other having the pyoverdine receptor at their surface is pushing the evolution of receptors and the co-evolution of non-ribosomal peptide synthetases (Smith et al. 2005; Tümmler and Cornelis 2005).

9.2.1
Biological Activities of Pyoverdines

Pyoverdine production was found to be of great importance for the capacity of *P. aeruginosa* to colonise mammalian hosts (Meyer et al. 1996; Handfield et al. 2000; Takase et al. 2000) and can also be considered as a signal molecule since it induces the expression of several virulence factors (exotoxin A, protease PrpL) via the receptor, FpvA, an anti-sigma factor, FpvR, and the sigma factor PvdS (Lamont et al. 2002; Visca et al. 2002; Beare et al. 2003).

A controversy exists concerning the importance of pyoverdines in the in vivo antagonism of fluorescent pseudomonads against plant root pathogenic fungi. Some authors found that production of pyoverdines contributes to the bio-control capacity of the fluorescent *Pseudomonas* (Becker and Cook 1988; Buyer and Leong 1986; Loper 1988; Keel et al. 1989; Loper and Buyer 1991; Lemanceau et al. 1992). Conversely, other studies failed to establish a link between production of pyoverdine and antagonism against phytopathogenic fungi (Paulitz and Loper 1991; Kraus and Loper 1992; Ongena et al. 1999). These seemingly contradictory results may be explained in terms of strain differences, production of other siderophores, and of regulation by pyoverdines of the production of other molecules (Lamont et al. 2002; Visca et al. 2002; Beare et al. 2003). Another possible explanation for these discrepant results is that pyoverdines with different peptide chains could have different biological effects on other microorganisms by acting as peptide antibiotics (Hancock and Chapple 1999). In this context it would be interesting to investigate thoroughly the antimicrobial activity of purified pyoverdines from different species of *Pseudomonas.*

9.2.2
Induction of Resistance to Plant Pathogens by Pyoverdines

In some instances pyoverdines (pseudobactins) have been shown to induce a resistance (Induced Systemic Resistance, ISR) to aerial phytopathogens when applied to roots, mimicking the effect of the colonization of the roots by the *Pseudomonas* rhizobacterium (Leeman et al. 1996; Meziane et al. 2005). For *P. putida* WCS358, Meziane et al. (2005) recently showed that pseudobactin is one of the components (together with LPS) that induce ISR against *Botrytis cinerea* in tomato plants. Interestingly, pseudobactin from the same strain does not induce resistance against Fusarium wilt of radish while the pseudobactin from *P. fluorescens* WCS374 does (Leeman et al. 1996).

9.3
Secondary Siderophores

Fluorescent pseudomonads are also able to produce siderophores other than pyoverdines, but their production is masked by pyoverdine (Cornelis and Matthijs 2002). Their detection is possible only in a pyoverdine-negative background, obtained after chemical or transposon mutagenesis, using the universal chrome azurol siderophore (CAS) detection assay of Schwyn and Neilands (Schwyn and Neilands 1987; Mossialos et al. 2000; Mirleau et al. 2000; Mercado-Blanco et al. 2001; Matthijs et al. 2004, 2007).

Table 9.1 gives a list of the known secondary siderophores identified so far in fluorescent pseudomonads while Fig. 9.1 shows the general strategy to detect the production of a secondary siderophore and to identify the genes for its biosynthesis and uptake.

9.3.1
Pyochelin and Other Thiazolines

Besides pyoverdine, *P. aeruginosa* produces another siderophore, pyochelin, with a lower affinity for iron (III) (Cox et al. 1981). The precursor is salicy-

Table 9.1. Secondary siderophores from fluorescent pseudomonads

Siderophore	Species	Characteristics	Reference
Pyochelin	*P. aeruginosa* *P. fluorescens* CHA0 *P. fluorescens* Pf-5	Binds other metals Redox-active	Cox et al. (1981) Visca et al. (1992) Britigan et al. (1997) Baysse et al. (2000) Duffy and Défago (2000) Paulsen et al. (2005)
Pseudomonine	*P. fluorescens* WCS374	Lower affinity siderophore	Mercado-Blanco et al. (2001)
Quinolobactin/ thioquinolobactin	*P. fluorescens* ATCC17400	Repressed by cognate pyoverdine Anti-*Pythium* activity	Mossialos et al. (2000) Matthijs et al. (2004, 2007)
PDTC	*P. stutzeri*[a] KC *P. putida*	Binds several metals Antimicrobial activity CCl_4 degradation	Stolworthy et al. (2001) Lewis et al. (2000) Lewis et al. (2004) Sebat et al. (2001)

[a] Species that does not produce pyoverdines

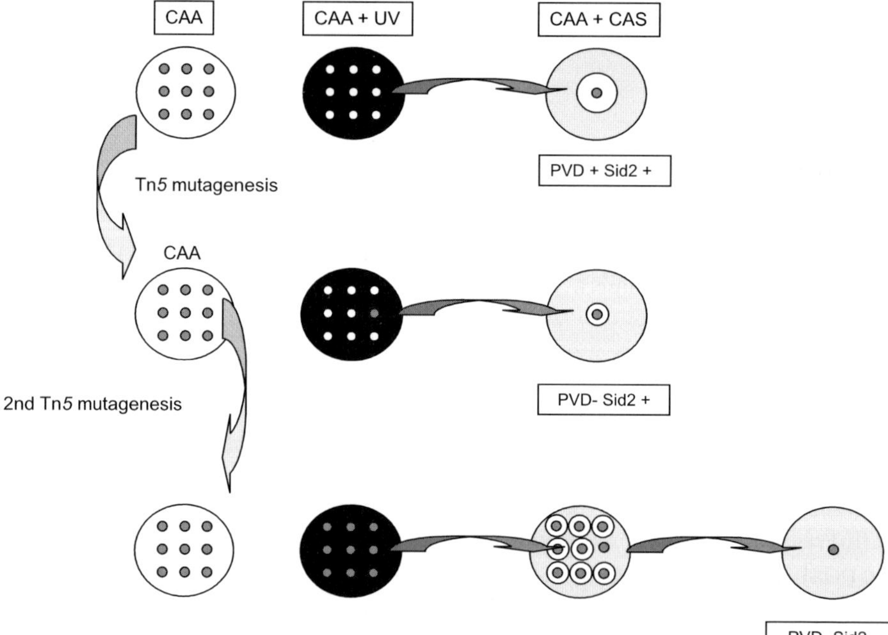

Fig. 9.1. General strategy to identify the presence of a secondary siderophore in fluorescent pseudomonads producing pyoverdine. Wild-type strains produce the yellow-green fluorescent pyoverdine siderophore in CAA medium, fluorescence being detected under UV. They also produce a large halo of discoloration on CAA plates containing chrome-azurol (CAS). After transposon mutagenesis, non-colored, non-fluorescent mutants can easily be selected. These mutants do not produce pyoverdine any more, but if their colonies still produce a halo on CAS, it can be reasonably assumed that they produce a secondary siderophore. This pyoverdine-negative, secondary siderophore-positive mutant can further be mutagenized with another Tn5 transposon, followed by selection for mutants that fail to produce a halo on CAS. DNA from these double-siderophore negative mutants can be isolated in order to identify the gene into which the second transposon inserted (Matthijs et al. 2004)

late, produced from chorismate with the products of the *pchAB* genes (Serino et al 1995, 1997). The complete biosynthetic pathway of pyochelin has been reviewed (Crosa and Walsh 2002). Visca et al. (1992) suggested that pyochelin could play a role in the acquisition of metals other than Fe(III), such as Co(II) and Mo(VI). Pyochelin can also form a 2:1 complex with vanadium (Baysse et al. 2000), which has a strong antibacterial activity against *P. aeruginosa*. It is also well documented that ferripyochelin can undergo a redox-cycle, especially in the presence of the *P. aeruginosa* phenazine pigment pyocyanin, resulting in the production of cell-damaging active oxygen species (Coffman et al. 1990; Britigan et al. 1997). A role for pyochelin (or its precursor salicylate) in combination with pyocyanin has also been suggested for the induction of resistance in tomato by the root-associated *P. aeruginosa* 7NSK2 against the aerial pathogen *Botrytis cinerea* (Audenaert et al. 2002). Pyochelin is produced by *P. fluorescens* CHA0 (Duffy and Défago 2000) and by *P. fluorescens* Pf5, the genome of which was recently published (Paulsen et al. 2005). Pyochelin is also produced by *Burkholderia cenocepacia* (Farmer and Thomas 2004).

Another siderophore derived from salicylic acid is pseudomonine, an isoxazolidone identified as being produced by two- *P. fluorescens* strains (Anthony et al. 1995; Mercado-Blanco et al. 2001).

9.3.2
Heteroaromatic Monothiocarboxylic Acids

Two siderophores have been described that belong to this group: pyridine-2,6-dithiocarboxylic acid (PDTC) and 8-hydroxy-4-methoxy-2-quinoline thiocarboxylic acid, which rapidly hydrolyzes to 8-hydroxy-4-methoxy-2-quinoline carboxylic acid, which is also called quinolobactin (Budzikiewicz 2003).

9.3.2.1
PDTC

Pyridine-2,6-dithiocarboxylic acid (PDTC) was first identified as the active component responsible for carbon tetrachloride degradation by *P. stutzeri* (Dybas et al. 1995). PDTC is able to form complexes with various metals and has been identified in the spent medium of different pseudomonads (reviewed in Budzikiewicz 2003). The PDTC stability constants for iron, cobalt, copper and nickel are very similar (Stolworthy et al. 2001). The siderophore activity of PDTC has now been clearly established (Lewis et al. 2004; Leach and Lewis 2006). Different metal-PDTC complexes have been found to have antimicrobial activities (Sebat et al. 2001).

9.3.2.2
Thioquinolobactin and Quinolobactin

Next to pyoverdine, *P. fluorescens* ATCC 17400 produces the secondary sidero-phore, quinolobactin or QB (Mossialos et al. 2000). Recently we described the genes and proposed a pathway for the synthesis of QB (Matthijs et al. 2004). Based on genome analysis predictions, the biosynthesis of the siderophore QB is likely to result from a combination of two pathways. The first pathway is involved in the production of xanthurenic acid through the catabolism of tryptophan, via a branch of the kynurenine pathway, while the second probably involves several enzymatic steps for the formation of thioquinolobactin (S-QB) from xanthurenic acid, some of them predicted to be similar to those described for the formation of the monothiocarboxylic acid PDTC (Fig. 9.2). From the

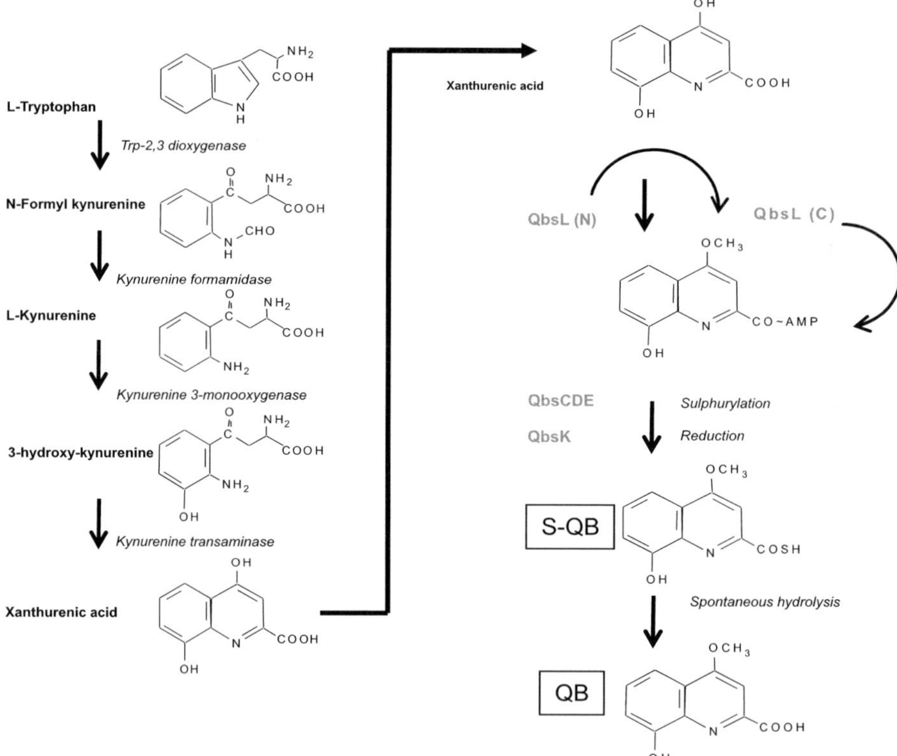

Fig. 9.2. Pathway for the biosynthesis of thio-quinolobactin (S-QB) as described by Matthijs et al. (2004). The degradation of tryptophan yields successively *N*-formyl-kynurenine, kynurenine, 3-hydroxykynurenine, and xanthurenice acid. Xanthurenic acid is then activated and methylated by the two-domain enzyme QbsL (AMP-ligase and methylase), followed by a sulfurylation which depends on the enzymes QbsCDE. QbsK is a reductase. the final product, S-QB, is unstable in solution and is hydrolyzed to yield QB (Matthijs et al. 2004)

pathway it can be inferred that 8-hydroxy-4-methoxy-2-quinoline thiocarboxylic acid (S-QB) is the end product, which gets rapidly hydrolysed in the culture medium to QB (Budzikiewicz 2003). This compound, named thioquinolobactin, could be isolated from culture supernatants of *P. fluorescens* ATCC 17400 and was shown to be rapidly hydrolyzed to quinolobactin (Matthijs et al. 2007). Furthermore, it was shown that thioquinolobactin, but not quinolobactin, had an antimicrobial activity against *Pythium* and some phytopathogenic fungi although both had a siderophore activity (Matthijs et al. 2007). In eukaryotes, the tryptophan-kynurenine pathway is the main way for aerobic tryptophan catabolism (Stone and Darlington 2002). Some metabolites derived from this pathway are neuro-active compounds, and have been implicated in neurological disorders, such as Alzheimer disease (Stone 2001; Stone and Darlington 2002; Guillemin and Brew 2002). Induction of tryptophan-2,3-dioxygenase and the resulting production of kynurenines are also known to modulate the immune response and induce apoptosis (reviewed by Grohmann et al. 2003). This pathway has long been considered to be specific for eukaryotic organisms only, but recently, several genes of the tryptophan-anthranilate branch of the kynurenine pathway were discovered in several bacterial genomes and shown to be functional (Kurnasov et al. 2003a, b).

An interesting aspect in the regulation of the synthesis of QB is its repression by pyoverdine (Mossialos et al. 2000). It seems that in *P. fluorescens* ATCC 17400 the synthesis of QB is the first way for this bacterium to deal with iron limitation. Although this siderophore has a lower affinity for iron than pyoverdine (Mossialos et al. 2000), it supports well the growth of pyoverdine-negative mutants in iron-limiting medium.

9.4
Conclusion

The aim of this chapter was to demonstrate that siderophores (not only from fluorescent pseudomonads) can have interesting properties aside from complexation of iron, and when in complex with other metals. One interesting research area would be to have a fresh look at the anti-microbial activity (anti-bacterial, anti-fungal) of different purified pyoverdines and siderophores.

A second neglected field of research is the search for new secondary siderophores in fluorescent pseudomonads. This is because the production of these siderophores is difficult to establish in a wild-type producing pyoverdine. It can be expected that by using a strategy such as described in Fig. 9.1 more genes for the biosynthesis of secondary metabolites with siderophore activity will be identified in the future. These siderophores could have important biotechnological applications since they can be endowed with biological activities such as antagonism against plant pathogens or as inducers of plant defence response.

References

Anthony U, Christophersen C, Nielse PH, Gram L, Petersen BO (1995) Pseudomonine, an isoxazolidone with siderophoric activity from *Pseudomonas fluorescens* AH2 isolated from lake Victoria Nile perch. J Nat Prod 58:1786–1789

Audenaert K, Pattery T, Cornelis P, Höfte M (2002) Induction of systemic resistance to *Botrytis cinerea* in tomato by *Pseudomonas aeruginosa* 7NSK2: role of salicylic acid, pyochelin, and pyocyanin. Mol Plant-Microbe Inter 15:1147–1156

Baysse C, Meyer JM, Plesiat P, Geoffroy V, Michel-Briand Y, Cornelis P (1999) Uptake of pyocin S3 occurs through the outer membrane ferripyoverdine type II receptor of *Pseudomonas aeruginosa*. J Bacteriol 181:3849–3851

Baysse C, De Vos D, Naudet Y, Vandermonde A, Ochsner U, Meyer JM, Budzikiewicz H, Fuchs R, Cornelis P (2000) Vanadium interferes with siderophore-mediated iron uptake in *Pseudomonas aeruginosa*. Microbiology 146:2425–2434

Beare PA, For RJ, Martin LW, Lamont IL (2003) Siderophore-mediated cell signalling in *Pseudomonas aeruginosa*: divergent pathways regulate virulence factor production and siderophore receptor synthesis. Mol Microbiol 47:195–207

Becker JO, Cook RJ (1988) Role of siderophores in suppression of *Pythium* species and the production of increased growth-response of wheat by fluorescent pseudomonads. Phytopathology 78:778–782

Britigan BE, Rasmussen GT, Cox CD (1997) Augmentation of oxidant injury to human pulmonary epithelial cells by the *Pseudomonas aeruginosa* siderophore pyochelin. Infect Immun 65:1071–1076

Budzikiewicz H (2003) Heteroaromatic monothiocarboxylic acids from *Pseudomonas* spp. Biodegradation 14:65–72

Buyer JS, Leong J (1986) Iron transport-mediated antagonism between plant growth-promoting and plant-deleterious *Pseudomonas* strains. J Biol Chem 261:791–794

Coffman TJ, Cox CD, Edeker BL, Britigan BE (1990) Possible role of bacterial siderophores in inflammation. J Clin Invest 86:1030–1037

Cornelis P, Matthijs S (2002) Diversity of siderophore-mediated iron uptake systems in fluorescent pseudomonads: not only pyoverdines. Environ Microbiol 4:787–798

Cox CD, Rinehart KL, Moore ML, Cook JC (1981) Pyochelin: novel structure of an iron chelating growth promoter for *Pseudomonas aeruginosa*. Proc Natl Acad Sci USA 78:4256–4260

Crosa JH, Walsh CT (2002) Genetics and assembly line enzymology of siderophore biosynthesis in bacteria. Microbiol Mol Biol Rev 66:223–249

Duffy BK, Défago G (2000) Controlling instability in *gacS-gacA* regulatory genes during inoculant production of *Pseudomonas fluorescens* biocontrol strains. Appl Environ Microbiol 66:3142–3150

Dybas MJ, Tatara GM, Criddle CS (1995) Localization and characterization of the carbon tetrachloride transformation activity of *Pseudomonas* sp. strain KC. Appl Environ Microbiol 61:758–762

Farmer KL, Thomas MS (2004) Isolation and characterization of *Burkholderia cenocepacia* mutants deficient in pyochelin production: pyochelin biosynthesis is sensitive to sulphur availability. J Bacteriol 186:270–277

Ghysels B, Thi Min Dieu B, Beatson SA, Pirnay JP, Ochsner UA, Vasil ML, Cornelis P (2004) FpvB, an alternative type I ferripyoverdine receptor of *Pseudomonas aeruginosa*. Microbiology 150:1671–1680

Goldberg J (2000) *Pseudomonas*: global bacteria. Trends Microbiol 8:55–57

Grohmann U, Fallarino F, Pucetti P (2003) Tolerance, DCs and tryptophan: much ado about IDO. Trends Immunol 24:242–248

Guillemin GJ, Brew BJ (2002) Implications of the kynurenine pathway and quinolinic acid in Alzheimer's disease. Redox Rep 7:199–206

Hancock RE, Chapple DS (1999) Peptide antibiotics. Antimicrob Chemother 43:1317–1323

Handfield M, Lehoux DE, Sanschagrin F, Mahan MJ, Woods DE, Levesque RC (2000) In vivo-induced genes in *Pseudomonas aeruginosa*. Infect Immun 68:2359–2362

Keel C, Voisard C, Berling CH, Kahr G, Défago G (1989) Iron sufficiency, a prerequisite for the suppression of tobacco black-root-rot by *Pseudomonas fluorescens* strain CHAO under gnotobiotic conditions. Phytopathology 79:584–589

Kraus J, Loper JE (1992) Lack of evidence for a role of antifungal metabolite production by *Pseudomonas fluorescens* Pf-5 in biological control of *Pythium* damping-off of cucumber. Phytopathology 82:264–271

Kurnasov O, Jablonski L, Polanuyer B, Dorrestein P, Begley T, Osterman A (2003a) Aerobic tryptophan degradation pathway in bacteria: novel kynurenine formamidase. FEMS Microbiol Lett 227:219–227

Kurnasov O, Goral V, Colabrov K, Gerdes S, Anantha S, Osterman A, Begley TP (2003b) NAD biosynthesis: identification of the tryptophan to quinolinate pathway in bacteria. Chem Biol 10:1195–1204

Lamont IL, Beare PA, Ochsner U, Vasil AI, Vasil ML (2002) Siderophore-mediated signaling mediates virulence factor production in *Pseudomonas aeruginosa*. Proc Natl Acad Sci USA 99:7072–7077

Lamont IL, Martin LW (2003) Identification and characterization of novel pyoverdine synthesis genes in *Pseudomonas aeruginosa*. Microbiology 149:833–842

Leach LH, Lewis TA (2006) Identification and characterization of *Pseudomonas* membrane transporters necessary for utilization of the siderophore pyridine-2,6-bis(thiocarboxylic acid) (PDTC). Microbiology 152:3157–3166

Leeman M, Den Ouden FM, Van Pelt JA, Dirkx FPM, Steijl H, Bakker PHAM, Schippers B (1996) Iron availability affects induction of systemic resistance to fusarium wilt of radish by *Pseudomonas fluorescens*. Phytopathology 86:149–154

Lemanceau P, Bakker PA, De Kogel WJ, Alabouvette C, Schippers B (1992) Effect of pseudobactin 358 production by *Pseudomonas putida* WCS358 on suppression of fusarium wilt of carnations by nonpathogenic *Fusarium oxysporum* Fo47. Appl Environ Microbiol 58:2978–2982

Lewis TA, Cortese MS, Sebat JL, Green TL, Lee CH, Crawford RL (2000) A *Pseudomonas stutzeri* gene cluster encoding the biosynthesis of the CCl_4-dechlorination agent pyridine-2,6-bis(thiocarboxylic acid). Environ Microbiol 2:407–416

Lewis TA, Leach L, Morales S, Austin PR, Hartwell HJ, Kaplan B, Forker C, Meyer, JM (2004) Physiological and molecular genetic evaluation of the dechlorination agent, pyridine-2,6-bis(monothiocarboxylic acid) (PDTC) as a secondary siderophore of *Pseudomonas*. Environ Microbiol 6:159–169

Loper JE (1988) Role of fluorescent siderophore production in biological control of *Pythium ultimum* by a *Pseudomonas fluorescens* strain. Phytopathology 78:166–172

Loper JE, Buyer JS (1991) Siderophores in microbial interactions on plant surfaces. Mol Plant-Microbe Interact 4:5–13

Matthijs S, Baysse C, Koedam N, Abbaspour-Tehrani K, Verheyden L, Budzikiewicz H, Schäfer M, Hoorelbeke B, Meyer JM, De Greve H, Cornelis P (2004) The *Pseudomonas* siderophore quinolobactin is synthezised from xanthurenic acid, an intermediate from the kynurenine pathway. Mol Microbiol 52:371–384

Matthijs S, Abbaspour-Therani K, Laus G, Jackson RW, Cooper RM, Cornelis P (2007) Thioquinolobactin, a *Pseudomonas* siderophore with antifungal and anti-*Pythium* activity. Environ Microbiol 9:425–434

Mercado-Blanco J, van der Drift KMGM, Olsson PE, Thomas-Oates JE, van Loon LC, Bakker PAHM (2001) Analysis of the *pmsCEAB* gene cluster involved in biosynthesis of salicylic acid and the siderophore pseudomonine in the biocontrol strain *Pseudomonas fluorescens* WCS374. J Bacteriol 183:1909–1920

Meyer JM (2000) Pyoverdines: pigments, siderophores and potential taxonomic markers of fluorescent *Pseudomonas* species. Arch Microbiol 174:135–142

Meyer JM, Neely A, Stintzi A, Georges C, Holder IA (1996) Pyoverdin is essential for virulence of *Pseudomonas aeruginosa*. Infect Immun 64:518–523

Meziane H, Van Der Sluis I, Van Loon LC, Höfte M, Bakker PA (2005) Determinants of *Pseudomonas putida* WCS358 involved in inducing systemic resistance in plants. Mol Plant Pathol 6:177–185

Michel-Briand Y, Baysse C (2002) The pyocins of *Pseudomonas aeruginosa*. Biochimie 84:499–510

Mirleau P, Delorme S, Philippot L, Meyer JM, Mazurier S, Lemanceau P (2000) Fitness in soil and rhizosphere of *Pseudomonas fluorescens* C7R12 compared with a C7R12 mutant affected in pyoverdine synthesis and uptake. FEMS Microbiol Ecol 34:35–44

Mossialos D, Meyer JM, Budzikiewicz H, Wolff U, Koedam N, Baysse C, Anjaiah V, Cornelis P (2000) Quinolobactin, a new siderophore of *Pseudomonas fluorescens* ATCC 17400 whose production is repressed by the cognate pyoverdine. Appl Env Microbiol 66:487–492

Mossialos D, Ochsner U, Baysse C, Chablain P, Pirnay JP, Koedam N, Budzikiewicz H, Uria-Fernandez D, Schäfer M, Ravel J, Cornelis P (2002) Identification of new, conserved, non-ribosomal peptide synthetases from fluorescent pseudomonads involved in the biosynthesis of the siderophore pyoverdine. Mol Microbiol 45:1673–1685

Ongena M, Daayf F, Jacques P, Thonart P, Benhamou N, Paulitz TC, Cornelis P, Koedam N, Bélanger RR (1999) Protection of cucumber against *Pythium* root rot by fluorescent pseudomonads: predominant role of induced resistance over siderophores and antibiosis. Plant Pathol 48:66–76

Ongena M, Jacques P, Delfosse P, Thonart P (2001) Unusual traits of the pyoverdin-mediated iron acquisition system in *Pseudomonas putida* BTP1. Biometals 15:1–13

Paulitz TC, Loper JE (1991) Lack of a role for fluorescent siderophore production in the biological control of *Pythium* damping-off of cucumber by a strain of *Pseudomonas putida*. Phytopathology 81:930–935

Paulsen IT, Press CM, Ravel J, Kobayashi DY, Myers GS, Mavrodi DV, DeBoy RT, Seshadri R, Ren Q, Madupu R, Dodson RJ, Durkin AS, Brinkac LM, Daugherty SC, Sullivan SA, Rosovitz MJ, Gwinn ML, Zhou L, Schneider DJ, Cartinhour SW, Nelson WC, Weidman J, Watkins K, Tran K, Khouri H, Pierson EA, Pierson LS III, Thomashow LS, Loper JE (2005) Complete genome sequence of the plant commensal *Pseudomonas fluorescens* Pf-5. Nat Biotechnol 23:873–878

Raaijmakers JM, Vandersluis I, Koster M, Bakker PAHM, Weisbeek PJ, Schippers B (1995) utilization of heterologous siderophores and rhizosphere competence of *Pseudomonas* spp. Can J Microbiol 41:126–135

Ravel J, Cornelis P (2003) Genomics of pyoverdine-mediated iron uptake in pseudomonads. Trends Microbiol 5:95–200

Schwyn B, Neilands JB (1987) Universal chemical assay for the detection and determination of siderophores. Anal Biochem 160:47–56

Sebat JL, Paszczynski AJ, Cortese MS, Crawford RL (2001) Antimicrobial properties of pyridine-2,6-dithiocarboxylic acid, a metal chelator produced by *Pseudomonas* spp. Appl Environ Microbiol 67:3934–3942

Serino L, Reimmann C, Baur H, Beyeler M, Visca P, Haas D (1995) Structural genes for salicylate biosynthesis from chorismate in *Pseudomonas aeruginosa*. Mol Gen Genet 249:217–228

Serino L, Reimmann C, Visca P, Beyeler M, Chiesa VD, Haas D (1997) Biosynthesis of pyochelin and dihydroaeruginoic acid requires the iron-regulated *pchDCBA* operon in *Pseudomonas aeruginosa*. J Bacteriol 179:248–257

Smith EE, Sims EH, Spencer DH, Kaul R, Olson MV (2005) Evidence for diversifying selection at the pyoverdine locus of *Pseudomonas aeruginosa*. J Bacteriol 187:2138–2147

Stolworthy JC, Paszczynski A, Korus R, Crawford RL (2001) Metal binding by pyridine-2,6-bis(monothiocarboxylic acid), a biochelator produced by *Pseudomonas stutzeri* and *Pseudomonas putida*. Biodegradation 12:411–418

Stone TW (2001) Kynurenines in the CNS: from endogenous obscurity to therapeutic importance. Prog Neurobiol 64:185–218

Stone TW, Darlington LG (2002) Endogenous kynurenines as targets for drug discovery and development. Nat Rev Drug Discov 1:609–620

Takase H, Nitanai H, Hoshino K, Otani T (2000) Impact of siderophore production on *Pseudomonas aeruginosa* infections in immunosuppressed mice. Infect Immun 68:1834–1839

Tümmler B, Cornelis P (2005) Pyoverdine receptor: a case of positive Darwinian selection in *Pseudomonas aeruginosa*. J Bacteriol 187:3289–3292

Visca P, Colotti G, Serino L, Versili D, Orsi N, Chiancone E (1992) Metal regulation of siderophore synthesis in *Pseudomonas aeruginosa* and functional effects of siderophore-metal complexes. Appl Environ Microbiol 58:2886–2893

Visca P, Leoni L, Wilson MJ, Lamont IL (2002) Iron transport and regulation, cell signalling and genomics: lessons from *Escherichia coli* and *Pseudomonas*. Mol Microbiol 45:1177–1190

10 Microbial Siderophores in Human and Plant Health-Care

S.B. Chincholkar, B.L. Chaudhari and M.R. Rane

10.1
Introduction

Modern biotechnology is based on foundation of bio-molecules. The world's scientific community is enamored of diversity of bio-molecules and their vivacious conduct. The way a particular molecule is employed is determined by virtue of its nature and the perspicacity of its employer. The present chapter is a product of a gentle survey for applications of siderophores and an effort to provoke a genuine reconsideration of siderophores for newer applications on modern horizons.

The history of iron nutrition of microorganisms dates back to the evolutionary period when anaerobic life came into existence and it is assumed that such forms of life were probably using ferrous ions. In due course of time, in the present context of an aerobic environment, the majority of microbes have opted to use the ferric instead of the ferrous form for their nutrition. The first few reports of microbial siderophores relate to the discovery of the growth factors for mycobacteria and fungi such as Pilobolus (Hesseltine et al. 1952). In 1952, "Neilands" who is now honored as the "Father of ferruginous facts" succeeded in isolating and purifying siderophores in crystalline form. Now, numerous organisms have been reported to produce siderophores in diverse molecular forms. Producers of siderophores include bacteria, fungi, yeast and plants. Techniques to explore siderophores slowly evolved to fascinate genuine intellectuals, explorers of the ferruginous world. Today, information exhibits knowledge on genetics of siderophores, biosynthetic pathways leading to production of siderophores, transport of desferri- and ferri- forms of siderophores, and reaction kinetics of iron solubilization by siderophore molecules with great details. Methods for siderophore analysis have evolved quite fast, leading to added accuracy and precision

School of Life Sciences, North Maharashtra University, Jalgaon, Post Box 80, Jalgaon 425001, Maharashtra, India, email: sudhirchincholkar@hotmail.com

Soil Biology, Volume 12
Microbial Siderophores
A. Varma, S.B. Chincholkar (Eds.)
© Springer-Verlag Berlin Heidelberg 2007

(Chincholkar et al. 2000). Jalal and van der Helm (1991) have elegantly reviewed methodologies for isolation and identification of fungal siderophores. Studies on cross-feeding of ferri-siderophores and ability of an organism to utilize non-self ferri-siderophores have also been made in equally great details (Poole and McKay 2003).

It is essential to provide information regarding "coining of term – siderophore". Chemically speaking, siderophores are low molecular weight metal chelating compounds secreted by microorganisms growing under low iron stress (Lankford 1973). In earlier literature, designations like siderochromes, sideramines, sideromycins and ionophores have been used. However, now these terms have been unanimously replaced by the term "siderophore" which means in Greek: *sideros* = iron and *phores* = bearer. The term siderophore is reserved for the iron-free ligands only (Neilands 1981). However, in some cases, the name had been given to iron complexed siderophore (ferrichrome, ferrioxamine). The corresponding iron-free form, therefore, needs the prefix desferri- or deferri- (desferricoprogen). Whereas, the siderophores which represent the iron free form (enterobactin, staphyloferrin, rhizoferrin) need the prefix ferri- or ferric- when iron is bound to it (Neilands 1981).

10.2
Microbial Siderophores for Living System/for Health

Extraordinary ability of siderophores to chelate and mobilize iron is a driving force to search its applications. Siderophores are being employed for effective iron nutrition of various organisms including human being; where as on the other hand, these molecules are being explored for reduction in iron overload under specific diseased conditions.

10.2.1
Applications in Health Sector

Iron is an inevitable element involved in proper functioning of various physiological activities among living systems viz micro-organisms, plants, animals and human beings. Its surfeit and deprivation leads to various diseased situations (Eisendle et al. 2006), which tempts siderophores to find some applications in medicine.

10.2.1.1
Siderophores as Medicaments

10.2.1.1.1
Iron Overload Conditions

In case of iron overload, iron can be removed from body by deferration. Several pathological situations arise as side effect of iron overload, e.g. x-linked sideroblastic anemia (XLSA) (Heller et al. 2004). Another example is Type 3 hemochromatosis, a rare autosomal recessive disorder due to mutations of the *TFR2* gene. There is a progressive hepatocellular iron accumulation from Rappaport's zone 1–3 and that iron loading in sinusoidal and portal macrophages occur only in a more advanced stage. This report says that, after removal of 26 g of iron by subcutaneous deferoxamine infusion a marked clinical improvement was possible (Riva et al. 2004). Eisendle et al. (2006) have reported that intracellular siderophore "Ferricrocin" is involved in iron storage, oxidative-stress resistance, germination, and sexual development in *Aspergillus nidulans.* Application of this siderophore to decrease iron overload in human seems to be an open area for further research.

10.2.1.1.2
Non-iron Overload Conditions

As studies indicate that iron deprivation is a useful therapeutic approach for malaria, investigations were made of the effect of the iron chelator desferrioxamine B (DFO B) on the clinical course of *Plasmodium falciparum* infection. Although some clinical benefits were observed, overall mortality was not observed as hydrophilic DFO displays poor penetration across a membrane. This failure, however, inspired further studies to employ synthetic siderophores to arrest uncontrolled proliferation of *P. falciparum* in human infected erythrocytes (Rotheneder et al. 2002). Voest et al. (2005) have revived the application of iron-chelating agents in non-iron overload conditions like rheumatoid arthritis, cardiac poisoning, reperfusion injury, solid tumors, hematologic malignancies, antiprotozoal activity (as discussed above), aluminium mediated pathological conditions like renal failure and Alzheimer diseases and so on. Almost all of the medical applications reported above involved use of deferoxamine. Table 10.1 shows few reported siderophores (dominantly deferoxamine) having potential as medicaments. However, deferoxamine application has three major drawbacks: i) oral inactivity, ii) non lipophilic nature, and iii) high cost. In this context, randomized, prospective trials are needed to evaluate the applicability of other known siderophores owing to its great structural diversity. Moreover, the finding of new siderophores is foreseen with a major breakthrough in the near future.

Table 10.1. Siderophores having potential as medicament

Pathological condition/symptom	Siderophore	Function of siderophore	Reference
Reperfusion injury	Exochelin	Prevent iron-mediated generation of hydroxyl radical (.OH) leading to reperfusion injury	Horwitz et al. 1989
X-linked sideroblastic anemia (XLSA)	Deferoxamine	Decrease the iron overload in body	Heller et al. 2004
Acute pancreatitis		Active free oxygen radical scavenger	Ateskan et al. 2003
Skin exposed to nitrogen mustard		Prevention of nuclear damage and for handling the burns	Karayilanoglu et al. 2003 Banin et al. 2003
Retrobulbar haematoma		Prevention degenerative changes in retrobulbar haematoma	Pelit et al. 2003
Iron-overload cardiomyopathy		Promotes survival and prevents electrocardiographic abnormalities	Obejero-Paz et al. 2003
Cancer		Markedly inhibiting the growth of aggressive tumors	Richardson 2002
Skeletal muscle ischemia		Enhances neovascularization and recovery of ischemic skeletal muscle	Chekanov et al. 2003
Microbial Infection		Inhibit *Trichomonas vaginalis* in vitro	Mahmoud 2002
Hyper-transfusion		Decrease iron overload	Link et al. 2003
Peritonitis in rat		Compliment antibiotic therapy	Soybir et al. 2002
Trypanosomiasis		Inhibition of bloodstream forms of *Trypanosoma brucei*	Breidbach et al. 2002
Pneumocystis carinii pneumonia		Inhibit *Pneumocystis carinii* by novel mechanism instead of iron-starvation	Clarkson et al. 2001

10.2.1.2
Siderophores as Trojan-Horse in Antibiotic Conjugates

The magic of antibiotics has started to wane due to indiscriminate administration of heavy doses and consequent development of resistant microbes. This has led to the development of blended preparations of various antibiotics. Considering the fact that most of the medicines introduced into the body on weight basis is in fact a nuisance to the body other than the target. Targeted drug delivery can increase precision with reliability and hence could decrease the abuse. Several targeted medicines with/without controlled delivery modes have reached the market. In most cases, target specific drug delivery is based on specific cell membrane proteins and its conjugation with antibody-like molecules.

Table 10.2. Siderophores conjugated with antibiotics

Siderophore type	Antibiotic	Target Organism	Reference
Catechol	Spiramycin 1 and Neospiramycin 2	Gram negative organism	Poras et al. 1998
Pyoverdin	Ampicillin	*Pseudomonas aeruginosa*	Kinzel et al. 1998
	Norfloxacin and Benzonaphthyridone	*Pseudomonas aeruginosa*	Hennard et al. 2001

Although siderophores have been studied for drug delivery, there has been no major achievement yet (Table 10.2). Pathogenic bacteria have a strict nutritional requirement for iron, and in vivo they must contend with the natural ability of the host to withhold free iron in the form of iron-protein complexes. Conjugation of drugs to siderophores uses the mechanism of active iron transport for pathogen directed drug delivery. In some studies it has been observed that linking of siderophores with antibiotics improves the penetration and hence increases the antibacterial activity of the antibiotics. It has been shown that a single siderophore can be used to deliver multiple drugs to target pathogenic microorganisms (Lu and Millar 1999). This has opened a new horizon in drug delivery.

Several natural iron-chelating antibiotics have been described in the literature (Prelog 1963; Rogers 1973). The antibiotic albomycin has been shown to be a linear peptide attached to a toxic thioribosyl unit, whose iron binding motifs are similar to siderophore ferrichrome. Albomycin is known to be transported in cells similar to ferrichrome via Fhu A OMP (Braun et al. 1983).

10.3
Iron Nutrition in Plants

Iron deficiency anemia (IDA) is a global problem threatening about one-third of the world population. The root cause of IDA is availability of mostly iron deficient foods, vegetables, and fruits and it's subsequent consumption (Beard and Stoltzfus 2000). As one-third of the global soils are calcareous in nature, soil iron becomes insoluble and the availability of soluble iron to plants even in iron rich soils is reduced (Chincholkar et al. 2000). This leads to the development of iron deficiency in the mass population due to the consumption of iron deficient plant products. Improving the proportion of this element in grains, seeds, fruits, and vegetables is the most suitable option to alleviate the problem. To improve the iron content of plant products, many strategies have been tried with some success (Lucca et al. 2002). Gramineous plants are known to produce phytosiderophores for iron assimilation in rhizosphere which usually fulfill its demand; however under certain conditions they fail to do so due to a low production profile and weak nature to chelate iron, which is evident from iron chlorosis and hence poor quality grains. In contrast, bacterial siderophores have been reported to be strong iron chelators in comparison with phytosiderophores and evidence is available indicating the acceptance of bacterial siderophore-iron complexes by plant roots (Yehuda et al. 1996). As depicted in Table 10.3, improvement in the growth and nutritional status of crops by application of siderophoregenic microorganisms is well established. One of the major factors involved in better plant growth is "iron nutrition" which is revealed through various growth effects. Evidence also suggests that the plants have an ability to accept iron derived from microbial siderophore-iron complexes in its biomass (Backer et al. 1985; Reid et al. 1984). This fact was confirmed at the molecular level by Castignetti and Smarrelli (1986) by proving that at least one plant enzyme, NADH: nitrate reductases, functions as ferri-siderophore reductases, indicating that plants accept iron available through iron-bacterial siderophore chelates. Thus, siderophore of microbial origin may play a key role in iron nutrition for all. Authors laboratory has contributed in observing, plant growth promotion by application of siderophoregenic microorganisms.

Table 10.3. Siderophoregenic microorganisms having PGPR activity

Siderophoregenic micro-organism	Plant	Growth effects	Reference
Pseudomonas sp	Maize	Increase in germination percentage, shoot and root length and dry weight	Sharma and Johri 2003
Bradyrhizobium japonicum	Soybean	Increase in percent germination, nodulation, chlorophyll, oil, protein content, and number of pods, shoot length, branches and root length	Khandelwal et al. 2002
Pseudomonas sp.	Maize	Better growth due to disease suppression	Pal et al. 2001
Proteus sp.	Mugbeans	Growth promotion with fungicidal activity	Barthakur 2000
Pseudomonas strain (fluorescent)	Pea nut	Antibiosis and disease suppression	Dileep and Kumar 2000
Penicillium chrysogenum[a]	Cucumber, maize	Increased chlorophyll content	Hordt et al. 2000
Kluyvera ascorbata	Tomato, Canola and Indian mustard	Decreased heavy metal toxicity	Burd et al. 2000
Pseudomonas B10	Potato	Increased growth	Buyer and Leong 1986
Pseudomonas aeruginosa	Groundnut[b]	Improved percent germination, root ramification, nodulation, height, foliage and chlorophyll content	Manwar 2001

[a] Ferreted siderophore mixture used under hydroponic conditions
[b] Increase in nutritional values of groundnut has also been reported

10.4
Role of Microbial Siderophores in Plant Health

Modern practices for suppressing root phytopathogens are largely based on use of synthetic pesticides. These chemical pesticides have proven to cause detrimental environmental effects. Its incorporation in the food chain poses hazards to macro- and micro-organisms in that ecosystem as well as to human health (Chincholkar et al. 2000). Hence, there is a demand and need to introduce new methods to supplement existing disease control strategies so as to achieve better disease control. Currently, among the disease control strategies, biological control is gaining attention over chemical agents (Chet and Inbar 1994). It provides an alternate means of reducing the incidence of plant diseases without those negative aspects of pesticide application (Papavizas 1985; Cook 1993). Table 10.4 shows the impact of biocontrol agents compared to chemical pesticides. Biocontrol of plant diseases involves the use of either a single organism or consortium of organisms known to produce secondary metabolite(s) and inhibit phytopathogens, thus reducing disease occurrence (Walsh et al. 2001).

Kloepper et al. (1980) were first to demonstrate the importance of siderophore synthesis as an integer of the mechanism of biological control. Production of siderophores by various strains with disease biocontrol ability have been studied and reviewed well in the literature (Cook et al. 1995; Handelsman and Stab 1996; Walsh et al. 2001; Whipps 2001; Chincholkar et al. 2000). In almost all studies, the role of siderophores is well accepted in iron nutrition, plant growth promotion and biological control of diseases. Researchers have speculated on a siderophore mediated biological control mechanism based on its role in competition for iron and induced systemic resistance.

Table 10.4. Comparative analysis of chemical pesticide and biocontrol agent

Parameter/impact	Chemical pesticide	Biocontrol agent
Effect	Immediate, could be for short term	Gradual, may last long
Accumulation in food chain	Observed	Not observed
Poising hazard	Very high possibility	Low possibility
Contribution to pollution	High and alarming	Not observed due to natural existence
Development of resistance	Usual phenomenon	Not yet reported
Economics of production	Expensive, could be cheap due to production at large scale	Cheap, could be expensive at low production scale

10.4.1
Competition for Iron

Iron, although an essential nutrient for most microorganisms, is extremely limiting due to its low solubility in an aerobic aqueous environment. Hence, these microbes excrete siderophores to fulfill the demand for iron for metabolic activities. A potent biocontrol agent produces strong siderophores, which chelate iron in the rhizosphere in competitive fashion and reduce its (iron's) availability to others in its niche. Moreover, it also possesses the ability to utilize foreign iron-siderophore complexes leading to further deprivation of iron for others. Considering the complexity of rhizosphere, experimental proof of this hypothesis is still controversial. Joshi et al. (2006) have described a nodule isolate capable of utilizing ferri-siderophore complexes of other rhizospheric isolates while siderophore produced by isolate was not utilized by other rhizospheric isolates.

10.4.2
Induced Systemic Resistance (ISR)

Some biocontrol agents induce a sustained change in plant, increasing its tolerance to infection by pathogen, a phenomenon known as induced systemic resistance. Disease resistance induction is an active plant defense process that is activated by biotic and abiotic inducers which depends on physical or chemical barriers of the host (Uknes et al. 1993). De Meyer (1999) has reported such a kind of systemic resistance induced by salicylic acid, a siderophore, produced at bean roots by a rhizobacterium *Pseudomonas aeruginosa*.

10.4.3
Indirect Contribution to Biological Control

Considering the fact that iron is an indispensable physiological factor, its absence lead to a reduction in production of various metabolites by micro-organisms including antibiotics and HCN, etc. However, siderophores aid its production by making iron available in the required quantity. Voisard et al. (1989) observed reduced control of black root rot of tobacco by Pvd– mutants of *P. fluorescence* CHA0. The loss of activity was attributed to loss of HCN synthesis due to its inability to synthesize siderophore hence deprived iron.

Table 10.5 illustrates various studies where siderophoregenic cultures have been successfully employed for demonstrating biocontrol activity against various plant pathogens.

Table 10.5. Recent examples of siderophoregenic microorganisms having biocontrol activity

Siderophoregenic organism	Pathogen controlled	Plant-disease	Reference
Rhizobium meliloti	*Macrophomina phaseolina*	Charcoal rot in groundnut	Arora et al. 2001
Pseudomonas fluorescens	*Rhizoctonia solani*	Rice sheath blight	Nagarajkumar et al. 2004
Pseudomonas sp.	*Colletotrichum dematium, Rhizoctonia solani* and *Sclerotium rolfsii*	Maize	Sharma and Johri 2003
Pseudomonas sp. EM85	*Macrophomina phaseolina, Fusarium moniliforme Fusarium graminearum*	Maize root diseases	Pal et al. 2001
Proteus sp.	*Fusarium oxysporum*	Mung beans	Barthakur 2000
Rhodotorula sp	*Botrytis cinerea*	Grey mould on a wide variety of host plants including numerous commercial crops	Calvente et al. 2001
Pseudomonas aeruginosa	*Macrophomina phaseolina, Fusarioum oxysporum*	-	Gupta et al. 1999
Pseudomonas aeruginosa	*Aspergillus niger, Aspergillus flavus, Aspergillus oryzae, Fusarium oxysporum, Sclerotium rolfsii* and *Alternaia alternata*	Groundnut	Manwar 2001

10.5
Conclusion

In the mid-twentieth century, siderophores were investigated and studied thoroughly using modern techniques; however, its dominant and single ability to chelate iron was ignored. In the 1990s, after understanding the chemical facts, its medical applications were proposed. Because of the interdisciplinary nature of siderophore production technology, various scientists were attracted to these fascinating molecules. Due to extraordinary metal chelating abilities, a few laboratories started working on sideorphore mediated uranium chelation.

In the twenty-first century, even though few reports are available on the biotechnological applications of siderophores, it is clear that they may have applications in many fields of human endeavors including health-care products, environment and industry. It is hoped that scientists from all disciplines will give prime attention towards these molecules to exploit them for betterment of human life.

References

Arora NK, Kang SC, Maheshwari DK (2001) Isolation of siderophore-producing strains of *Rhizobium meliloti* and their biocontrol potential against *Macrophomina phaseolina* that causes charcoal rot of groundnut. Curr Sci 81:673–677

Ateskan U, Mas MR, Yasar M, Deveci S, Babaoglu E, Comert B, Mas NN, Doruk H, Tasci I, Ozkomur ME, Kocar IH (2003) Deferoxamine and meropenem combination therapy in experimental acute pancreatitis. Pancreas 27:247–252

Backer JO, Messens E, Hedges RW (1985) The influence of agrobactin on the uptake of ferric iron by plants, FEMS Microbial Ecol 31:171–175

Banin E, Morad Y, Berenshtein E, Obolensky A, Yahalom C, Goldich J, Adibelli FM, Zuniga G, DeAnda M, Pe'er J, Chevion M (2003) Injury induced by chemical warfare agents: characterization and treatment of ocular tissues exposed to nitrogen mustard. Invest Ophthalmol Vis Sci 44:2966–2972

Barthakur M (2000) Plant growth promotion and fungicidal activity in a siderophore-producing strain of Proteus sp. Folia Microbiol 45:539–543

Beard J, Stoltzfus R (eds) (2000) Iron-deficiency anemia: reexamining the nature and magnitude of the public health problem, Proceedings of the WHO–INACG Conference, Belmont, MD, USA, 2000. J Nutr (2001), Supplement 131:2S-II

Braun V, Gunthner K, Hantke K, Zimmermann L (1983) Intracellular activation of albomycin in *Escherichia coli* and *Salmonella typhymurium*. J Bacteriol 156:308–315

Breidbach T, Scory S, Krauth-Siegel RL, Steverding D (2002) Growth inhibition of bloodstream forms of *Trypanosoma brucei* by the iron chelator deferoxamine. Int J Parasitol 32:473–479

Burd GI, Dixon DG, Glick BR (2000) Plant growth-promoting bacteria that decrease heavy metal toxicity in plants. Can J Microbiol 46:237–245

Buyer JS, Leong J (1986) Iron transport-mediated antagonism between plant growth-promoting and plant-deleterious *Pseudomonas* strains. J Biol Chem 261:791–794

Calvente V, de Orellano ME, Sansone G, Benuzzi D, Sanz de Tosetti MI (2001) Effect of nitrogen source and pH on siderophore production by Rhodotorula strains and their application to biocontrol of phytopathogenic moulds. J Ind Microbiol Biotechnol 26:226–229

Castignetti D, Smarrelli J (1986) Siderophores, the iron nutrition of plants and nitrate reductases. FEBS Lett 209:147–151

Chekanov VS, Nikolaychik V, Maternowski MA, Mehran R, Leon MB, Adamian M, Moses J, Dangas G, Kipshidze N, Akhtar M (2003) Deferoxamine enhances neovascularization and recovery of ischemic skeletal muscle in an experimental sheep model. Ann Thorac Surg 75:184–189

Chet I, Inbar J (1994) Biological control of fungal pathogens. Appl Biochem Biotechnol 48:37–43

Chincholkar SB, Chaudhari BL, Talegaonkar SK, Kothari RM (2000) Microbial iron chelators: a sustainable tool for the biocontrol of plant disease. In: Upadhyay RK, Mukerji KG, Chamola PC (eds) Biocontrol potential and its exploitation in sustainable agriculture. Kluwer Academic Publishers, pp 49–70

Clarkson AB Jr, Turkel-Parrella D, Williams JH, Chen LC, Gordon T, Merali S (2001) Action of deferoxamine against *Pneumocystis carinii*. Antimicrob Agents Chemother 45:3560–3565

Cook RJ (1993) Making greater use of introduced microorganisms for biological control of plant pathogens. Ann Rev Phytopathol 31:53–80

Cook RJ, Thomashow LS, Weller DM, Fujimoto D, Mazzola M, Bangera G, Kim D (1995) Molecular mechanisms of defense by rhizobacteria against root disease. Proc Natl Acad Sci USA 92:4197–4201

De Meyer G, Capieau K, Audenaert K, Buchala A, Metraux JP, Hofte M (1999) Nanogram amounts of salicylic acid produced by the rhizobacterium *Pseudomonas aeruginosa* 7NSK2 activate the systemic acquired resistance pathway in bean. Mol Plant Microbe Interact12:450–458

Dileep C, Kumar BS (2000) Influence of metham sodium on suppression of collar rot disease of peanut, in vitro antibiosis, siderophore production and root colonization by a fluorescent pseudomonad strain FPO4. Indian J Exp Biol 38:1245–1250

Eisendle M, Schrettl M, Kragl C, Muller D, Illmer P, Haas H (2006) The intracellular siderophore ferricrocin is involved in iron storage, oxidative-stress resistance, germination, and sexual development in *Aspergillus nidulans*. Eukaryot Cell 5:1596–1603

Gupta CP, Sharma A, Dubey RC, Maheshwari DK (1999) *Pseudomonas aeruginosa (GRC1)* as a strong antagonist of *Macrophomina phaseolina* and *Fusarium oxysporum*. Cytobios 99:183–189

Handelsman J, Stabb EV (1996) Biocontrol of soilborne plant pathogens. Plant Cell 8:1855–1869

Heller T, Hochstetter V, Basler M, Borck V (2004) Vitamin B6-sensitive hereditary sideroblastic anemia. Dtsch Med Wochenschr 129:141–144

Hennard C, Truong QC, Desnottes JF, Paris JM, Moreau NJ, Abdallah MA (2001) Synthesis and activities of pyoverdin-quinolone adducts: a prospective approach to a specific therapy against *Pseudomonas aeruginosa*. J Med Chem 44:2139–2151

Hesseltine CW, Pidacks C, Whiteball AR, Bononos N, Hutchings BL, Williams JH (1952) Coprogen, a new growth factor for coprophillic fungi. J Am Chem Soc 74:1362

Hordt W, Romheld V, Winkelmann G (2000) Fusarinines and dimerum acid, mono- and dihydroxamate siderophores from Penicillium chrysogenum, improve iron utilization by strategy I and strategy II plants. Biometals 13:37–46

Horwitz LD, Sherman NA, Kong Y, Pike AW, Gobin J, Fennessey PV, Horwitz MA (1989) Lipophilic siderophores of *Mycobacterium tuberculosis* prevent cardiac reperfusion injury. Proc Natl Acad Sci USA 95:5263–5268

Jalal MAF, van der Helm D (1991) Isolation and spectroscopic identification of fungal siderophores. In: Winkelmann G (ed) Handbook of microbial iron chelators. CRC Press, Boca Raton, pp 235–236

Joshi FR, Kholiya SP, Archana G, Desai AJ (2006) Siderophore cross-utilization amongst nodule isolates of the cowpea miscellany group and its effect on plant growth in the presence of antagonistic organisms. Microbiol Res [Epub ahead of print]

Karayilanoglu T, Gunhan O, Kenar L, Kurt B (2003) The protective and therapeutic effects of zinc chloride and desferrioxamine on skin exposed to nitrogen mustard. Mil Med 168:614–617

Khandelwal SR, Manwar AV, Chaudhari BL, Chincholkar SB (2002) Siderophoregenic bradyrhizobia boost yield of soybean. Appl Biochem Biotechnol 102/103:155–168

Kinzel O, Tappe R, Gerus I, Budzikiewicz H (1998) The synthesis and antibacterial activity of two pyoverdin-ampicillin conjugates, entering *Pseudomonas aeruginosa* via the pyoverdin-mediated iron uptake pathway J Antibiot (Tokyo) 51:499–507

Kloepper JW, Leong J, Tientze M, Schroth MN (1980) Pseudomonas siderophores: a mechanism explaining disease suppressive soils. Curr Microbiol 4:317–320

Lankford CE (1973) Bacterial assimilation of iron. Crit Rev Microbiol 2:273

Link G, Ponka P, Konijn AM, Breuer W, Cabantchik ZI, Hershko C (2003) Effects of combined chelation treatment with pyridoxal isonicotinoyl hydrazone analogs and deferoxamine in hypertransfused rats and in iron-loaded rat heart cells. Blood 101:4172–4179

Lu Y, Miller MJ (1999) Syntheses and studies of multiwarhead Siderophore-5-fluorouridine conjugates. Bioorg Med Chem 7:3025–3038

Lucca P, Hurrell R, Potrykus I (2002) Fighting iron deficiency anemia with iron-rich rice. J Am Coll Nutr 21:184S–190S

Mahmoud MS (2002) Effect of the iron chelator deferoxamine on *Trichomonas vaginalis* in vitro. J Egypt Soc Parasitol 32:691–704

Manwar AV (2001) Application of microbial iron chelators (siderophores) for improving yield of groundnut. Ph.D. Thesis, North Maharashtra University, Jalgaon

Nagarajkumar M, Bhaskaran R, Velazhahan R (2004) Involvement of secondary metabolites and extracellular lytic enzymes produced by *Pseudomonas fluorescens* in inhibition of *Rhizoctonia solani*, the rice sheath blight pathogen. Microbiol Res 159:73–81

Neilands JB (1981) Microbial iron compounds. Ann Rev Biochem 50: 715–731

Obejero-Paz CA, Yang T, Dong WQ, Levy MN, Brittenham GM, Kuryshev YA, Brown AM (2003) Deferoxamine promotes survival and prevents electrocardiographic abnormalities in the gerbil model of iron-overload cardiomyopathy. J Lab Clin Med 141:121–130

Pal KK, Tilak KV, Saxena AK, Dey R, Singh CS (2001) Suppression of maize root diseases caused by *Macrophomina phaseolina, Fusarium moniliforme* and *Fusarium graminearum* by plant growth promoting rhizobacteria. Microbiol Res 156:209–223

Papavizas GC (1985) Trichoderma and Gliocladium: biology, ecology and the potential for biocontrol. Ann Rev Phytopathol 23:23–54

Pelit A, Haciyakupoglu G, Zorludemir S, Mete U, Daglioglu K, Kaya M (2003) Preventative effect of deferoxamine on degenerative changes in the optic nerve in experimental retro-bulbar haematoma. Clin Experiment Ophthalmol 31:66–72

Poole K, McKay GA (2003) Iron acquisition and its control in *Pseudomonas aeruginosa*: many roads lead to Rome. Front Biosci 8:661–686

Poras H, Kunesch G, Barriere JC, Berthaud N, Andremont A (1998) Synthesis and in vitro antibacterial activity of catechol-spiramycin conjugates. Antibiot 51:786–794

Prelog V (1963) Iron containing antibiotics and microbial growth factors. Pure Appl Chem 6:327–332

Reid CPP, Crowley DE, Powell PE, Kim HJ, Szaniszlo PJ (1984) Utilization of iron by oat when supplied as ferrated hydroxamate siderophore or as ferrated synthetic chelate. J Plant Nutr 7:437–447

Richardson DR (2002) Therapeutic potential of iron chelators in cancer therapy. Adv Exp Med Biol 509:231–249

Riva A, Mariani R, Bovo G, Pelucchi S, Arosio C, Salvioni A, Vergani A, Piperno A (2004) Type 3 hemochromatosis and beta-thalassemia trait. Eur J Haematol 72:370–374

Rogers HJ (1973) Iron-binding catechols and virulence of *Escherichia coli*. Infect Immun 7:445–456

Rotheneder A, Fritsche G, Heinisch L, Mollmann U, Heggemann S, Larcher C, Weiss G (2002) Effect of synthetic siderophores on proliferation of *Plasmodium falciparum* in infected human erythrocytes, Antim Ag Ch 46:2010–2013

Sharma A, Johri BN (2003) Growth promoting influence of siderophore-producing Pseudomonas strains GRP3A and PRS9 in maize (Zea mays L.) under iron limiting conditions. Microbiol Res 158:243–248

Soybir N, Soybir G, Lice H, Dolay K, Ozseker A, Koksoy F (2002) The effects of desferrioxamine and vitamin E as supplements to antibiotics in the treatment of peritonitis in rats J R Coll Surg Edinb 47:700–704

Uknes S, Winter AM, Delaney T, Vernooij B, Morse A, Friedrich L, Nye G, Potter S, Ward E, Ryals J (1993) Biological induction of systemic acquired resistance in *Arabidopsis*. Mol. Plant-Microbe Interact 6:692–698

Voest EE, Vreugdentil G, Marx JJM (2005) Iron chelating agents in non-iron-overload conditions, Ann Intern Med 120:490–510

Voisard C, Keel C, Haas D, Defago G (1989) Cyanide production by Pseudomonas fluorescens helps suppress black root rot of tobacco under gnotobiotic conditions, EMBO J 8:351–358

Walsh UF, Morrisey JP, O'Gara F (2001) Pseudomonas for biocontrol of phytopathogens: from functional genomics to commercial exploitation. Curr Opin Biotechnol 12:289–295

Whipps JM (2001) Microbial interactions and biocontrol in the rhizosphere, J Exp Bot 52:487–511

Yehuda Z, Shenker M, Romheld V, Marschner H, Hadar Y, Chen Y (1996) The role of ligand exchange in the uptake of iron from microbial siderophores by gramineous plants. Plant Physiol 112:1273–1280

11 Biotechnological Production of Siderophores

María Elena Díaz de Villegas

11.1
Introduction

The increasing introduction of biotechnological products in agriculture allows the achievement of higher yields in almost every present-day commercial crop, leading at the same time to a higher quality and the generation of less aggressive wastes, thereby minimizing ecological damage (Cohen et al. 1998; Liebman and Gallaundt 1997). Biotechnology may be used to develop environmentally safe and economically sound alternatives to chemical pesticides, simultaneously providing efficient methods for pest control.

Newer biotechnological products are currently being developed through the stimulation of plant self-defense by the application of plant-beneficial bacteria for biological control disease (bio pesticides) (Thomashow and Weller 1995) and finally, aiming to plant growth regulation (bio fertilizers) (Montesinos et al. 2002). The commercial interest and user acceptance for biological products depend on the development of low-cost, stable products able to provide a consistent efficacy.

In this context, biological control has gained importance in recent years, especially the products constituted by *Pseudomonas* spp. due to its possibilities to colonize plant roots and to produce siderophores and antimicrobial metabolites, which inhibit the growth of pathogens (Buysens et al. 1996; Fujimoto et al. 1995; Vidhyasekaran and Muthamilan 1999; Nagarajkumar et al. 2004; Walsh et al. 2001). It has been suggested that siderophores are antagonistic with plant pathogens by sequestering iron from the environment, thereby restricting the growth of harmful microorganisms (Leeman et al. 1996; Loper and Buyer 1991).

Pseudomonas' siderophores have also been involved in the induction of systemic resistance (ISR) in plants (Leeman et al. 1996; Maurhofer 1994), e.g.

Cuban Institute for Research on Sugar Cane by Products (ICIDCA),
Vía Blanca No 804 y Carretera Central, San Miguel Del Padrón, Ciudad de La Habana, Cuba,
email: mariaelena.diaz@icidca.edu.cu

Soil Biology, Volume 12
Microbial Siderophores
A. Varma, S.B. Chincholkar (Eds.)
© Springer-Verlag Berlin Heidelberg 2007

the enhancement of the capacity of self-defense against a broad spectrum of pathogens, triggered by non-pathogenic plant growth-promoting rhizobacteria (Buysens et al. 1996).

Nowadays much research is being devoted to the development of *Pseudomonas* inoculants and other biological products constituted by active metabolites such as antibiotics and siderophores as bio-control agents (Mark et al. 2006). *Pseudomonas* spp. have been efficiently used for pest bio-control in the past, and at present time there are several commercial products already in the market (Wilson 1997); nevertheless, the applications of purified siderophores as bacteriostatic or fungistatic agents in combination with other antibacterial factors will certainly raise great interest in the near future (Raaska et al. 1993).

Our research group has developed a biological product GLUTICID, which is an antifungal product constituted by antimicrobial metabolites such as siderophore pyoverdine and salicylic acid produced by *Pseudomonas aeruginosa* PSS, very effective against *Paeronospora tabacina* in tobacco culture, *Alternaria solani* in tomato and *Pseudoperonospora cubensis* in cucumber (Villa and Díaz de Villegas 1996; Villa et al. 2002, 2004).

Several factors can be mentioned that affect the efficacy of siderophores as control agents against plant pathogens, the most important among them being (Glick and Bashan 1997):

- Type of microorganism
- Target phytopathogen
- Medium composition

A better understanding of the culture medium and environmental factors of siderophores production by *Pseudomonas* spp is needed if new and efficient products are going to be developed. This review focuses on the influence of these factors in the production of siderophores.

11.2
Genus *Pseudomonas*

According to Todar 2004, the general characteristics of the *Pseudomonas* genus are: Gram-negative, rod-shaped, strictly aerobic bacteria, motile by polar flagella, chemo-organotrophic metabolism, catalase-positive, usually oxidase-positive. No organic growth factors are required. Diffusible and/or insoluble pigments may be produced.

Fluorescent species that belong to Group I include *P. aeruginosa*, *P. fluorescens*, *P. syringae*, and *P. cichorii*. This Group I also includes several non-fluorescent species such as *P. stutzeri* and *P. mendocina*.

A common characteristic of fluorescent pseudomonads is the production of pigments that fluoresce under short wave-length (254-nm) ultraviolet light, particularly after growth under conditions of iron limitation. Some of these pigments and/or their derivatives are known to play a role as siderophores in the iron uptake systems of the bacteria and hence their production is markedly enhanced under conditions of iron deficiency.

They have a strict metabolism with oxygen as final electron acceptor, although in some cases NO_3 can play that role. They could degrade an ample group of organic compounds as aromatic compounds and halogen derivatives (Kerster et al. 1996).

Because of their catabolic versatility, their excellent root-colonizing abilities, and their capacity to produce a wide range of antifungal metabolites, the soil-borne fluorescent pseudomonads have received particular attention as efficient biological control agents (Boruah and Kumar 2002; Nautiyal et al. 2003).

Pseudomonas spp. produce several siderophores such as pyoverdine, pyochelin, azotobactin, salicylic acid (Budzikiewicz 1997; Dave and Dube 2000). Recently Mercado-Blanco et al. (2001) reported a siderophore pseudomonine which chelates iron and other metals. All these siderophores contribute to disease suppression through the competition for iron (De Meyer and Höfte 1997; Loper and Henkels 1997).

11.3
Siderophores Production in Liquid Culture

Liquid culture constituted a feasible alternative for the production of different metabolites as siderophores due to its potentialities in the regulation and optimization of the parameters that affect the biosynthesis of these pigments. Fermentation conditions, such as dissolved oxygen, carbon source, pH or temperature, were controlled to allow the highest production of siderophores and to prevent the production of phytotoxic metabolites (Slininger and Shea-Wilbur 1995; Slininger et al. 1997, 1998).

Research published concerning the impact of liquid culture conditions (carbon and nitrogen sources, carbon-to-nitrogen ratio, nutrients, temperature, pH, and dissolved oxygen) (Sabra et al. 2002), microbial physiology (growth state) and metabolites on the qualities of the biocontrol products is scarce, yet these relationships are fundamental to design production processes (Slininger et al. 2003).

Siderophores production by strains of *Pseudomonas* spp. is mainly influenced by the following environmental factors:

- Iron concentration (Budzikiewicz 1993; Laine et al. 1996; Díaz de Villegas et al. 2002)
- Concentration and nature of carbon and nitrogen source (Park et al. 1988; Duffy and Défago 1999; Albesa et al. 1985; Díaz de Villegas 1999)
- Phosphate concentration (Défago and Haas 1990; Barbahaiya and Rao 1985)
- pH (Loper and Henkels 1997; Manninen and Mattila-Sandholm 1994)

- Temperature (Digat and Mattar 1990)
- Oxygen transfer (Kim et al. 2003)

Among all these factors, the most important is the concentration of Fe(III).

11.3.1
Iron Influence

Iron is an essential element for microbial metabolism and plays an important role in membrane-bound electron transport chains and in cytoplasmic redox enzymes. Despite its relative abundance in nature, in aerobic conditions in neutral and alkaline pH, it is present in its trivalent form Fe(III) as oxyhydroxides, which are not assimilated by microorganisms (Loper and Henkels 1997; Elliot and Huang 1979). To fulfill their iron requirement, microbes have developed a specific system to scavenge the last traces of iron from such an environment, low molecular weight chelators called siderophores.

Siderophores are iron-specific compounds which are secreted under low iron stress and which capture iron from the environment (Budzikiewicz 1993).

The function of siderophores is to solubilize iron atoms that might be available from the surroundings and to transport them into the cell. As less iron is available, a greater amount of siderophores will be produced. They are produced as free ligands (the desferri form) and become a complex with the iron available in the medium to give the ferri-complex. This complex is then taken up by cellular proteins that are in the outer cell-membrane (Höfte et al. 1993; Weller 1988; Neilands 1982).

The influence of iron in the production of siderophores has been the subject of several papers in recent years (Ochsner et al.2002; Braud et al. 2006). Iron concentration is the most important factor, since they are produced under iron-limiting conditions (Budzikiewicz 1993; Vasil and Ochsner 1999; Visca et al. 1993). Regarding the iron availability, it is important to take into account the environment in which the microorganism is growing, because in an intensively aerated liquid medium, iron exists mainly in the form of Fe(III) which is extremely insoluble at neutral pH. Thus the increase of oxygen pressure in the culture broth can reduce iron availability (Kim et al. 2003).

With laboratory culture media, the amount of siderophores produced by iron-sufficient growth can be as little as 0.1% of that produced under iron-deficient conditions. It is therefore necessary to remove even the smallest trace of iron from the culture medium if a successful production of siderophores is required (Messenger and Ratledge 1985).

Siderophores production is associated with iron concentration in the medium. Studies carried out with different *Pseudomonas* strains confirm this affirmation. *Pseudomonas fluorescens* (VTE94558), *Pseudomonas fluorescens* (VTT/ELT 116) and *Pseudomonas chlororaphis* (VTT-E-94557) at concentration of Fe(III) of 100 µM suppress the production of siderophores (Laine et al. 1996).

Iron concentrations of about 10 µM are considered high enough and generally result in excellent cell-mass accumulation with only modest yields of siderophores (Neilands 1984); nevertheless, *Pseudomonas fluorescens* 94 produces siderophores at an iron concentration of 50 µM (Manninen and Mattila-Sandholm 1994).

On the other hand, *Pseudomonas syringae* B30ID gradually represses siderophore production at Fe(III) concentration from 1 to 10 µM although low concentrations (between 0 and 1 µM) similarly induce siderophore production (Bultreys and Gheysen 2000).

According to these results, in our research we found that although cell growth of *Pseudomonas aeruginosa* PSS reached a maximum of siderophore concentration at values above 10 µM Fe(III), their biosynthesis was lowered at this concentration, since cell growth and siderophore production are inversely proportional under such conditions (Díaz de Villegas et al. 2002).

Pseudomonas aeruginosa NCCB 2452 and ATCC 15692 produce the siderophores pyoverdine and pyochelin. The concentration of both them was increased in iron-limited culture although the level of pyochelin was 10 times lower than that of pyoverdine and the total pioverdine is a close function of biomass concentration (Kim et al. 2003).

An alternative to produce pyoverdine in media containing Fe(III) by *Pseudomonas* spp is by the use of mutants. One example is the overproduction of pyoverdine by Fur mutants of *Pseudomonas aeruginosa* strain PAO1 and Fe 10 reported by Johnová et al. (2001). These authors obtained a pyoverdin overproduction in both iron-depleted and iron-supplemented Casamino acid (CA) media.

Pyochelin, one of the siderophore produced by *Pseudomonas aeruginosa*, binds other transition metals such as: Cu(II), Co(II), Mo(VI) and Ni(II) with significant affinity. Pyochelin synthesis is repressed by Fe(III) and Co(II) at 10 µM or Mo(VI), Ni(II) and Cu(II) at 100 µM, in this case the functional effects of Fe(III) can be mimicked by other transition metals, while pyoverdine synthesis were affected only by Fe(III) (Visca et al. 1992).

11.3.2
Nature of Nitrogen and Carbon Source

A critical step in the development of an economical fermentation process is the selection and optimization of the carbon and nitrogen source.

Nitrogen is a major component of important cellular elements including proteins, nucleic acid and cell wall and has been demonstrated its role in the regulation of a multitude of primary and secondary cellular processes and is often the most expensive component of production medium.

It is likely that all bacteria can utilize ammonia (as the NH_4^+ ion) which reflects its central role in nitrogen metabolism as the form in which nitrogen is incorporated into organic cell components. Amino acids are particularly good sources of nitrogen, generally inducing an increased growth rate; however, economic considerations undoubtedly make use of NH_4 salts more attractive in most industrial processes (Dunn 1985). Cellular nitrogen has to follows a pathway that goes through glutamine, glutamate or both in order to be assimilated, unless the latter was supplied in the culture broth and directly incorporated (Huntr 1985).

Glutamic acid as sole source of carbon and nitrogen also improves the production of siderophores (Casida 1992). In our experience *Pseudomonas aeruginosa* PSS produces siderophores in parallel to growth with an specific growth rate (μ) of $0.064 \ h^{-1}$ in a glutamic minimum medium with glutamic acid as sole carbon source in a concentration as low as $1 \ g \ L^{-1}$. A maximum siderophore value of 140 μM was reached after 25 h. A high siderophores productivity (P μM $L^{-1} \ h^{-1}$) of 5.75 was obtained due to the presence of glutamic acid in a culture medium that induces siderophores production, as other amino acids (Díaz de Villegas et al. 2002).

For siderophores production, different nitrogen sources have been reported. Among them are the supplements of other amino acids (Albesa et al. 1985). Strains of *Pseudomonas syringae* produce high levels of siderophore with almost all the 20 amino acid in standard proteins when used as the sole source of both carbon and nitrogen (Bultreys and Gheysen 2000).

Another amino acid commonly present in such media is asparagine, which is used as carbon and nitrogen source by almost all fluorescent *Pseudomonas* (Palleroni 1984). Solid and liquid culture media containing asparagines are reproducible and highly effective for the induced production of siderophores by strains of *Pseudomonas syringae* (Bultreys and Gheysen 2000). This amino acid is usually combined with sucrose in the medium known as sucrose-asparagine (SA), for the production of siderophores (Laine et al. 1996; Morris et al. 1992).

Carbon is a major constituent of cellular material, and in addition to this role it is frequently the energy source. It constituted approximately 50% of the dry weight of bacteria. *Pseudomonas* species have very simple nutritional requirements; they can use more than 150 different organic compounds as carbon source (Todar 2004).

For the production of siderophores in particular, it is possible to use almost any organic substrate, as was previously reported by Meyer and Abdallah (1978). Glycerol has been widely used as the carbon source in different media (Nowak-Thompson and Gould 1994), among them the King´s Medium B (King et al. 1954). In this medium it is possible to demonstrate the production of fluorescent yellow-green pigments.

On the other hand, Duffy and Défago (1999) reported that the production of pyochelin, a siderophore produced by *Pseudomonas fluorescens* CHAO, did not show a significant increase when glycerol was used alone or in combination with minerals; however, salicylic acid production exhibited a significant increase with that substrate.

Another carbon source used in siderophore production is sodium succinate in a medium with salts known as succinate-medium (Meyer and Abdallah 1978; Johnová et al. 2001; Boruah and Kumar 2002; Sharma and Johri 2003).

In our laboratory, when *Pseudomonas aeruginosa* PSS was cultured in such a medium, the highest siderophore concentration achieved was close to 60 μM.

The specific growth rate was 0.07 h^{-1} (Diaz de Villegas 1999), lower than that previously reported by Champomier-Verges et al. (1996) with *Pseudomonas aeruginosa* PA01 where a maximum growth (OD$_{650}$) of 0.650 h^{-1} was obtained, suggesting that succinate was used by *Pseudomonas aeruginosa* PSS more efficiently in siderophore synthesis than in growth, presumably due to the influence of succinate in the synthesis of this metabolite. This proposal is based on the structure of pyoverdine in which the 3-amino moiety of the chromophore is substituted with various acyl groups derived from succinate, malate or α-ketoglutarate (Demange et al. 1987; Linget et al. 1992).

Some sugars such as sucrose (Laine et al. 1996; Morris et al. 1992) and glucose (Sabra et al. 2000) are also appealing because they are easy and economical to provide during submerged cultures. Laine et al. (1996) reported that *Pseudomonas* strains grew well in sucrose-l-asparagine (SA) medium and produced siderophores in this low-iron medium, although differences in production efficiency between strains were detected.

Duffy and Défago (1999) reported that glucose used alone and in combination with minerals generally increased the production of siderophore pyochelin.

Our research showed that for *Pseudomonas aeruginosa* PSS grown on glucose minimum media, the siderophore production achieved was 180 μM at the end of the growth period, while the specific growth rate (μ) was 0.1394 h^{-1} (Díaz de Villegas et al. 2002).

11.3.3
Behavior of pH During Siderophore Production

During the production of siderophores by *Pseudomonas* strains, marked changes in the pH of the culture medium were observed in populations grown in different media.

In a medium containing asparagine, glucose and salts (GASN medium) in Petri dishes containing agar blocks, the pH decreased from 7 to 4.6 after one day, increased to 6.6 on the second day and rose to 7.5 on the third day. The increase in pH was accompanied by an abrupt increase in siderophore concentration. Weak bacterial growth and siderophore production occurred in cultures grown in asparagine minimal medium contained no glucose (Bultreys and Gheysen 2000). These authors also reported the alkalization of asparagine minimal medium (ASN-M-medium), where asparagine was adjusted to 5–7 before autoclaving. The level of siderophore production was relatively high compared with the amount of bacterial growth.

When *Pseudomonas aeruginosa* PSS was cultured in succinate medium, with sodium succinate as carbon and energy source, the pH only shifted from 7 to 7.5 despite a significant siderophore production took place during the growth period (Díaz de Villegas et al. 2002). In a medium containing glucose, urea and potassium phosphate during the growth period, the pH decreased from 7 to 5.5

in 18 h in accordance with siderophore concentration, which was lowered from 80 μM to 50 μM. After that time, the pH medium shifted from 5.5 to 7, which correlated well with a high level of siderophore concentration of 180 μM at the end of the growth period. A further result in a glutamic acid medium was that the pH increased from 7 to 8.5 during the growth period in accordance with the siderophore concentration (Díaz de Villegas 2002), which suggests that alkalinity in the medium is important to avoid siderophores destruction as has been pointed out by Budzikiewicz (1993).

11.3.4
Influence of Culture Temperature on Siderophore Production

Temperature is a very important environmental factor for siderophore production by *Pseudomonas* strains. All the known species of *Pseudomonas,* generally grow quite well at 28–30 °C (Todar 2004) while siderophores synthesis of some fluorescent *Pseudomonas* is inhibited above 33 °C (Loper and Schroth 1986). In agreement with this, Meyer et al. (1996) reported that *Pseudomonas aeruginosa* ATCC 15692 (PAO1 strain) in succinate medium show a drastic loss of its ability to produce pyoverdine when the cells were grown at temperatures ranging from 40 to 43 °C. However, growth itself was unaffected at this temperature range. Neither pyoverdine nor pyochelin was detectable in the growth supernatants obtained from iron-starved cultures incubated at 43 °C. In this respect, *P. aeruginosa* behaved similar to other bacteria which lack siderophore production capacity while grown at sub-lethal temperatures.

On the other hand, Digat and Mattar (1990) found that strains of fluorescent *Pseudomonas putida* from tropical origin, grew normally at 35 °C. At this temperature, most strains from temperate countries grew quite slowly or did not grow at all, and without production of siderophores with few exceptions, while tropical strains did.

11.4
Conclusion

- Concentration of Fe (III) is the most important factor in the production of siderophores. Higher concentration of siderophore is produced under lower concentration of Fe(III).
- Amino acids, particularly glutamic acid and asparagine, improve the production of siderophores.
- It is possible to use almost any organic substrate, such as glycerol, sodium succinate, glucose and sucrose as the carbon source.

- During the production of siderophore the pH changes but at the end of the culture generally the pH of the medium increases to alkaline values.
- The production of siderophores is inhibited above 33 °C.

References

Albesa I, Barberes LI, Pajaro MC, Craso AJ (1985) Pyoverdine production by *Pseudomonas* fluorescentes in synthetic media with various sources of nitrogen J Gen Microb 131:3251–3254

Barbahaiya HB, Rao KK (1985) Production of pyoverdine, the fluorescent pigment of *Pseudomonas aeruginosa* PAO1. FEMS Microbiol Lett 27:233–235

Boruah HPD, Kumar BSD (2002) Biological activity of secondary metabolites produced by strain of *Pseudomonas fluorescens*. Folia Microbiol 47:359–363

Braud A, Jézéquel K, Léger MA, Lebeau T (2006) Siderophore production by using free and immobilized cells of two pseudomonads cultivated in a medium enriched with Fe and/ or toxic metals (Cr, Hg, Pb). Biotechnol Bioeng 94:1080–1088

Budzikiewicz H (1993) Secondary metabolites from fluorescent pseudomonads. FEMS Microbiol Rev 104:209–228

Budzikiewicz H (1997) Siderophores of fluorescent pseudomonads. Z Naturforsch 52C:713–720

Bultreys A, Gheysen D (2000) Production and comparison of peptide siderophores from strains of distantly related pathovars of *Pseudomonas syringae* and *Pseudomonas viridiflava* LMG 2352. Appl Environ Microbiol 66:325–331

Buysens S, Heunges K, Poppe J, Höfte M (1996) Involvement of pyochelin and pyoverdine in suppression of phythium-induced damping-off of tomato by *Pseudomonas aeruginosa* 7NSK2. Appl Environ Microbiol 62:865–871

Casida LE Jr (1992) Competitive ability and survival in soil of *Pseudomonas* strain 679-2 a dominant, nonobligate bacterial predator of bacteria. Appl Environ Microbiol 58:32–37

Champomier-Verges MC, Stintzi A, Meyer JM (1996) Acquisition of iron by the non- siderophore-producing *Pseudomonas fragi*. Microbiology 142:1191–1199

Cohen JI, Falconi C, Komen J (1998) Strategic decisions for agricultural biotechnology: synthesis of four policy seminars. ISNAR Briefing Paper 38:1–11. ISSN 1021-2310

Demange P, Wenderbaum S, Bateman A, Dell A, Abdallah MA (1987) Bacterial siderophores: structure and physicochemical properties of pyoverdins and related compounds. In: Winkelman G, Helm DVD, Neilands JB (eds) Iron transport in microbes, plants and animals. VCH, Weinkeim, pp 167–187

Dave BP, Dube HC (2000) Detection and chemical characterization of siderophores of rhizobacterial fluorescent pseudomonads. Indian Phytopathol 53:97–98

De Meyer G, Höfte M (1997) Salicylic acid produced by the rhizobacterium *Pseudomonas aeruginosa* 7NSK2 induces resistance to leaf infection by Botrytis cinerea on bean. Phytopathology 87:588–593

Défago G, Haas D (1990) Pseudomonads as antagonists of soilborne plant pathogens. Modes of action and genetic analysis. In: Bollag JM, Stotzky YG (eds) Soil biochemistry, vol 6. Marcel Dekker, New York Basel, pp 249–291

Díaz de Villegas ME (1999) Evaluation of siderophore production by *Pseudomonas* spp. MSc Thesis. Faculty of Biology, Havana University (in Spanish)

Díaz de Villegas ME, Villa P, Frías A (2002) Evaluation of the siderophores production by *Pseudomonas aeruginosa* PSS. Rev Latinoam Microb 44:112–117

Digat B, Mattar J (1990) Effects of temperature on growth and siderophore production of Pseudomonas fluorescents-putida. Symbiosis 9:203–207

Duffy BK, Défago G (1999) Environmental factors modulating antibiotic and siderophore biosynthesis by *Pseudomonas fluorescens* biocontrol strains. Appl Environ Microbiol 65:2429–2438

Dunn GM (1985) Nutritional requirements of microorganism. In: Moo-Youg M (ed) Comprehensive biotechnology, vol 1. Pergamon Press, pp 113–125

Elliot HA, Huang CP (1979) The effect of complex formation on the adsorption characteristic of heavy metals. Environ Int 2:145–152

Fujimoto DK, Weller DM, Thomashow LS (1995) Role of secondary metabolites in root disease suppression. In: Inderjit KMM, Dakshini, Einhellig FA (eds) Allelopathy, organisms, processes and applications. ACS. Symposium Series 582, American Chemical Society, Washington, DC, pp 330–347

Glick BR, Bashan Y (1997) Genetic manipulation of plant growth-promoting bacteria to enhance biocontrol of phytopathogens. Biotechnol Adv 15:353–378

Höfte M, Buysens S, Koldam N, Cornelis P (1993) Zinc affects siderophore-mediated high affinity iron uptake systems in the rhizosphere *Pseudomonas aeruginosa* 7NSK2. BioMetals 6:85–91

Huntr IS (1985) Assimilation of nitrogen. In: Moo-Young M (ed) Comprehensive biotechnology, vol 1. Pergamon Press, pp 141–155

Johnová A. Dobisová M, Abdallah MA, Kyslik P (2001) Overproduction of pyoverdins by Fur mutants of *Pseudomonas aeruginosa* strains PAO1 and Fe10 in stirred bioreactors. Biotechnol Lett 23:1759–1763

Kerster K, Ludwig W, Vancanneyt M, De Vos P, Gillis M, Schleifer K-H (1996) Recent changes in the classification of the *Pseudomonads*: an overview. Sys Appl Microbiol 19:465–477

Kim EJ, Sabra W, Zeng AP (2003) Iron deficiency leads to inhibition of oxygen transfer and enhanced formation of virulence factors in cultures of *Pseudomonas aeruginosa* PAO1. Microbiology 149:2627–2634

King EO, Ward MK, Raney DE (1954) Two simple media for the demostration of pyocyanin and fluorescin. J Lab Clin Med 44:301–307

Laine MH, Karwoski MT, Raaska L, Mattila-Sandholm T (1996) Antimicrobial activity of *Pseudomonas* spp. against food poisoning bacteria and moulds. Lett Appl Microbiol 22:214–218

Leeman M, Den Ouden FM, Van Pelt JA, Dirkx FPM, Steijil H, Bakker PAHM, Schippers B (1996) Iron availability affects induction of systemic resistance to Fusarium wilt of radish in commercial greenhouse trials by seed treatment with Pseudomonas fluorescens WCS 374. Phytopathology 85:149–155

Liebman M, Gallandt ER (1997) Many little hammers: ecological approaches for management of crop-weed interactions In: Jackson LE (ed) Ecology in agriculture. Academic Press, San Diego CA, pp 291–343

Linget C, Slylianou DG, Dell A, Wolff RE, Piémont Y, Abdallah MA (1992) Bacterial siderophores: the structure of a desferribactin produced by *Pseudomonas fluorescens* ATTC 13525. Tetrahedron Lett 33:3851–3854

Loper JE, Schroth MN (1986) Importance of siderophores in microbial interactions in the Rhizosphere. In: Swinburne TR (ed) Iron, siderophores and plant diseases. Plenum Press, New York, pp 85–98.

Loper JE, Buyer JS (1991) Siderophores in microbial interactions on plant surfaces. Mol Plant-Microbe Interact 4:5–13

Loper JE, Henkels MD (1997) Availability of iron to *Pseudomonas fluorescens* in rhizosphere and bulk soil evaluated with an ice nucleation reporter. Gene Appl Environ Microbiol 99–105

Manninen E, Mattila-Sandholm T (1994) Methods for the detection of *Pseudomonas* siderophores. J Microbiol Meth 19:223–234

Mark GL, Morrissey JP, Higgins P, O'Gara F (2006) Molecular-based strategies to exploit
Pseudomonas biocontrol strains for environmental biotechnology applications. FEMS
Microbiol Ecol 56:167–177

Maurhofer M, Hase C, Meuwly P, Métraux JP, Défago G (1994) Induction of systemic resistance of tobacco to tobacco necrosis virus by the root-colonizing Pseudomonas fluorescens strain CHAO: influence of gacA gene and the pyoverdine production. Phytopathology 84:139–146

Mercado-Blanco J, van der Drift KMGM, Olsson PE, Thomas-Oates JE, van Loon LC, Bakker
PAHM (2001) Analysis of the *pmsCEAB* gene cluster involved in biosynthesis of salicylic
acid and the siderophore Pseudomonine in the biocontrol strain *Pseudomonas fluorescens* WCS374. J Bacteriol 183:1909–1920

Messenger AJM, Ratledge C (1985) Siderophores. In: Moo-Youg M (ed) Comprehensive biotechnology, vol 3. Pergamon Press, pp 275–294

Meyer JM, Abdallah MA (1978) The fluorescent pigment of *Pseudomonas fluorescens*. Biosynthesis, purification and physicochemical properties. J Gen Microbiol 107:319–328

Meyer JM, Neely A, Stintzi A, Georges C, Holder IA (1996) Pyoverdin is essential for virulence of *Pseudomonas aeruginosa*. Infect Immun 64:518–523

Montesinos E, Bonaterra A, Badosa E, Francãç J, Alemany J, Llorente I, Moragrega C (2002)
Plant-microbe interactions and the new biotechnological methods of plant disease control. Int Microbiol 5:169–175

Morris J, O´Sullivan DJ, Koster M, Leong J, Weisbeek PJ, O´Gara F (1992) Characterization of fluorescent siderophore-mediated iron uptake in *Pseudomonas* sp.strain M114:
evidence for the existence of an additional ferric siderophore receptor. Appl Environ
Microbiol 58:630–635

Nagarajkumar M, Bhaskaran R, Velazhahan R (2004) Involvement of secondary metabolites
and extracellular lytic enzymes produced by Pseudomonas fluorescens in inhibition of
Rhizoctonia solani , the rice sheath blight pathogen. Microbiol Res 159:73–81

Nautiyal CS, Johri JK, Singh HB (2003) Survival of the rhizosphere-competent biocontrol
strain Pseudomonas fluorescens NBR 12650 in the soil and phytosphere. Can J Microbiol 48:588–601

Neilands JB (1982) Microbial envelope proteins related to iron. Annu Rev Microbiol
36:285–309

Neilands JB (1984) Methodology of siderophores. Struct Bond 58:1–24

Nowak-Thompson B, Gould SJ (1994) A simple assay for fluorescent siderophores produced
by Pseudomonas species and an efficient isolation of pseudobactin. BioMetals 7:20–24

Ochsner UA, Wilderman PJ, Vasil AI, Vasil ML (2002) GeneChip® expression analysis of the
iron starvation response in *Pseudomonas aeruginosa*: identification of novel pyoverdine
biosynthesis genes. Mol Microbiol 45:1277

Palleroni NJ (1984) Family I Pseudomonadaceae. In: Krieg NR, Holt JG (eds) Bergey´s Manual of systematic bacteriology, vol 1. The Williams and Wilkins Co Baltimore, Md, pp
141–199

Park CS, Paulitz TC, Baker R (1988) Biocontrol of *Fusarium* wilt of cucumber resulting from
interactions between *Pseudomonas putida* and non pathogenic isolates off *Fusarium oxysporum*. Phytopathology 78:190–194

Raaska L, Viikari L, Mattila-Sandholm T (1993) Detection of siderophores in growing cultures of *Pseudomonas spp.*. J Ind Microbiol 11:181–186

Sabra W, Zeng A-P, Lünsdorf H, Deckwer W-D (2000) Effect of oxygen on formation and
structure of *Azotobacter vinelandii* alginate and its role in protecting nitrogenase. Appl
Environ Microbiol 66:4037–4044

Sabra W, Kim EJ, Zeng AP (2002) Physiological responses of Pseudomonas aeruginosa PAO1 to oxidative stress in controlled microaerobic and aerobic cultures. Microbiology 148:3195–3202

Slininger PJ, Shea-Wilbur MA (1995) Liquid-culture pH, temperature, and carbon (not nitrogen) source regulate phenazine productivity of the take-all biocontrol agent *Pseudomonas fluorescens* 2-79. Appl Microbiol Biotechnol 43:794–800

Slininger PJ, VanCauwenberge JE, Shea-Wilbur MA, Burkhead KD, Schisler DA, Bothast RJ (1997) Reduction of phenazine-1-carboxylic acid accumulation in growth cultures of the biocontrol agent *Pseudomonas fluorescens* 2-79 eliminates phytotoxic effects of wheat seed inocula without sacrifice to take-all suppressiveness. In A. Ogoshi, K. Kobayashi, Y. Homma, F. Kodama, N.Kondo & S. Akino (eds.), Plant growth-promoting rhizobacteria present status and future prospects. The 4th PGPR International Workshop Organizing Committee, Faculty of Agriculture, Hokkaido University, Lab of Plant Pathology, Sapporo, Japan, , p. 464–467

Slininger PJ, VanCauwenberge JE, Shea-Wilbur MA, Bothast RJ (1998) Impact of liquid culture physiology, environment, and metabolites on biocontrol agent qualities- *Pseudomonas fluorescens* 2-79 versus wheat take-all, p. 199-221. In: Boland GJ, Kuykendall LD (eds) Plant-microbe interactions and biological control. Marcel Dekker, New York, pp 329–353

Slininger P, Behle RW, Jackson MA, Schisler DA (2003) Discovery and development of biological agents to control crop pest. Neotrop Entomol 32:183–195

Sharma A, Johri BN (2003) Combat of iron-deprivation through aplant growth promoting fluorescent *Pseudomonas* strain GRP3A in mung bean (*Vigna radiate* L. Wilzeck). Microbial Res 158:77–81

Todar K (2004) Pseudomonas and its relatives. Todar's online textbook of bacteriology. www.textbookofbacteriology.net

Thomashow LS, Weller DM (1995) Current concepts in the use of introduced bacteria for biological disease control: mechanisms and antifungal metabolites. In: Stacey G, Keen N (eds) Plant microbe interactions, vol 1. Chapman and Hall, New York, pp 187–235

Vasil ML, Ochsner UA (1999) The response of *Pseudomonas aeruginosa* to iron: genetics, biochemistry and virulence. Mol Microbiol 34:399–413

Vidhyasekaran P, Muthamilan M (1999) Evaluation of powder formulation of Pseudomonas fluorescens Pf1 for control of rice sheath blight. Biocontrol Sci Technol 9:67–74

Villa P, Bell A, Frías A, Díaz de Villegas ME, Martínez J, Gutiérrez I, Torres P, Redondo D, Hernández Y, Stefanova M, Alfonso I, Budzikiewicz H (2004) Biosynthesis of metabolites from *Pseudomonas aeruginosa* strain PSS for the control of phytopathogenic fungus. Memorias Encuentro Latinoamericano y del Caribe de Biotecnología Agrícola. Taller 3 21-25 de Junio 2004, Boca Chica, Rep. Dominicana (in Spanish)

Villa P, Díaz de Villegas ME (1996) Potentialities of a biological product from *Pseudomonas* sp. strain PSS for the control of fungus and weeds. Sobre los Deriv 30:6-12 (in Spanish)

Villa P, Díaz de Villegas ME, Stefanova M, Michelena G, Rodríguez JA, Gutiérrez I, Frías A (2002) Procedimiento de obtención de metabolitos antifúngicos de *Pseudomonas aeruginosa* PSS por vía biotecnológica Patente cubana No 22 805

Visca P, Colotti G, Serino L, Verzili D, Orsi N, Chiancone E (1992) Metal regulation of siderophore synthesis in pseudomonas aeruginosa and functional effects of siderophore- metal complexes. Appl Environ Microbiol 58:2886–2893

Visca P, Ciervo A, Sanfilippo V, Orsi N (1993) Iron-regulated salicylate synthesis by Pseudomonas spp. J Gen Microbiol 139:1995–2001

Walsh UF, Morrissey JP, O'Gara F (2001) *Pseudomonas* for biocontrol of phytopathogens: from functional genomics to commercial exploitation. Curr Opin Biotechnol 12:289–295

Weller DM (1988) Biological control of soilborne plant pathogens in the rhizosphere with bacteria. Ann Rev Phytopathol 26:379–407
Wilson M (1997) Biocontrol of aerial plant diseases in agriculture and horticulture: current approaches and future prospects. J Ind Microbiol Biotechnol 19:188–191

12 Microbial Siderophore: A State of Art

S.B. Chincholkar, B.L. Chaudhari, and M.R. Rane

12.1
Introduction

Iron is ubiquitous in almost all environments of this world. During the period of origin of life this iron was available in soluble form as the atmosphere was reducing. In fact, an abundance of iron might have led to its involvement in a wide range of metabolic reactions and physiological functions such as respiratory chain and transport of oxygen. Later release of oxygen during photosynthetic reactions resulted in an oxidizing atmosphere making iron almost insoluble. The solubility product of iron under aerobic conditions approaches as low as 10^{-38} (Castignetti and Smarrelli 1986). However, iron is physiologically indispensable. Plants being auto-lithotrophs developed various mechanisms to collect insoluble iron from rhizosphere. A type of mechanism exists among non-gramineous monocots and all dicots, which create a localized reducing environment in rhizosphere by secretion of protons (in the form of organic acids) complemented with membrane-associated reductases. A specialized mechanism for iron uptake is observed in gramineous plants which, via roots, release iron chelating non-proteinogenic amino acids called phytosiderophores (such as mugineic acid by barley, distichonic acid by beer barley, avenic acid A by oats, deoxymugineic acid by wheat, hydroxymugineic acid by rye and nicotinamine by tobacco). Some plants like barley are able to take up ferriphytosiderophores 100–1000 times faster than other ferri-chelators (Castignetti and Smarrelli 1986).

Microorganisms being unicellular are in intimate association and hence depend directly on a variable environment to meet its requirements. Almost all aerobic micro-organisms have evolved siderophore based specialized systems to overcome thermodynamic and kinetic barriers involved in sequestration and mobilization of soluble iron. One of the first few reports on microbial sider-

School of Life Sciences, North Maharashtra University, Jalgaon, Post Box 80, Jalgaon 425001, Maharashtra, India, email: sudhirchincholkar@hotmail.com

Soil Biology, Volume 12
Microbial Siderophores
A. Varma, S.B. Chincholkar (Eds.)
© Springer-Verlag Berlin Heidelberg 2007

ophores was aimed at searching for the growth factors for mycobacteria and fungi (Hesseltine et al. 1952). After isolation of siderophores in crystalline form by "Neilands: father of ferruginous facts" in 1952, numerous organisms have been reported to produce siderophore in diverse molecular forms. Baukhalfa et al. (2002) estimated that almost 500 siderophores have been reported. Although this seems to be a liberal estimate it may not be far from the truth as we were able to enlist 105 siderophores with a limited literature survey (http://groups.msn.com/siderophores). Considering the fact that many investigations/reports are lost in the huge amount of literature over a period, we invite readers/researchers to suggest additions, deletions and corrections in our list. Through discussion with compilers, (siderophores@groups.msn.com) new reports on siderophore will be added to update the list. Siderophores have been well studied in the last seven decades and information exists on methodologies of siderophore analysis (Rane et al. 2004; Jalal and van der Helm 1991), genetics (Lehoux et al. 2000), biosynthetic pathways of its synthesis (Meyer and Abdallah 1978), transport of desferri- and ferri- forms of siderophores (Faraldo-Gómez et al. 2002; Faraldo-Gómez and Sansom 2003), and reaction kinetics of iron solubilization by siderophore (Baukhalfa and Crumbliss 2002; Kraemer 2004) etc. Siderophores have received wide applications in medicine, agriculture, environment and industry (Chincholkar et al. 2005).

That there are exceptions to the rule that life forms require iron is a miraculous feature of nature. Likewise, the best studied eukaryotic model micro-organism *Saccharomyces cerevisiae* lacks the ability to synthesize siderophores although it can utilize siderophores produced by other species via "reductive and non-reductive iron assimilation" (Eissendle et al. 2003). Pandey et al. (1994) studied 23 strains of lactic acid bacilli for their ability to produce siderophores. The growth of several lactobacilli tested was unaffected by a deficiency of iron and no direct effect due to iron chelation by synthetic iron chelator was observed. Hence, it was confirmed that these 23 strains of lactic acid bacilli do not require iron! Hubmacher et al. (2002) reported that *Lactobacillus plantarum* and *Borelia burgdorferi* are known to modulate their metabolism and enzymes in an iron independent way. It is interesting to note that although some organisms were originally thought not to be producing siderophore, they were later found to produce it; for example Reeves et al. (1983) reported that Legionella sp. do not synthesize siderophores. According to them the absence of siderophore has also been noted in *Yersinia* sp. and *Neisseria gonorrhoeae*; however, siderophores of these organisms (Brem et al. 2001; Perry et al. 1999; Chambers et al. 1996; Yancey and Finkelstein 1981) including *Liegionella* sp. (Liles et al. 2000) have been reported in later years. Thus, it is clear that detection of siderophore is critical and a defined protocol needs to be followed to make sure whether particular organisms produce siderophore or not.

Fermentative production of siderophore is tricky due to its dependence on critical amounts of iron. Moreover, several factors influence the production of siderophore. The present review is directed to elaborate this issue based on experiments performed in our laboratory.

12.2
Fermentative Production of Siderophores

Bacteria as well as fungi can be used for siderophore production by a process generally called fermentation. Fermentative production of siderophore essentially involves preparation and use of an iron controlled medium, availability of specific nutrients and provision of proper physico-chemical conditions, which cumulatively tempt microbes to synthesize siderophore.

12.3
Media Preparation

Production of siderophore for the most part revolves around the iron content of the medium. Biosynthesis and overproduction of siderophore is expected only in the culture, which is stressed for iron. Even a small addition of ferric sulfate inhibits siderophore production and promotes culture density. It is, however, clear that complete absence of iron in the medium is not recommendable as a trace of the same is essential for physiological growth of organisms and induction of siderophore biosynthetic machinery. It is absolutely necessary to use iron-free glassware, water and media. In the case of glassware, being silicates, iron may be available from deep within the glass. Hence, to avoid this, regular treatment of containers using concentrated nitric acid (other acid solutions can also be used) for 4–5 h is required. Such treatment is mostly necessary for new glassware before use. Subsequently, two to three rinses with de-ionized distilled water generally remove traces of acids. Most workers prefer to use ion-exchange resin treated water for rinsing the glassware and subsequent preparation of media. Water should not be stored in plastic containers for more than one week after which bacterial growth may reach 10^4 cfu/ml. In the case of in-house distillation facility, it needs to be ensured that water does not travel through iron/stainless steel pipes. Moreover, regular visual examination for corroding metal surfaces and dead insects/rodents is needed to avoid the problem of iron contamination in water. Nowadays many types of equipment are available for obtaining ultra-pure water employing reverse osmosis and ion exchange resins. Conductivity checking of water is one of the easy processes to ensure the quality of water (Jalal and van der Helm 1991).

Siderophore production in a minimal synthetic medium excluding iron salts is most suitable as complex media components usually carry excess iron. However, with some organisms, a complex medium is required for growth wherein media treatments (defer ration) have been employed to overcome the iron content of this media/iron-rich ingredient of media (Patil 1998). Deferration can be achieved by vigorously shaking the medium with a solution of iron/metal chelators highly soluble in water-immiscible solvents like chloroform (Lankford 1973), precipitation of iron in the form of salt by addition of salts like magnesium carbonate, and passing the solution of media/ingredient through active resin-polymers, e.g. Chelex 100 (Sokol 1984). Each method has advantages and disadvantages and hence the choice of method depends solely on the trial-error and availability of chemicals. When needed, the iron chelator ethylene-di(o-hydroxyphenylacetic acid) (EDDHA) is added to the medium at concentrations from 1 to 20 mM to chelate contaminating iron or to ensure iron-limited conditions. Pre-incubation of medium in the cold for 48 h prior to use to promote iron chelation by EDDHA is recommended (Oger et al. 2001). After the process of deferration, it is recommended to add 0.1 micromoles of $FeSO_4$ (Ong et al. 1979).

12.4
Siderophore Production by Bacteria

Siderophore production by marine *Pseudomonas aeruginosa* ID 4365 in a medium containing varying concentrations of $FeCl_3$ (Fig. 12.1) revealed that, in the absence of iron (exogenous), the organism produced the maximum amount of siderophores, while gradual increase in iron content of the medium decreased siderophore production. On the other hand, the reverse situation was observed in the case of growth. Siderophore production by *Pseudomonas* was repressed when supplemented with 20 µM and above of iron (Manwar 2001). Similar observations have been reported by Persmark et al. (1990).

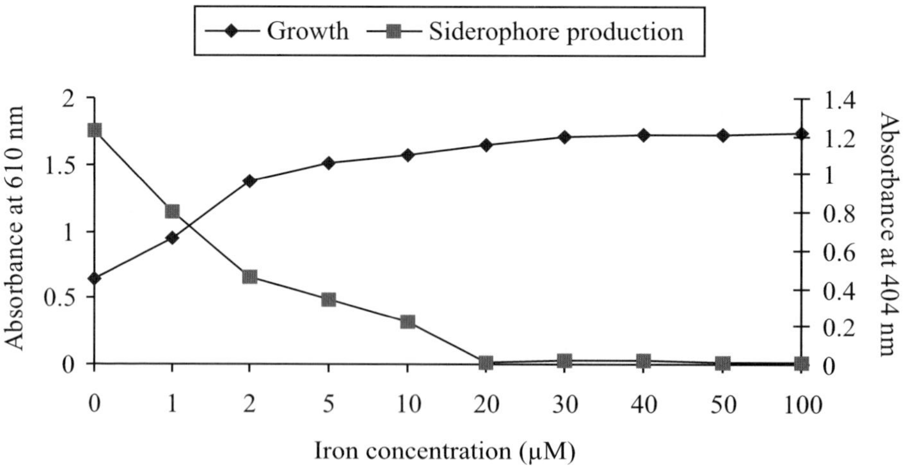

Fig. 12.1. Influence of iron (exogenous) on growth and siderophore production by *Pseudomonas aeruginosa ID 4365* (Manwar 2001)

Just a properly iron stressed medium is not enough for siderophore production; several other factors also influence it which can be observed in several microbes. If one considers the case of a versatile microorganism *Pseudomonas*, it is known to produce various siderophores, namely pyochelin (Cox 1982), salicylic acid (De Meyer et al. 1999), cepabactin (Meyer et al. 1989), corrugatin (Risse et al. 1998), ferribactin (Maurer et al. 1968), Ferrocins A, B, C and D (Katayama et al. 1993), Ornibactin (Stephan et al. 1993), pyridine-2-6-di-monothyocarboxylic acid (Ockels et al. 1978), Quinolobactin (Mossialos et al. 2000) and more than 50 types of pyoverdin (also known as Pseudobactin) type of siderophores. It has an innate capability to withstand and flourish under a wide range of microenvironments. In addition to iron concentration (Budzikiewicz 1993), siderophore production by *Pseudomonas* sp. is influenced by the nature of the carbon and nitrogen sources (Manwar 2001; Park et al. 1988), phosphate concentration (Barbhaiya and Rao 1985), pH (Manninen and Mattila-Sandholm 1994), temperature (Manwar 2001), oxygen transfer (Kim et al. 2003), heavy metals and amino acids (Manwar 2001). The mechanism of iron regulated siderophore genesis in enteric bacteria and pseudomonads is known where transcription of iron-regulated gene is under the negative control of *fur* (ferric uptake regulation) protein (repressor) with Fe^{2+} as an essential co-repressor (De Lorenzo et al. 1987).

Manwar (2001) employed succinate medium (SM), succinate medium with reduced phosphate, i.e. Barbhaiya and Rao medium (BRM), Cas amino acid medium (CAA), Philson and Linas medium (PLM), and succinate medium in artificial sea water (ASW), Succinate medium in tap water (SMT) for siderophore production. Analysis showed that BRM is the best medium and 82.56% decolorization of CAS by 40 h incubated broth was observed. The range of suitable media based on the amount of CAS decolorization was determined as BRM>SMT>SM>PLM>ASW>CAA. Recent work in our laboratory revealed that phosphate has the ability to induce siderophore production in iron rich media like PPMD meant for phenazine production (unpublished data). Succinic acid content in the medium had great influence on the production of siderophores by *Pseudomonas aeruginosa* ID 4365 wherein 0.2% concentration was optimum for high production of siderophores. Increase in its concentration enhanced the growth of bacteria but not siderophore. Hydrogen ion concentration in the medium had high impact on siderophore production. Medium pH 6.0 was optimum and siderophores were not produced at all below pH 4.0, and it gradually decreased after increasing alkalinity where it ceased at pH 10.0. The major siderophore of *Pseudomonas aeruginosa* is pyoverdine, which contains a peptide chain. Hence, fermentation directed by an exogenous supply of amino acid had significant influence on the formation of pyoverdine. Accordingly, serine, ornithine, lysine and arginine stimulated it to exhibit earlier and more production of siderophore whereas glycine, tyrosin and threonine showed no effect on production of this metabolite. When aminoacids were used as sole nitrogen and carbon source, it was observed that tyrosin>alanine>lysine>ornithine supported growth of organism where as serine>lysine>ornithine>alanine supported fluorescence production. In addition to succinic acid, butyric, citric and aspartic acids also supported production of siderophore by *Pseudomonas aeruginosa*; however, citric and aspartic acids delayed the production. Heavy metals also have a profound effect on siderophore production. The addition of Co, Cd and Zn to the media enhanced production of siderophores whereas Mo and Mn hindered the same. Braud et al. (2006) found that in a culture medium enriched with Fe and/or toxic metals (Cr, Hg and Pb), siderophore production of immobilized cells (in Ca-alginate beads) of *Pseudomonas aeruginosa* and *Pseudomonas fluorescence* was higher than for free cells, the possible mechanism being that Ca-alginate adsorbs soluble iron. Another report on similar lines (Karamanoli and Lindow 2006) states that leaf washings from 16 of the 52 plant species as well as tannic acid solutions stimulated pyoverdine synthesis in the presence of iron. The influence of temperature on pyoverdin production is also well recorded; however, it is felt that the optimum temperature for siderophore production by each Pseudomonas strain depends on the origin of the strain. Rhizospheric pseudomonads secreted higher siderophores between 28 and 30 °C whereas soil isolates of tropical origin do so well in range of 34–37 °C (Digat and Mattar 1990). Meyer et al. (1996) have also reported that *Pseudomonas aeruginosa* ATCC 15692 (PAO1) strain in succinate medium showed drastic loss of its ability to produce pyoverdin when grown above 40 °C.

Naphade (2002), who studied exopolysaccharide production by *Enterobacter cloacae* in detail at the authors laboratory, has also observed siderophore production by this organism. Siderophore production was tried in BRM, Brow and Dilworth medium, YEM broth and SM. Maximum siderophores were produced in SM after 36 h on a CAS test based comparison. This is contradictory to studies on siderophore of Pseudomonas, mentioned previously where phosphate deficient BRM media was most suitable. This suggests that the effect of phosphate concentration on siderophore production is species dependent.

12.5
Siderophore Production by Fungi

Curvularia lunata is a fascinating fungus exploited by authors for transformation of several steroids and an antibiotic (Sukhodolskaya et al. 1996; Chincholkar et al. 1993). Chaudhari (1998) has isolated four siderophores namely neocoprogen I, neocoprogen II, coprogen and ferricrocin from deferrated sucrose medium. When the media was inoculated with spore suspension and incubated on a rotary shaker at 220 rpm, high siderophore production was observed in 120 h. During logarithmic growth of the organism, there was gradual and slow increase in concentration of siderophore; however, after initiation of the stationary phase there was an abrupt increase in siderophore concentration. Addition of exogenous iron inhibited production of siderophore at concentration of 50 µM. Siderophore production was optimum at pH 6 while its secretion dropped at the rate of 20% per unit pH deviation on both sides. Due to the presence of an ornithine moiety in ferricrocin and coprogen structures, it was thought that directed fermentation by addition of individual amino acid would improve production of siderophores. Some amino acids, glycine>d,l-aspartic acid>l-glutamic acid increased the production of siderophores; ornithine and serine had no effect where as threonine hindered siderophore production. Although further research on this aspect is necessary, most probably larger/complex amino acids, which repress anabolism of complex amino acids initially, were utilized as nutrient source by organism, whereas small amino acids which easily took part in amino acid anabolism, increased levels of de novo synthesis of siderophores.

Patil (1998) from the author's laboratory studied siderophore production by *Cunninghamella blakesleeana* in iron deficient growth (sucrose) medium. An aliquot of spore suspension was inoculated and incubated at 28 °C at 220 rpm on rotary shaker for five days. Quantitative measurement revealed production up to 0.105 g/L crude siderophores. Further analysis showed that *C. blakesleeana* produced two siderophores namely ferrichrycin and a coprogen family member. An experimental set to determine effect of iron on siderophore production revealed that maximum siderophore was at zero concentration of iron (no exogenous addition of iron) after 168 h and it gradually decreased with increasing concentration of iron. An iron concentration of 50 μM led to negligible production of siderophores.

12.6
Conclusions

- For siderophore secretion there is need of specific nutrients and conditions which may vary from case to case, but iron-controlled conditions remain unchanged and play a vital role in its secretion.
- Iron is indispensable for almost all organisms with few exceptions.
- Plants and microorganisms have developed special mechanisms to obtain iron from the environment which involves synthesis of various siderophores.
- Study of the microbial system is required to channel the formation process to produce solely mono-siderophore but not the mixture and determine its aptitude in iron acquisition and anything else.

References

Barbhaiya HB, Rao KK (1985) Production of pyoverdine, the fluorescent pigment of *Pseudomonas aeruginosa* PAO1. FEMS Microbiol Lett 27:233–235
Boukhalfa H, Crumbliss A (2002) Chemical aspects of siderophore mediated iron transport. BioMetals 15:325–339

Baukhalfa HJ, Lack SD, Hersman RL, Neu MP (2002) Siderophore production and facilitated uptake of iron and plutonium in *P. putida* AIP Conf Proc 673(1):343–344

Braud A, Jezequel K, Leger MA, Lebeau T (2006) Siderophore production by using free and immobilized cells of two pseudomonads cultivated in a medium enriched with Fe and/or toxic metals (Cr, Hg, Pb). Biotechnol Bioeng 94:1080–1088

Brem D, Pelludat C, Rakin A, Jacobi CA, Heesemann J (2001) Functional analysis of yersiniabactin transport genes of *Yersinia enterocolitica*. Microbiology 147(5):1115–1127

Budzikiewicz W (1993) Secondary metabolites from fluorescent Pseudomonads. FEMS Microbiol Rev 204:209–228

Castignetti D, Smarrelli J (1986) Siderophores, the iron nutrition of plants and nitrate reductases. FEBS Lett 209:147–151

Chambers CE, McIntyre DD, Mouck M, Sokol PA (1996) Physical and structural characterization of yersiniophore, a siderophore produced by clinical isolates of *Yersinia enterocolitica*. Biometals 9:157–167

Chaudhari BL (1998) Studies on siderophores of *Curvularia lunata* NCIM 716. PhD Thesis, North Maharashtra University, Jalgaon

Chincholkar SB, Sukhodolskaya GV, Baklashova TG, Koscheyenko KA (1993) Characteristics of the 11β-hydroxylation of steroid compounds by *Curvularia lunata* VKM F-644 mycelium, in the presence of β-cyclodextrin. Appl Biochem Microbiol 28:517–524

Chincholkar SB, Rane MR, Chaudhari BL (2005) Siderophores: their biotechnological applications. In: Podila GK, Varma A (eds) Biotechnological applications: microbes; Microbiology Series, IK International, New Delhi, pp 177–198

Cox CD (1982) Effect of pyochelin on the virulence of *Pseudomonas aeruginosa*. Infect Immun 36:17–23

De Lorenzo V, Wee S, Herrero M, Page WJ (1987) Operator sequences of the aerobactin operon of plasmid Col IV K30 binding the ferric uptake regulation (*fur*) repressor. J Bacteriol 169:2624–2630

De Meyer G, Capieau K, Audenaert K, Buchala A, Metraux JP, Hofte M (1999) Nanogram amounts of salicylic acid produced by the rhizobacterium *Pseudomonas aeruginosa* 7NSK2 activate the systemic acquired resistance pathway in bean. Mol Plant Microbe Interact 12:450–458

Digat B, Mattar J (1990) Effects of temperature on growth and siderophore production of *Pseudomonas fluorescence-putida*. Symbiosis 9:203–207

Eissendle M, Oberegger H, Zadra I, Hass H (2003) The siderophore system is essential for viability of *Aspegillus nidulus*: functional analysis of two genes encoding l-ornithine *N*-monooxygenase (*Sid* A) and a non-ribosomal peptide synthesis (*Sid* C); Mol Microbiol 49:359–375

Faraldo-Gómez JD, Sansom MSP (2003) Acquisition of iron-siderophores in Gram-negative bacteria. Nat Rev Mol Cell Biol 4:105–116

Faraldo-Gómez JD, Smith GR, Sansom MSP (2002) Molecular dynamics simulations of the bacterial outer membrane protein FhuA: a comparative study of the ferrichrome-free and bound states; Biophys J 85:1406–1420

Hesseltine CW, Pidacks C, Whiteball AR, Bononos N, Hutchings BL, Williams JH (1952) Coprogen, a new growth factor for coprophillic fungi. J Am Chem Soc 74:1362

Hubmacher D, Matzanke BF, Anemuller S (2002) Investigations of iron uptake in *Halobacterium salinarum*. Biochem Soc Transac 30:710–712

Jalal MAF, van der Helm D (1991) Isolation and spectroscopic identification of fungal siderophores. Winkelmann G (ed) CRC Press, Boca Raton, Florida, pp 235–269

Karamanoli K, Lindow SE (2006) Disruption of *N*-acyl homoserine lactone-mediated cell signaling and iron acquisition in epiphytic bacteria by leaf surface compounds. Appl Environ Microbiol [Epub ahead of print]

Katayama N, Nozaki Y, Okonogi K, Harada S, Ono H (1993) Ferrocins, new iron-containing peptide antibiotics produced by bacteria taxonomy, fermentation and biological activity. J Antibiot 46:65–70

Kim EJ, Sabra W, Zeng AP (2003) Iron deficiency leads to inhibition of oxygen transfer and enhanced formation of virulence factors in cultures of *Pseudomonas aeruginosa* PAO1. Microbiology 149:2627–2634

Kraemer SM (2004) Iron oxide dissolution and solubility in presence of siderophores. Aquat Sci 66:3–18

Lankford CE (1973) Bacterial assimilation of iron. Crit Rev Microbiol 2:273–331

Lehoux DE, Sanschagrin F, Levesque RC (2000) Genomics of the 35-kb pvd locus and analysis of novel *pvd*IJK genes implicated in pyoverdine biosynthesis in *Pseudomonas aeruginosa* FEMS. Microbiol Lett 190:141–146

Liles MR, Scheel TA, Cianciotto NP (2000) Discovery of a nonclassical siderophore, legiobactin, produced by strains of *Legionella pneumophila*. J Bacteriol 182:749–757

Manninen E, Mattila-Sandholm T (1994) Methods for the detection of Pseudomonas siderophores. J Microbiol Meth 19:223–234

Manwar AV (2001) Application of microbial iron chelators (Siderophores) for improving yield of groundnut. PhD Thesis, North Maharashtra University, Jalgaon

Maurer B, Muller A, Keller-Schierlein W, Zahner H (1968) Metabolic products of microorganisms 61 Ferribactin, a siderochrome from *Pseudomonas fluorescens* Migula. Arch Mikrobiol 60:326–339

Meyer JM, Abdallah MA (1978) The fluorescent pigment of Pseudomonas fluorescens: biosynthesis, purification and physicochemical properties. J Gen Microbiol 107:319–328

Meyer JM, Hohnadel D, Halle F (1989) Cepabactin from *Pseudomonas cepacia*, a new type of siderophores. J Gen Microbiol 135:1479–1487

Meyer JM, Neely A, Stintzi A, Georges C, Holder IA (1996) Pyoverdin is essential for virulence of *Pseudomonas aeruginosa*. Infect Immun 64:518–523

Mossialos D, Meyer JM, Budzikiewicz H, Wolff U, Koedam N, Baysse C, Anjaiah V, Cornelis P (2000) Quinolobactin, a new siderophore of *Pseudomonas fluorescens* ATCC 17400, the production of which is repressed by the cognate pyoverdine. Appl Environ Microbiol 66:487–492

Naphade BS (2002) Studies on the application of Siderophoregenic E. cloacae as a biocontrol agent. PhD Thesis, North Maharashtra University, Jalgaon

Ockels W, Ro¨mer A, Budzikiewicz H (1978) An Fe(III) complex of pyridine-2,6-di-(monothiocarboxylic acid)—a novel bacterial metabolic product. Tetrahedron Lett 1978:3341–3342

Oger R, Lopez M, Farrand SK (2001) Iron-binding compounds from Agrobacterium Spp: biological control strain *Agrobacterium rhizogenes* K84 produces a hydroxamate siderophores. Appl Environ Microbiol 67:654–664

Ong SA, Peterson T, Neilands JB (1979) Agrobactin, a siderophore from *Agrobacterium tumefaciens*. J Biol Chem 254:1860–1865

Pandey A, Bringel F, Meyer JM (1994) Iron requirement and search for siderophores in lactic acid bacteria; Appl Microbiol Biotech 40:735–739

Park CS, Paulitz TC, Baker R (1988) Biocontrol of Fusarium wilt of cucumber resulting from interactions between *Pseudomonas putida* and non pathogenic isolates off *Fusarium oxysporum*. Phytopathology 78:190–194

Patil BB (1998) Studies on the role of iron metabolism in biotransformation with special reference to hydroxylation of progesterone and oxidation of rifamycin B. PhD Thesis, North Maharashtra University, Jalgaon

Perry RD, Balbo PB, Jones HA, Fetherston JD, DeMoll E (1999) Yersiniabactin from *Yersinia pestis*: biochemical characterization of the siderophore and its role in iron transport and regulation. Microbiology 145:1181–1190

Persmark M, Frejd T, Mattiasson B (1990) Purification, characterization and structure of pseudobactin 589 A, a siderophore from a plant growth promoting Pseudomonas. Biochem. 29:7348–7356

Rane MR, Naphade BS, Sayyed RZ, Chincholkar SB (2004) Methods for microbial iron chelator (siderophore) analysis. In: Podila GK, Varma A (eds) Basic research and applications of Mycorrhizae; Microbiology Series. IK International, New Delhi, 475–492

Reeves M, Pine L, Neilands JB, Ballows A (1983) Absence of siderophore activity in Legionella sp. grown in iron deficient media. J Bacteriol 154:324–329

Risse D, Beiderbeck H, Taraz K, Budzikiewicz H, Gustine D (1998) Corrugatin, a lipopeptide siderophore from *Pseudomonas corrugate*. Z Naturforsch 53:295–304

Sokol PA (1984) Production of the ferripyochelin outer membrane receptor by Pseudomonas species. FEMS Microbiol Lett 23:313–317

Stephan H, Freund S, Beck W, Jung G, Winkelmann G (1993) Ornibactins – a new family of siderophores from *Pseudomonas*. BioMetals 6:93–100

Sukhodolskaya GV, Gulevskaya SA, Chincholkar SB, Koscheyenko KA, Joshi AK, Bihari V, Basu SK (1996) Steroid 11 β-hydroxylase activity of Curvularia lunata VKM-F-644 and its stabilization factors. Appl Biochem Microbiol 32:63–71

Yancey RJ, Finkelstein RA (1981) Siderophore production by pathogenic *Neissria* sp. Infect Immun 32:600–608

Subject Index